Advances in Ecophysiology of Root Systems-Environment Interaction

Advances in Ecophysiology of Root Systems-Environment Interaction

Editor

Lorenzo Rossi

MDPI

Basel • Beijing • Wuhan • Barcelona • Belgrade • Novi Sad • Cluj • Manchester

Editor
Lorenzo Rossi
University of Florida
Fort Pierce, FL
USA

Editorial Office
MDPI
St. Alban-Anlage 66
4052 Basel, Switzerland

This is a reprint of articles from the Special Issue published online in the open access journal *Plants* (ISSN 2223-7747) (available at: https://www.mdpi.com/journal/plants/special_issues/29L7JB6Z92).

For citation purposes, cite each article independently as indicated on the article page online and as indicated below:

Lastname, A.A.; Lastname, B.B. Article Title. *Journal Name* **Year**, *Volume Number*, Page Range.

ISBN 978-3-7258-1017-8 (Hbk)
ISBN 978-3-7258-1018-5 (PDF)
doi.org/10.3390/books978-3-7258-1018-5

Cover image courtesy of Lorenzo Rossi

Contents

About the Editor

Lorenzo Rossi

Lorenzo Rossi is an Assistant Professor affiliated with the Horticultural Sciences Department at the University of Florida, stationed at the Indian River Research and Education Center in Fort Pierce, FL. With a primary focus on plant root biology, Lorenzo Rossi conducts field and greenhouse trials to study the root system architecture of citrus, as well as specialty crops such as peaches and olive trees. In addition to his research endeavors, he actively engages in teaching, overseeing courses such as Advanced Horticultural Physiology and Root and Rhizosphere Ecology. As an active member of the American Society of Horticultural Sciences, he is nationally and internationally recognized. Lorenzo Rossi develops and delivers seminars and field days to students and professionals, bringing his wealth of experience and expertise to the classroom, and, in turn, nurturing the next generation of horticultural scientists. Before joining the University of Florida, he completed two postdoctoral positions: one at Texas A&M University and another at North Carolina State University. Originally from Italy, he earned his Ph.D. in Plant Biology from Scuola Superiore Sant'Anna in Pisa, Italy, the country's most prestigious scientific university.

plants

MDPI

Article

Interactive Effect of Cultivars, Crop Years and Rootstocks on the Biochemical Traits of *Prunus persica* (L.) Batsch Fruits

Roberto Ciccoritti [1], Rossella Manganiello [2], Francesca Antonucci [2,*] and Danilo Ceccarelli [1]

1 Consiglio per la ricerca in agricoltura e l'analisi dell'economia agraria (CREA)—Centro di ricerca Olivicoltura, Frutticoltura e Agrumicoltura—Via di Fioranello 52, 00134 Rome, Italy; roberto.ciccoritti@crea.gov.it (R.C.); danilo.ceccarelli@crea.gov.it (D.C.)

2 Consiglio per la ricerca in agricoltura e l'analisi dell'economia agraria (CREA)—Centro di ricerca Ingegneria e Trasformazioni agroalimentari—Via della Pascolare 16, Monterotondo, 00015 Rome, Italy; rossella.manganiello@crea.gov.it

* Correspondence: francesca.antonucci@crea.gov.it

Abstract: Peach fruit is one of the most economically widespread temperate fruits, whose productivity, and nutritional and sensory qualities are determined by interactions among several environmental and genetic factors, rootstocks, agronomic practices and pedo-climatic conditions. In recent years, climate change has prompted peach breeding programs to use specific rootstocks that are well adapted to unusual soil and climate characteristics, thus improving the plant's adaptability and fruit quality. The aim of this work was to assess the biochemical and nutraceutical profile of two different peach cultivars, considering their growth on different rootstocks over three crop years. An analysis was carried out evaluating the interactive effect of all factors (i.e., cultivars, crop years and rootstocks) revealing the advantages or disadvantages on growth of the different rootstocks. Soluble solids content, titratable acidity, total polyphenols, total monomeric anthocyanins and antioxidant activity in fruit skin and pulp were analyzed. An analysis of variance was performed to assess the differences between the two cultivars considering the effect of rootstock (one way) and crop years, rootstocks and their interaction (two ways). In addition, two principal component analyses were performed separately on the phytochemical traits of the two cultivars to visualize the distributions of the five peach rootstocks during the three crop years. The results showed that fruit quality parameters are strongly dependent on cultivars, rootstocks and climatic conditions. All these aspects could be useful for the choice of rootstock in relation to agronomic management, making this study a valuable tool for choosing the best rootstock, considering simultaneously more factors affecting peaches' biochemical and nutraceutical profile.

Keywords: climatic effects; phytochemicals; multivariate statistics; nutraceutical plant profile; grafting

Citation: Ciccoritti, R.; Manganiello, R.; Antonucci, F.; Ceccarelli, D. Interactive Effect of Cultivars, Crop Years and Rootstocks on the Biochemical Traits of *Prunus persica* (L.) Batsch Fruits. *Plants* **2023**, *12*, 2325. https://doi.org/10.3390/plants12122325

Academic Editor: Lorenzo Rossi

Received: 19 May 2023
Revised: 7 June 2023
Accepted: 13 June 2023
Published: 15 June 2023

1. Introduction

Peach [*Prunus persica* (L.) Batsch] fruits represent one of the most widespread stone fruits, due to its good values from an energetic dietetic, nutritional and nutraceutical point of view. The peach fruit represents an important source of antioxidants, especially phenols, vitamin C and carotenoids, which are present in greater quantities, especially in the peel, although this part is not appreciated by consumers [1,2]. The antioxidant capacity due to the presence of phenolic compounds in these fruits is strongly influenced by the genotype [3]. In recent years, the consumption of peaches has decreased significantly globally, mainly due to dissatisfaction among consumers, who find fruits on the market to be mostly tasteless and of poor consistency, as they are harvested before they are fully ripe so that they can have a longer shelf-life. The main field factors that can influence fruit quality are the genotype, rootstock, orchard cultivation systems, harvest time and agro-climatic conditions. The lower quality is also due to bad post-harvest handling and storage [2].

From this perspective, peach and nectarine breeding programs are developing new genotypes that meet consumer expectations, and are trying to strike the right balance between quality and maturity at market harvest time. In addition, climatic changes in recent years have prompted peach breeding programs to improve the adaptability of peach trees to different soil and climatic conditions and cultivation systems to support high production standards, increase consumption and maintain a sustainable and profitable industry. For example, by using specific rootstocks that are adapted to unusual soil or water stress characteristics, fruit quality can be improved while also expanding ripening seasons [4]. In general, rootstocks influence many vegetative and reproductive traits of plants, including tree size, water and mineral requirements, climatic adaptation of flower buds, flowering and ripening times, yield and fruit quality [5]. The nutritional quality of the fruits is closely related to the interaction of the rootstock with water and nutrient availability in the soil [6]. Rootstocks provide a cultural tool for peach growers to increase productivity and improve efficiency via better tree survival, controlled tree vigor and increased fruit size, yield and quality. Thus, the choice of rootstock becomes as economically important as the scion cultivar whenever peach trees must be grown on soils with high bulk density, coarse texture (sand), parasitic nematodes, root rot fungal pathogens, high pH or other orchard replant problems [7].

Weather and climatic conditions and the growing environment, such as, for example, canopy irradiation, vigor management and carbon supply, also strongly influence fruit quality, especially exocarp metabolic profiles [8]. High CO_2 concentration, high temperature and limited water availability, as a result of climate change, have a negative impact on flowering and production [9].

In this scenario, the aim of the present study was to assess the biochemical and nutraceutical profile of two different peach cultivars, considering their growth on different rootstocks over three crop years. This analysis was carried out by evaluating the interactive effect of all factors (i.e., cultivars, crop years and rootstocks), revealing the advantages or disadvantages about growth of the different rootstocks.

2. Materials and Methods

2.1. Plant Material and Climatic Data Collection

This study was performed during three crop years (2011, 2017 and 2019) at the experimental farm of Centro di ricerca Olivicoltura, Frutticoltura e Agrumicoltura of Rome—Consiglio per la ricerca in agricoltura e l'analisi dell'economia agraria (CREA) (Central Italy, 41.8000° N, 12.5690° E, alt. 86 m a.s.l.).

Two different peach cultivars, grafted on five different rootstocks, grown with the same agronomic techniques (i.e., fertilization, irrigation and pest control), were considered. For the chemical and pomological analyses, fruits were harvested at consumption maturity [10].

The *P. persica* cultivars used were Ghiaccio-1*, characterized by their total lack of pigment and their white pulp, fruit with long storage capability, and sweet and aromatic test coupled to a rustic tree, and Romestar*, characterized by deep yellow pigmented fruit pulp with red pigment around the stone [11]. Finally, rootstocks (i.e., GF677, Cadaman® Avimag, Barrier®, Isthara® and GxN22 (Felinem)) were chosen in relation to the different characteristics, as reported in Table 1.

Climatic data (i.e., daily mean temperature, daily mean rainfall, daily mean relative humidity and the average total hourly solar radiation) during the three years were acquired from the Agenzia Regionale per lo Sviluppo e l'Innovazione dell'Agricoltura del Lazio (ARSIAL) weather station (RM17SIE) positioned at Marino (Rome).

Three sets of samples were collected from each cultivar (i.e., Ghiaccio-1* and Romestar*) for each rootstock for each of the three years. For the quality traits analyses, three replicates of fresh pitted fruits (about 1.5 kg) from each set were randomly sampled. For phytochemical traits determination, the samples were stored at −80 °C until the analyses.

Table 1. Rootstocks origin and agronomic characteristics. Modified from [12–14].

Rootstocks	Origin	Characteristics
GF677	*Prunus persica* (L.) Batsch x *Prunus amygdalus* Batsch	Induces high tree vigor, rapid entry into production, high production yields (both dry and irrigated). Superior adaptability even to difficult soils, tolerating active lime levels up to 12%. Tolerates water deficiency and stumpiness fairly well. Poor resistance to Agrobacterium, nematodes and Armillaria.
Cadaman® Avimag	*Prunus davidiana* (Carrière) Franch. x *Prunus persica* (L.) Batsch	Medium-to-high vigor (similar to GF677) with good growth rate. Suitable for fresh, poor and even tendentially asphyctic soils. Induces slight earliness of maturity and increase in size. Resistant to some types of nematodes. Presents some polloniferous aptitude.
Barrier®	*Prunus davidiana* (Carrière) Franch. x *Prunus persica* (L.) Batsch	Slightly lower vigor than GF677. Good anchorage, good-to-high productivity, induces better fruit size and coloration, slightly delays both flowering and ripening. Suitable for replanting. Resistant to some nematodes, asphyctic soils and chlorosants.
Isthara®	(*Prunus cerasifera* Ehrh. x *Prunus salicina* Lindl.) x (*Prunus cerasifera* Ehrh. x *Prunus persica* (L.) Batsch)	Medium vigor, with lower development than "franco" seedling rootstocks, but with good vegetative renewal and decent adaptability. Moderate suckering activity. Induces early fruiting, high yields and good-sized fruit. It is susceptible to Armillaria, but resistant to some nematodes.
GxN22 (Felinem)	*Prunus amygdalus* Batsch x *Prunus persica* (L.) Batsch	In grafted cultivars induces vigor and productivity similar to GF-677 or Hansen 536. High resistance to the main root nematode species that attack *Prunus*. Adapts well to calciferous soil. Resistant toward chlorosis.

2.2. Chemicals

All reagents were of analytical High-Performance Liquid Chromatography (HPLC)-grade (Merk Life Science S.r.l, Milan, Italy). Folin–Ciocalteau reagent, 2,2-diphenyl-1-picrylhydrazyl (DPPH), (±)-6-hydroxy-2,5,7,8-tetramethylchromane-2-carboxylic acid (commonly called Trolox) (T), sodium carbonate and gallic acid (GA) and cyanidin 3-O-glucoside (CG) were purchased from Sigma-Aldrich (St. Louis, MO, USA). Milli-Q water (Millipore, Bedford, MA, USA) passed through 0.45 nylon membrane filters (Pall Corporation, Ann Arbor, MI, USA) was used for the study.

2.3. Analytical Methods

2.3.1. Pomological Traits Determination

Pitted fresh fruits (about 500 g per sample) were homogenized, and titratable acidity (TA) and pH were determined, according to Ceccarelli et al. [10], on 10 g aliquots diluted to 50 mL with distilled water using an automatic titration system (785 DPM Titrino, Metrohm Ldt, Herisau, Switzerland). TA content was expressed as mEq L^{-1} of NaOH 0.1 M. Soluble solids content (SSC) was determined on fruit juice with a digital refractometer (Refracto 30PX, Mettler Toledo, Greifensee, Switzerland) and expressed as g 100 g^{-1} FW (Brix degrees).

2.3.2. Phytochemical Content, Total Monomeric Anthocyanin and Antioxidant Activity Extraction

Defrosted samples (about 5 g) were homogenized with a blender (Ultra-Turrax T25, IKA Labortechnik, Staufen, Germany) in 25 mL of methanol solution (methanol/water 70/30 *v*/*v*), adding 5 mM HCl to determine total phenolic content (TPC), total monomeric anthocyanin (TMA) and antioxidant activity (AA). According to Ceccarelli et al. [10], the extraction was carried out under shaking in a thermostatic bath at 37 °C for 2 h, and then were centrifuged (centrifuge mod. 4239R, ALC International—Milan, Italy) at 8000× *g* for 15 min at 5 °C, recovering the supernatant.

2.3.3. Total Phenolic Content Determination

The TPC was determined using the Folin–Ciocalteu method as reported by Ceccarelli et al. [10]. Results were calculated and expressed in mg of gallic acid equivalent (GAE) 100 g^{-1} FW. Determination of GAE was performed using the GA standard curve (0.025–0.5 mg mL^{-1}). Briefly, 0.4 mL of extract, were added to 16.0 mL water, 2.0 mL of Folin–Ciocalteu phenol reagent and 6.0 mL of 1M sodium carbonate, and the final volume was adjusted to 25 mL with the same solution used for the extraction. Samples were read at 760 nm after 2 h using an Evolution 300 UV–Vis Spectrophotometer (Thermo Electron Scientific Instruments, Madison, WI, USA). The samples were analyzed in triplicate.

2.3.4. Total Monomeric Anthocyanins Determination

TMA were determined using the pH differential method as described by Giusti and Wrolstad [15], using the UV–Vis spectrophotometer at 510 and 700 nm. Results were calculated and expressed as mg cyanidin 3-O-glucoside equivalents (CGE) 100 g^{-1} FW. All samples were analyzed in triplicate.

2.3.5. Antioxidant Activity Evaluation

The AA of the extracts was determined using the DPPH (2,2-diphenyl-1-picrylhydrazyl) method as described by Ceccarelli et al. [10]. In this procedure, 1.5 mL of DPPH solution was added to 1.5 mL of fruit extract, and after 15 min, absorbance at 515 nm was determined using the UV–Vis Spectrophotometer. The percent inhibition activity of fruit extract was calculated as:

$$[(A_0 - A_1)/A_0]\cdot 100$$

where A_0 was the control absorbance and A_1 the extract absorbance. Trolox (0.5–10 µg mL^{-1}) was used as the reference compound and AA was expressed as µg of Trolox equivalent (TE) mg^{-1} FW. All samples were analyzed in triplicate.

2.4. Statistical Analysis

All the statistical analyses described below were carried out on the mean of three replicates for each cultivar. In addition, all the rootstocks were considered for the analyses (Past v. 4.02). Firstly, a one-way analysis of variance (ANOVA) was performed on the whole dataset to evaluate the differences between the two cultivars (i.e., Ghiaccio-1* and Romestar*) considering the effect of rootstock on phytochemical composition. Then, a two-way ANOVA considering the effects of crop years (Y), rootstock (R) and their interaction (Y × R), followed by a post hoc Tukey test, was carried out. In addition, two Principal Component Analyses (PCA) were performed (considering the two cultivars separately) on the mean data of SSC, TA, TPC, TMA and AA to visualize the distribution of the five peach rootstocks during the three crop years.

3. Results

3.1. Climatic Condition

The hourly variation in weather and climate parameters (i.e., average daily temperature, average daily precipitation, average daily relative humidity and average total solar radiation) recorded during the experimental crop years (2011, 2017 and 2019) is shown in Figure 1.

The monthly average air temperature ranged from 6.0 °C (January 2017) to 27.4 °C (August 2017). The highest average temperature was recorded in August 2017 (28.2 °C), and the lowest in January 2017 (5.5 °C). In winter, minimum temperatures never fell below 5 °C, while maximum temperatures in summer averaged around 25–26 °C. The average air temperature during the growing season (II trimester) was around 19 °C for the 2011–2017 crop years and 17.6 °C for 2019. The average solar radiation was very high in the second and third quarters of 2017, reaching a maximum value of 1240 kJ m^{-2} in June, while lower and similar values were reported during the 2011 and 2019 crop years. The rainfall distribution during the growing season was higher in 2019; in particular, May 2019 was an extremely rainy month (average value 235 mm), while it never rained in the following month. In the same period of the 2011 and 2017 crop years, the average total amount of rainfall was very low, varying between 30 and 80 mm. Finally, the highest average monthly relative humidity was recorded in November 2019 (86.7%), corresponding to the maximum rainfall (516 mm), while the lowest average relative humidity was recorded in July 2017 (51.6%).

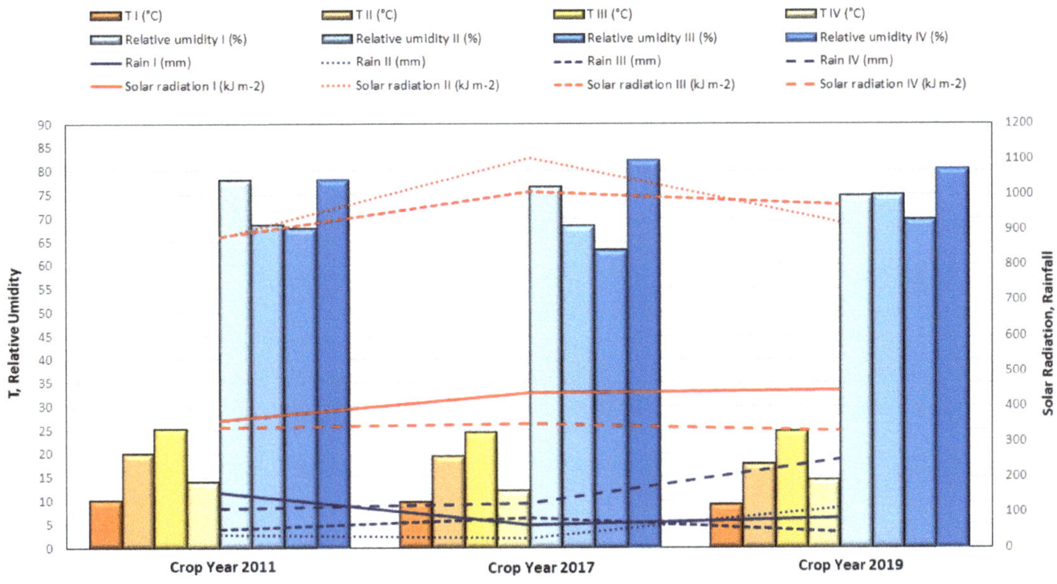

Figure 1. Means of the daily climatic data (i.e., air temperature at 2 m (T, °C), relative humidity at 2 m (%), rainfall (mm) and solar radiation (kJ m^{-2}) during the four trimesters of the 2011, 2017 and 2019 crop years. The first quarter (I) refers to the months of January to March, the second (II) to the months of April to June, the third (III) to the months of July to September and the fourth (IV) to the months of October to December.

3.2. Peach Cultivars Biochemical Variability

The one-way ANOVA to evaluate the rootstock effect between the two cultivars (i.e., Ghiaccio-1* and Romestar*) showed that the two cultivars were significantly ($p < 0.05$) different for all investigated parameters, except for antioxidant activity. Figure 2 shows the box plots displaying these values.

For Ghiaccio-1*, the SSC and the TA were higher than those of Romestar*, when considering all the five different rootstocks together (15 $\pm$ 1 g/100 g FW and 50 $\pm$ 10 mEq L^{-1}, respectively). Considering the peel bioactive compounds (TPC and TMA), the highest values were observed in Romestar*, whereas no significant differences were found in AA (Figure 2). No significant differences among TPC, TMA or AA were found between the two cultivars in flesh fruits.

3.3. Peach Years and Rootstock Biochemical Variability

The results of the two-way ANOVA reported in Table 2 show significant effects ($p < 0.001$) of crop years (Y), rootstock (R) and their interaction (Y $\times$ R) for all investigated phytochemicals in Ghiaccio-1*.

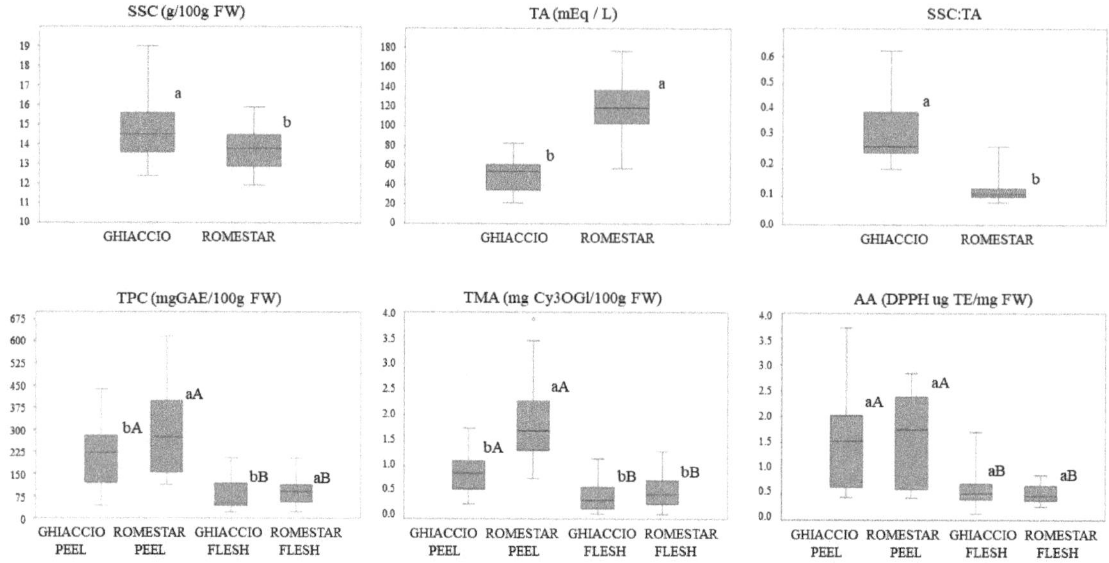

Figure 2. Box plots extracted from the one-way analysis of variance (ANOVA) examining soluble solid content (SSC), titratable acidity (TA), soluble solid content and titratable acidity ratio (SSC: TA) of the entire pitted fresh fruits, and the total phenolic compound (TPC), total monomeric anthocyanins (TMA) and antioxidant activity (AA) of the peel and flesh of the two cultivars Ghiaccio-1* (Ghiaccio) and Romestar* during the three crop years 2011, 2017 and 2019. Values belonging to the same traits without letters in common are statistically different according to LSD ($p \leq 0.05$). Lower-case letters are used for intragroup difference evaluation, upper-case letters are applied for intergroup evaluation (peel vs. flesh).

Y (Table 2A) was the main factor that affected all parameters, except for flesh AA, which was strongly influenced by the interaction Y × R. In particular, the highest TA value values of the Y mean data were observed in 2017 and 2019(61 ± 12 and 55 ± 5 mEq L^{-1}, respectively), while the lowest was detected in 2011 (32 ± 6 mEq L^{-1}).

A low TA variability was observed among the rootstocks, ranging from 54 ± 6 (GxN22) to 46 ± 5 mEq L^{-1} in the Isthara®. The highest SSC amount was found in 2017, ranging from 14.1 to 17.9 g 100 g^{-1} FW (Table 2B). Among the rootstocks, the mean highest SSC was found in GxN22 (mean 16 ± 1 g 100 g^{-1} FW), followed by Barrier® (mean 15 ± 1 g 100 g^{-1} FW) and by Cadaman® and Isthara® (14.2 ± 0.9 g 100 g^{-1} FW). In addition, slight differences in SSC:TA ratio were found among Isthara® and the others (Table 2B).

Significant differences ($p < 0.001$) were observed among the Y and between the two cultivars, showing the highest TPC in 2017 both in peel and flesh (307 ± 80 and 142 ± 36 mg GAE 100 g^{-1} FW, respectively). The highest TMA values were found in peel (1.2 ± 0.3 mg Cy3OGl 100 g^{-1} FW) in 2017 and in flesh in 2019 (2.2 ± 0.8 mg Cy3OGl 100 g^{-1} FW; Table 2B). Finally, the mean highest AA in peel was found in 2011, whereas no significant differences were found among the years in flesh.

Among the studied rootstocks, Barrier®, Cadaman® and Isthara® showed an interesting phytochemical profile characterized by the highest TPC, TMA and AA values. The rootstocks GxN22 and GF677 showed low TPC, TMA and AA values both in peel and flesh (Table 2B).

Regarding Romestar*, the two-way ANOVA revealed similar effects of Y, R and Y × R for all the parameters. Additionally, for this cultivar, the Y was the main discriminant factor except for SSC and SSC:TA ratio, which were strongly affected by Y × R (Table 3A).

Table 2. Results of the two-way analysis of variance (ANOVA) of pitted fruit, flesh and peel of the Ghiaccio-1* cultivar: (**A**) two-way ANOVA performed on the mean squares of the five rootstocks (R) for three crop years (Y); (**B**) mean values, for R and Y, of titratable acidity (TA), soluble solids content (SSC), SSC:TA, total phenolic content (TPC), total monomeric anthocyanins (TMA) and antioxidant activity (AA). [1] Degrees of freedom (DF); * significant ($p < 0.05$); ** significant ($p < 0.001$). Different letters in the same column indicate significant differences.

			Ghiaccio-1*								
	(A)			Pitted Fruit			Peel			Flesh	
	Source	D.F. [1]	TA	SSC	SSC:TA	TPC	TMA	AA	TPC	TMA	AA
	Crop Years (Y)	2	15,254.1 **	79.32 **	0.68 **	776,527 **	7.7 **	25.5 **	201,693 **	2.7 **	0.96 **
	Rootstock (R)	4	661.6 **	28.67 **	0.02 ns	70,126.8 **	1.5 **	22.5 **	11,088 **	0.7 **	1.66 **
	Y × R	5	3210.3 *	64.22 **	0.03 ns	130,864 **	2.1 **	15.2 **	11,878 **	0.8 **	6.91 **
	Within	75	42.8 ns	44.56 ns	0.19 ns	62,892 ns	0.9 ns	4.5 ns	13,910 ns	1.6 ns	0.52 ns
	Total	89	19,877.4 ns	219.79 ns	0.92 ns	1.1×10^6 ns	12.2 ns	67.6 ns	238,570 ns	5.8 ns	10.05 ns
	(B)		TA	SSC	SSC:TA	TPC	TMA	AA	TPC	TMA	AA
ROOTSTOCKS	Barrier®		48 ± 8 ab	15.0 ± 1.1 ab	0.4 ± 0.1 a	235 ± 27 a	1.0 ± 0.1 a	1.4 ± 0.1 bc	96 ± 17 a	0.5 ± 0.1 a	0.5 ± 0.2 a
	Cadaman®		51 ± 7 a	14.2 ± 0.9 b	0.3 ± 0.1 a	223 ± 20 a	0.9 ± 0.1 a	2.5 ± 0.2 a	76 ± 16 ab	0.3 ± 0.1 b	0.6 ± 0.1 a
	GF677		48.0 ± 6 ab	14.8 ± 1.0 ab	0.3 ± 0.1 a	155 ± 17 b	0.7 ± 0.1 b	1.2 ± 0.1 c	67 ± 10 b	0.3 ± 0.1 b	0.5 ± 0.2 a
	GXN22		54 ± 6 a	15.9 ± 1.3 a	0.3 ± 0.1 a	204 ± 26 ab	0.9 ± 0.2 ab	1.2 ± 0.1 c	68 ± 8 b	0.4 ± 0.1 ab	0.5 ± 0.2 a
	Isthara®		46 ± 5 b	14.2 ± 0.8 b	0.3 ± 0.1 a	219 ± 12 a	1.0 ± 0.2 a	1.6 ± 0.2 b	88 ± 12 a	0.6 ± 0.2 a	0.6 ± 0.1 a
			TA	SSC	SSC:TA	TPC	TMA	AA	TPC	TMA	AA
YEARS	2011		31.5 ± 5.5 b	13.8 ± 0.8 bc	0.45 ± 0.08 a	90 ± 27 b	0.6 ± 0.1 b	1.5 ± 0.4 a	43.8 ± 12.0 b	0.5 ± 0.3 b	0.5 ± 0.1 a
	2017		60.8 ± 11.9 a	16.0 ± 1.9 a	0.27 ± 0.05 b	307 ± 80 a	1.2 ± 0.3 a	1.0 ± 0.1 b	142 ± 36 a	0.6 ± 0.1 b	0.6 ± 0.2 a
	2019		55.4 ± 5.1 a	14.6 ± 0.9 b	0.26 ± 0.03 b	223 ± 66 ab	0.9 ± 0.3 ab	0.22 ± 0.08 c	49 ± 20 b	2.2 ± 0.8 a	0.7 ± 0.1 a

Table 3. Results of the two-way analysis of variance (ANOVA) of pitted fruit, flesh and peel of the Romestar* cultivar: (**A**) two-way analysis of variance (ANOVA) performed on the mean squares of the five rootstocks (R) for three crop years (Y); (**B**) mean values, for the R and Y, of titratable acidity (TA), soluble solids content (SSC), SSC:TA, total phenolic content (TPC), total monomeric anthocyanins (TMA) and antioxidant activity (AA). [1] Degrees of freedom (DF); * significant ($p < 0.001$). Different letters in the same column indicate significant differences.

			Romestar*								
(A)			Pitted Fruits			Peel			Flesh		
	Source	D.F. [1]	TA	SSC	SSC:TA	TPC	TMA	AA	TPC	TMA	AA
	Crop Years (Y)	2	20,291.8 *	6.9 *	0.04 *	1.1×10^6 *	29.3 *	56.4 *	132,714 *	5.74 *	1.77 *
	Rootstock (R)	4	12,499.2 *	22.6 *	0.05 *	72,425.7 *	2.6 *	1.3 *	16,196.2 *	0.56 *	0.14 *
	Y × R	5	10,201.6 *	36.3 *	0.07 *	383,575 *	10.5 *	1.1 *	20,998.6 *	1.94 *	0.24 *
	Within	75	8779.4 ns	35.0 ns	0.01 ns	81,072.5 ns	5.3 ns	2.7 ns	2.1×10^5 ns	0.56 ns	0.32 ns
	Total	89	51,772.1 ns	100.9 ns	0.16 ns	1.6×10^6 ns	47.7 ns	61.5 ns	1.9×10^5 ns	8.81 ns	2.45
	(B)		TA	SSC	SSC:TA	TPC	TMA	AA	TPC	TMA	AA
ROOTSTOCKS	Barrier®		119 ± 9 a	14.3 ± 0.8 a	0.12 ± 0.02 ab	263 ± 18 b	1.7 ± 0.4 a	1.4 ± 0.2 a	101 ± 17 a	0.6 ± 0.2 a	0.5 ± 0.1 a
	Cadaman®		123 ± 11 a	13.2 ± 0.9 b	0.11 ± 0.02 ab	276 ± 11 b	1.9 ± 0.3 a	1.5 ± 0.3 a	81 ± 7 b	0.4 ± 0.1 a	0.5 ± 0.1 a
	GF677		131 ± 10 a	13.5 ± 0.8 b	0.10 ± 0.01 ab	282 ± 13 b	1.9 ± 0.3 a	1.5 ± 0.3 a	80 ± 6 b	0.5 ± 0.1 a	0.5 ± 0.1 a
	GxN22		115 ± 5 ab	13.2 ± 0.9 b	0.12 ± 0.02 ab	315 ± 18 a	2.0 ± 0.4 a	1.6 ± 0.2 a	81 ± 5 b	0.7 ± 0.2 a	0.5 ± 0.1 a
	Isthara®		96 ± 15 b	14.5 ± 0.9 a	0.17 ± 0.07 a	314 ± 19 a	1.8 ± 0.2 a	1.7 ± 0.3 a	113 ± 12 a	0.5 ± 0.1 a	0.6 ± 0.1 a
			TA	SSC	SSC:TA	TPC	TMA	AA	TPC	TMA	AA
YEARS	2011		95.9 ± 20.8 b	13.5 ± 1.2 b	0.15 ± 0.06 a	145 ± 20 b	1.3 ± 0.3 b	0.6 ± 0.3 b	76 ± 15 b	0.8 ± 0.3 a	0.7 ± 0.1 a
	2017		125.4 ± 24.1 a	14.2 ± 1.0 a	0.12 ± 0.04 a	311 ± 73 ab	2.7 ± 0.6 a	1.7 ± 0.4 a	144 ± 37 a	0.6 ± 0.1 a	0.37 ± 0.04 b
	2019		127.9 ± 15.5 a	13.6 ± 0.9 b	0.11 ± 0.02 a	412 ± 116 a	2.4 ± 0.4 a	1.7 ± 0.5 a	57 ± 13 b	0.5 ± 0.1 ab	0.23 ± 0.03 c

In particular, for 2017 and 2019, the highest TA values were equal to 125 ± 24 and 128 ± 16 mEq L^{-1}, respectively, while the lowest was observed for 2011 (96 ± 21 mEq L^{-1}; Table 3B).

High TA variability was observed among the rootstocks, which ranged from 131 ± 10 for GF677 to 96 ± 15 mEq L^{-1} for Isthara®. The highest SSC was found in 2017, ranging from 13.2 to 15.2 g 100 g^{-1} FW (Table 2B). Among the rootstocks, the highest mean SSC was found in Isthara® (14.5 ± 0.9 g 100 g^{-1} FW), followed by Barrier® (14.3 ± 0.8 g 100 g^{-1} FW) and by Cadaman® and GxN22 (13.2 ± 0.9 g 100 g^{-1} FW). Finally, slight differences in SSC:TA ratio were found between Isthara® and all the others (Table 3B).

The years 2019 and 2017 showed the highest peel values of TPC, TMA and AA, whereas the lowest were observed in 2011 (Table 3B). The flesh showed the highest value of TPC (144 ± 37 mg GAE 100 g^{-1} FW) during 2017, and of AA during 2011 (0.7 ± 0.1 µg g^{-1} TE FW).

As regards rootstocks, only the TPC values showed significant differences. In detail, Isthara® showed the highest TPC value both in its peel and flesh.

3.4. Chemometric Elaboration

To visualize the distribution of the five peach rootstocks during the three crop years according to the phytochemical traits (i.e., SSC, TA, TPC, TMA and AA), two separate PCAs were performed for the two cultivars. Figure 3 shows the scatter plot of Ghiaccio-1* considering the five different rootstocks. The first two principal components (PC1 and PC2) represent an explained variance equal to 58.3% and 28.3%, respectively. The PC1 was positively correlated with SSC, TPC and TMA. Meanwhile, the PC2 was positively correlated with AA and TA. It is possible to observe well-defined clusters grouping the samples in three main classes (Figure 3). The first one was placed in the first and fourth quadrant and included all samples grown during 2017, characterized by the highest SSC, TPC and TMA values. The second one was located in the first and second quadrant, and included all the samples grown during 2019, and was characterized by the highest AA. Finally, the third group was located in the third quadrant, and presented the highest SSC:TA ratio values.

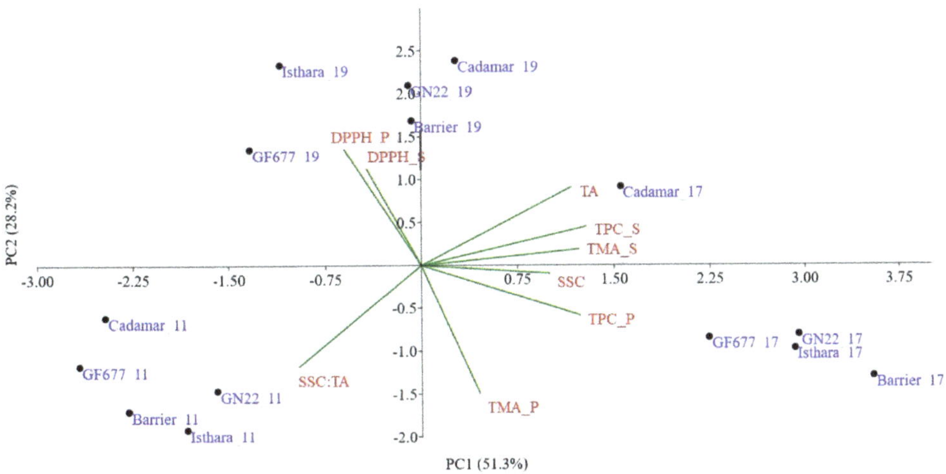

Figure 3. Principal Component Analysis (PC1 vs. PC2) of Ghiaccio-1* samples performed on the five rootstocks (i.e., Cadaman®, GF677, GxN22, Isthara® and Barrier®) analyzed in 2019 (black dots), 2017 (red dots) and 2011 (blue dots). The scatter plot reports: solid soluble content (SSC), titratable acidity (TA), total phenolic content (TPC), total monomeric anthocyanins (TMA) and antioxidant activity (AA) in peel (S) and flesh (P).

The PC1 in Figure 4 was positively correlated with TMA, SSC:TA ratio and AA, while the PC2 was positively correlated with TPC and SSC. Additionally, for this cultivar, the PCA reported three well-defined clusters representing the three crop years. In each group, the samples were spread in relation to the rootstocks, which influenced the fruits' biochemical composition.

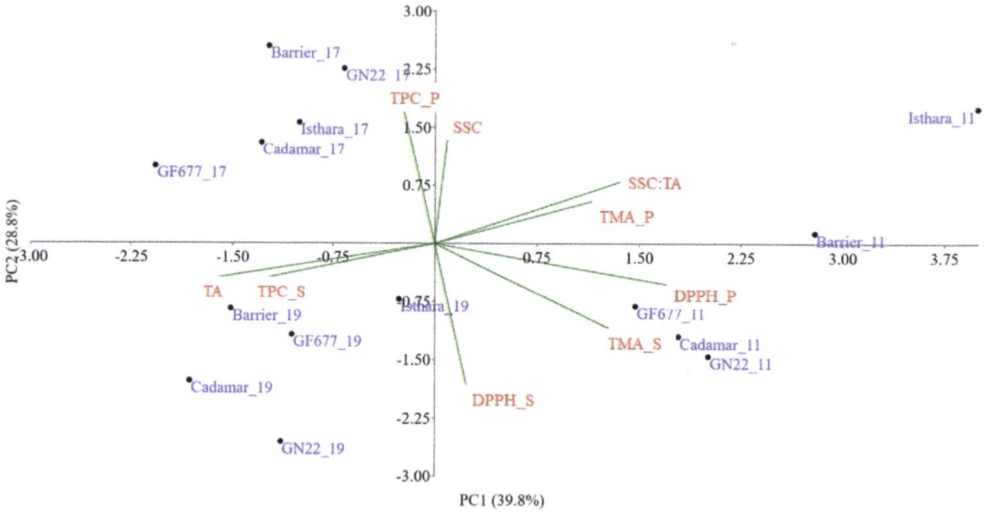

Figure 4. Principal Component Analysis (PC1 vs. PC2) of Romestar* samples performed on the five rootstocks (i.e., Cadaman®, GF677, GxN22, Isthara® and Barrier®) analyzed in 2019 (black dots), 2017 (red dots) and 2011 (blue dots). The scatter plot reports: solid soluble content (SSC), titratable acidity (TA), total phenolic content (TPC), total monomeric anthocyanins (TMA) and antioxidant activity (AA) in peel (S) and flesh (P).

4. Discussion

The aim of this study was the valuation of the advantages of five different rootstocks through the characterization of the biochemical and nutraceutical profiles of two different peach cultivars (i.e., Ghiaccio-1* and Romestar*). Several authors [4,16–19] have underlined the key role of rootstock in determining the quality of production and the nutraceutical characteristics of fruits. Numerous of these specified that some peach rootstocks increased the yield, size and quality of commercial peach [5,8,16,20,21]. To strengthen the purpose of this research, three crop years were considered. This, coupled with chemometric elaboration, allowed the evaluation of the effects of climatic conditions on the fruits' qualitative traits, defining the effects of rootstock during the experiment. Generally, as reported in this work, the fruit quality parameters were found to be strongly dependent on the cultivars, rootstocks and climatic conditions (Figure 1; Tables 2 and 3). This has been confirmed by many studies present in the literature [9,16,22–24].

As reported in Tables 2 and 3, Ghiaccio-1* showed the highest SSC and the lowest TA values with respect to Romestar*. Its high SCC:TA ratio, its acid levels and its soluble solid concentration could make it very suitable for consumers [25,26]. Generally, significant differences in TPC and TMA were found in this study between the two cultivars with respect to AA, probably due to their specific chemical composition.

The multivariate statistical PCA was applied to evaluate the similarity of phytochemical parameters between the different rootstock types during the three crop years. Figure 3 shows that, for Ghiaccio-1*, Cadaman® rootstock was located the furthest from the others. All the samples grafted on Cadaman® were located in the positive side of PC2, highlighting high values of TA and AA, peel TPC and TMA, and suggesting high adaptability to the environment. Finally, the Barrier® rootstock appears to induce higher sweetness, low acidity and higher TPC content in Ghiaccio-1*.

On the other hand, Figure 4 shows that, for Romeestar*, the best performance in terms of quality biochemical traits was obtained by using the Isthara® rootstock.

In addition, climatic conditions affect peach quality parameters and chemical composition differently. Moreover, they could influence the phytochemical profile in different ways. For example, water stress in the final stages of growth in plum fruits causes a significant decrease in size, but accelerates maturation and SSC level [27]. Higher precipitation has been found to be related to higher TPC values, as was observed for Romestar* in 2019. The high humidity registered during harvest time in this year could be an additional factor contributing to higher phenols content, possibly due to minor abiotic stress which could have influenced transpiration and photosynthetic activities [28]. In addition, the abundant solar radiation in 2017 contributed to an increase in the amount of the bioactive compounds TPC and TMA in Ghiaccio-1*, thanks to the greater development of pigments, especially in the peel. This was in agreement with the study of Solovchenko and Schmitz-Eiberger [29], which claimed that specific spectral light properties of solar radiation and temperatures are important for the regulation of antioxidant biosynthesis. This research highlighted that the parameter most influenced by climatic conditions is the cultivar.

The results of this work demonstrated that the main bioactive compounds and antioxidant activity are significantly influenced by rootstock, even if it is not possible to define a regular trend. Indeed, rootstocks of similar vigor (Table 1) produced fruits with very different nutritional characteristics, indicating that the effects of climatic condition and grafted varieties could strongly affect the peach quality, underlining the fact that biochemical composition is strictly related to the interaction of rootstock water composition and nutrient soil availability [30].

5. Conclusions

Peach fruit quality is highly dependent on cultivars and growing years. Moreover, this is closely related to the choice of rootstock on which the plants are grafted, which is becoming an important parameter to consider in new plantings. In fact, rootstocks are valuable tools used by growers to improve the efficiency of cultivation, such as by increasing plant survival under different soil and climate conditions. In addition, it can control their vigor and increase productivity, with effects on fruit yield and phytochemical characteristics. All these aspects could be useful for the choice of rootstock in relation to the agronomic management, making this study a valuable tool for choosing the best rootstock while simultaneously considering more factors affecting the fruit's qualitative parameters.

Author Contributions: Conceptualization, R.C., D.C. and F.A.; methodology, R.C., D.C. and F.A.; software, R.C., R.M. and F.A.; validation, R.C., D.C., R.M. and F.A.; formal analysis, R.C. and D.C.; investigation, R.C., R.M, D.C. and F.A.; resources, D.C.; data curation, R.C., R.M. and F.A.; writing—original draft preparation, R.C., R.M., D.C. and F.A.; writing—review and editing, R.C., R.M., D.C. and F.A.; visualization, R.C. and F.A.; supervision, R.C. and F.A.; project administration, D.C.; funding acquisition, D.C. All authors have read and agreed to the published version of the manuscript.

Funding: This research received no external funding.

Data Availability Statement: The data used in the current study are contained within the article.

Acknowledgments: The authors thank Luigia Principio, Carolina Talento and Roberto Ciorba for their technical contribution.

Conflicts of Interest: The authors declare no conflict of interest.

References

1. Zhao, X.; Zhang, W.; Yin, X.; Su, M.; Sun, C.; Li, X.; Chen, K. Phenolic Composition and Antioxidant Properties of Different Peach [*Prunus persica* (L.) Batsch] Cultivars in China. *Int. J. Mol. Sci.* **2015**, *16*, 5762–5778. [CrossRef]
2. Minas, I.S.; Tanou, G.; Molassiotis, A. Environmental and orchard bases of peach fruit quality. *Sci. Hortic.* **2018**, *235*, 307–322. [CrossRef]
3. Petruccelli, R.; Bonetti, A.; Ciaccheri, L.; Ieri, F.; Ganino, T.; Faraloni, C. Evaluation of the Fruit Quality and Phytochemical Compounds in Peach and Nectarine Cultivars. *Plants* **2023**, *12*, 1618. [CrossRef]
4. Giorgi, M.; Capocasa, F.; Scalzo, J.; Murri, G.; Battino, M.; Mezzetti, B. The rootstock effects on plant adaptability, production, fruit quality, and nutrition in the peach (cv. "Suncrest"). *Sci. Hortic.* **2005**, *107*, 36–42. [CrossRef]
5. Minas, I.S.; Anthony, B.M.; Pieper, J.R.; Sterle, D.G. Large-scale and accurate non-destructive visual to near infrared spectroscopy-based assessment of the effect of rootstock on peach fruit internal quality. *Eur. J. Agron.* **2023**, *143*, 126706. [CrossRef]
6. Alfaro, J.M.; Bermejo, A.; Navarro, P.; Quiñones, A.; Salvador, A. Effect of rootstock on Citrus fruit quality: A review. *Food Rev. Int.* **2021**, 1–19. [CrossRef]
7. Reighard, G.L. Peach, plum and apricot rootstocks for the 21st century. *Asp. Appl. Biol.* **2013**, *119*, 59–66.
8. Anthony, B.M.; Minas, I.S. Redefining the impact of preharvest factors on peach fruit quality development and metabolism: A review. *Sci. Hortic.* **2022**, *297*, 110919. [CrossRef]
9. Jiménez, S.; Fattahi, M.; Bedis, K.; Nasrolahpour-Moghadam, S.; Irigoyen, J.J.; Gogorcena, Y. Interactional effects of climate change factors on the water status, photosynthetic rate, and metabolic regulation in peach. *Front. Plant Sci.* **2020**, *11*, 43. [CrossRef]
10. Ceccarelli, D.; Antonucci, F.; Costa, C.; Talento, C.; Ciccoritti, R. An artificial class modelling approach to identify the most largely diffused cultivars of sweet cherry (*Prunus avium* L.) in Italy. *Food Chem.* **2020**, *333*, 127515. [CrossRef]
11. Conte, L.; Fantechi, P.; Insero, O.; Liverani, L.; Nicotra, A. *Monografia di Cultivar di Pesco, Nettarine e Percoche*; Ministero delle Politiche Agricole e Forestali: Rome, Italy; Istituto Sperimentale per la Frutticoltura: Rome, Italy, 2004; pp. 46, 75.
12. Beckman, T.G. Register of New Fruit and Nut Cultivars (Almond Rootstock). *Hortic. Sci.* **2008**, *43*, 1321–1322.
13. Fideghelli, C.; Loreti, F.; Ancarani, V.; Fei, C.; Godini, A.; Grandi, M.; Liverani, A.; Lugli, S.; Massai, R.; Palasciano, M.; et al. *Monografia dei Portinnesti dei Fruttiferi*; Progetto Finalizzato MIPAAF Regioni "Liste di orientamento varietale dei fruttiferi"; Ministero delle Risorse Agricole, Alimentari e Forestali: Rome, Italy, 2009; p. 239.
14. Felipe, A.J. 'Felinem', 'Garnem', and 'Monegro' almond×peach hybrid rootstocks. *Hortic. Sci.* **2009**, *44*, 196–197. [CrossRef]
15. Giusti, M.M.; Wrolstad, R.E. Characterization and measurement of anthocyanins by UV-visible spectroscopy. *Curr. Protoc. food Anal. Chem.* **2001**, *1*, F1–F2. [CrossRef]
16. Remorini, D.; Tavarini, S.; Degl'Innocenti, E.; Loreti, F.; Massai, R.; Guidi, L. Effect of rootstocks and harvesting time on the nutritional quality of peel and flesh of peach fruits. *Food Chem.* **2008**, *110*, 361–367. [CrossRef]
17. Gil, M.; Tomas-Barberan, A.T.; Hess-Pierce, B.; Kader, A.A. Antioxidant capacities, phenolic compounds, carotenoids and vitamin C content of nectarine and plum cultivars from California. *J. Agric. Food Chem.* **2002**, *50*, 4976–4982. [CrossRef]
18. Font i Forcada, C.; Gogorcena, Y.; Moreno, M.Á. Agronomical parameters, sugar profile and antioxidant compounds of "Catherine" peach cultivar influenced by different plum rootstocks. *Int. J. Mol. Sci.* **2014**, *15*, 2237–2254. [CrossRef]
19. Drogoudi, P.D.; Tsipouridis, C.G. Effects of Cultivar and Rootstock on the Antioxidant Content and Physical Characters of Clingstone Peaches. *Sci. Hortic.* **2007**, *115*, 34–39. [CrossRef]
20. Mestre, L.; Reig, G.; Betrán, J.A.; Moreno, M.Á. Influence of plum rootstocks on agronomic performance, leaf mineral nutrition and fruit quality of 'Catherina' peach cultivar in heavy-calcareous soil conditions. *Span. J. Agric. Res.* **2017**, *15*, e0901. [CrossRef]
21. Gullo, G.; Motisi, A.; Zappia, R.; Dattola, A.; Diamanti, J.; Mezzetti, B. Rootstock and fruit canopy position affect peach [*Prunus persica* (L.) Batsch] (cv. Rich May) plant productivity and fruit sensorial and nutritional quality. *Food Chem.* **2014**, *153*, 234–242. [CrossRef]
22. Ceccarelli, D.; Talento, C.; Sartori, A.; Terlizzi, M.; Caboni, E.; Carbone, K. Comparative characterization of fruit quality, phenols and antioxidant activity of de-pigmented "Ghiaccio-1*" and white flesh peaches. *Adv. Hortic. Sci.* **2016**, *30*, 175–182.
23. De Salvador, F.R.; Proietti, G.; Tomasone, R.; Cedrola, C. Field performance of several hybrid rootstocks with six peach cultivars. In Proceedings of the X International Symposium on Integrating Canopy, Rootstock and Environmental Physiology in Orchard Systems, Stellenbosch, South Africa, 3–6 December 2012; pp. 577–583.
24. Cipriani, G.; Terlizzi, M.; Bevilacqua, D.; Di Cintio, A.; Rosato, T.; Sartori, A. Peach breeding programme for new cultivars and different traits-pomological and phenological data analysis with a ranking method. In Proceedings of the II International Symposium on Horticulture in Europe, Angers, France, 1–5 July 2012; pp. 695–702.
25. Crisosto, C.H.; Day, K.R.; Crisosto, G.M.; Garner, D. Quality attributes of white flesh peaches and nectarines grown under California conditions. *J. Am. Pomol. Soc.* **2001**, *55*, 45–51.
26. Crisosto, C.H.; Crisosto, G.M. Relationship between ripe soluble solids concentration (RSSC) and consumer acceptance of high and low acid melting flesh peach and nectarine (*Prunus persica* (L.) Batsch) cultivars. *Postharvest Biol. Technol.* **2005**, *38*, 239–246. [CrossRef]
27. Intrigliolo, D.S.; Castel, J.R. Response of plum trees to deficit irrigation under two crop levels: Tree growth, yield and fruit quality. *Irrig. Sci.* **2010**, *28*, 525–534. [CrossRef]
28. Sahamishirazi, S.; Moehring, J.; Claupein, W.; Graeff-Hoenninger, S. Quality assessment of 178 varieties of plum regarding phenolic, anthocyanin and sugar content. *Food Chem.* **2017**, *214*, 694–701. [CrossRef]

29. Solovchenko, A.; Schmitz-Eiberger, M. Significance of skin flavonoids for UV-B protection in apple fruits. *J. Exp. Bot.* **2003**, *54*, 1977–1984. [CrossRef]
30. Tavarini, S.; Gil, M.I.; Tomas-Barberan, F.A.; Buendia, B.; Remorini, D.; Massai, R.; Degl'Innocenti, E.; Guidi, L. Effects of water stress and rootstocks on fruit phenolic composition and physical/chemical quality in Suncrest peach. *Ann. Appl. Biol.* **2011**, *158*, 226–233. [CrossRef]

plants

MDPI

Article

Influence of Planting Density on Sweet Potato Storage Root Formation by Regulating Carbohydrate and Lignin Metabolism

Qinggan Liang [1,2,3,†], Hongrong Chen [2,†], Hailong Chang [2], Yi Liu [1,3], Qinnan Wang [2], Jiantao Wu [2], Yonghua Liu [1,3], Sunjeet Kumar [1,3], Yue Chen [1,3], Yanli Chen [1,3,*] and Guopeng Zhu [1,3,*]

1 Key Laboratory of Quality Regulation of Tropical Horticultural Crop in Hainan Province, School of Horticulture, Hainan University, Haikou 570228, China
2 Institute of Nanfan & Seed Industry, Guangdong Academy of Sciences, Guangzhou 510310, China
3 Hainan Yazhou Bay Seed Laboratory, Sanya Nanfan Research Institute of Hainan University, Sanya 572025, China
* Correspondence: chenyanli@hainanu.edu.cn (Y.C.); zhuguopeng@hainanu.edu.cn (G.Z.)
† These authors contributed equally to this work.

Abstract: An appropriate planting density could realize the maximum yield potential of crops, but the mechanism of sweet potato storage root formation in response to planting density is still rarely investigated. Four planting densities, namely D15, D20, D25, and D30, were set for 2-year and two-site field experiments to investigate the carbohydrate and lignin metabolism in potential storage roots and its relationship with the storage root number, yield, and commercial characteristics at the harvest period. The results showed that an appropriate planting density (D20 treatment) stimulated cambium cell differentiation, which increased carbohydrate accumulation and inhibited lignin biosynthesis in potential storage roots. At canopy closure, the D20 treatment produced more storage roots, particularly developing ones. It increased the yield by 10.18–19.73% compared with the control D25 treatment and improved the commercial features by decreasing the storage root length/diameter ratio and increasing the storage root weight uniformity. This study provides a theoretical basis for the high-value production of sweet potato.

Keywords: sweet potato; carbohydrate; lignin; storage root number; yield

Citation: Liang, Q.; Chen, H.; Chang, H.; Liu, Y.; Wang, Q.; Wu, J.; Liu, Y.; Kumar, S.; Chen, Y.; Chen, Y.; et al. Influence of Planting Density on Sweet Potato Storage Root Formation by Regulating Carbohydrate and Lignin Metabolism. *Plants* **2023**, *12*, 2039. https://doi.org/10.3390/plants12102039

Academic Editor: Lorenzo Rossi

Received: 22 February 2023
Revised: 14 May 2023
Accepted: 15 May 2023
Published: 19 May 2023

1. Introduction

Sweet potato (*Ipomoea batatas* L.) is the seventh most important crop around the world for its yield and cultivated area, and China is the largest sweet potato producer worldwide [1]. The yield of sweet potatoes is determined by the average storage root number per plant and the individual storage root weight [2]. The average storage root number per plant significantly contributes to the yield [3]. Improving the average storage root number per plant could lead to a good appearance and improve the commercial characteristics and yield of sweet potato storage roots [4]. Sweet potato storage root formation is a vital process determined by the degree of primary cambium development and stele cell lignification in adventitious roots [5]. The first clear sign of storage roots is the formation of secondary cambial cells (anomalous cambium) encircling the adventitious root's primary and secondary xylem elements [6]. Moreover, cambial cell proliferation forms starch-accumulating parenchyma cells in the root's vascular cylinder, accompanied by massive starch accumulation [7]. Simultaneously, carbohydrates provide energy for cambium cell development. The lignin content in potential storage roots is a common physiological indicator of sweet potato for the calculation of lignification [8]. The upregulation of carbohydrate metabolism and down-regulation of lignin biosynthesis usually facilitate sweet potato storage root formation [9]. Sucrose synthase (SuSy) catalyzes the reversible cleavage of sucrose into fructose and either uridine diphosphate glucose (UDP-G) or adenosine diphosphate glucose

(ADP-Glc) [10]. SPS catalyzes the conversion of fructose-6-phosphate and uridine diphosphateglucose (UDP-glucose) into sucrose-6-phosphate [11]. AGPase catalyzes the first step of starch biosynthesis by producing ADP-Glc and pyrophosphate (PPi) from Glc-1-P and ATP [12]. SSS acts to elongate linear chains, SBE promotes chain branching, and GBSS is an enzyme that is responsible for the elongation of amylose chains [13]. According to Du [14] and Si [15], the SuSy and SPS activity in young roots has great potential to promote storage root formation and increase the storage root number in sweet potato.

Meanwhile, increased AGPase, SSS, SBE, and GBSS activity promoted starch synthesis and deposition in the sink, resulting in sink bulking [16]. Lignin biosynthesis was initiated with the deamination of phenylalanine by phenylalanine ammonialyase (PAL), followed by a series of reactions that involved numerous enzymes, such as cinnamate 4-hydroxylase (C4H), 4-coumarate: CoA ligase (4CL), p-hydroxycinnamoyl-CoA: quinate shikimate p-hydroxycinnamoyltransferase (HCT), caffeoyl-CoAO-methyltransferase (CCoAOMT), and cinnamyl alcohol dehydrogenase (CAD) [17]. C4H and 4CL are two enzymes involved in phenylpropane synthesis, and HCT catalyzes the reactions by converting coumaroyl-CoA into coumaroylshikimate/quinate and caffeoyl shikimate/quinate into caffeoyl CoA. CCoAOMT further catalyzes the reaction by methylating caffeoyl CoA to feruloyl CoA, and CAD catalyzes the final step of lignin biosynthesis by converting the corresponding cinnamyl aldehydes into cinnamyl alcohols [18–20]. The downregulation of gene expression in lignin biosynthesis could reduce the lignin content in the plant tissue, including the root and stem [8,21]. Ibkn1, Ibkn2, and Ibkn3 are members of the class I knotted 1-like (KNOX-box) gene family, and they are expressed in primary cambium cells [8,15]. They regulate cell proliferation and differentiation in potential storage roots [22,23], and their expression in storage roots was two-fold higher than that in fiber roots [8,15].

Plant root system development is significantly affected by the planting density. The root architecture's response to planting density is mainly manifested in the alternation of the root length, root diameter, root biomass, and root number. In addition, this root system enables the plantlet to adapt to space and source competition by adjusting nutrition and water absorption [24]. A high planting density could cause crop logging by limiting lignin biosynthesis, decrease the root size, and limit the activity of root absorption for shoot development, ultimately causing a yield decline [24–26]. Under a low planting density, the source competition between individual plants could be alleviated, but the development between shoots and roots could lose its balance, which also limits yield formation [27]. Planting density is an important factor in regulating yield formation in the crop lifespan. However, the planting density of sweet potatoes is low in most areas of China at approximately 40,000–50,000 plants ha^{-1} (70–80 cm row space and 25–30 cm plant space). Therefore, increased planting density is a vital cropping measure to realize the maximum sweet potato yield potential [28,29]. An appropriate planting density is hypothesized to coordinate the relationship between shoot and root development and promote sweet potato storage root formation by stimulating carbohydrate accumulation and limiting lignin biosynthesis, ultimately increasing the storage root number. In the present study, two widely cultivated sweet potato cultivars, Yanshu 25 (YS-25) and Pushu 32 (PS-32), were used for 2-year and two-site field experiments. Four planting densities were set (D15, 15 cm plant distance and 83,280 plants ha^{-1}; D20, 20 cm plant distance and 62,520 plants ha^{-1}; D25, 25 cm plant distance and 50,025 plants ha^{-1}; and D30, 30 cm plant distance and 41,640 plants ha^{-1}; the row distance was set at 80 cm for all) to investigate carbohydrate and lignin metabolism in potential storage roots, cambium development, root morphology, the source–sink relationship during the storage root formation period, and the yield and its components. The storage roots' commercial characteristics in the harvest period were also investigated. This work presents a theoretical foundation for improved sweet potato productivity and storage root commercial quality under a suitable planting density in China.

2. Results

2.1. *Sweet Potato Storage Root Yields, Yield Components, and Appearance Quality*

The results of the 2-year field experiment showed that YS-25 and PS-32 elicited the same effect of the planting density on the storage root yield, the yield and its components, and the appearance quality (Table 1). The storage root diameter and average storage root weight significantly increased with the decrease in planting density. Furthermore, the storage root yield significantly increased under a higher planting density (D15 and D20; $p < 0.05$) compared with the control D25 treatment. The yield increased by 1.56–5.45% and 3.46–9.87% under the D15 treatment and by 10.93–19.73% and 10.18–14.13% under the D20 treatment in 2021 and 2022, respectively. In addition, the average number of storage roots per plant increased initially and subsequently decreased as the planting density decreased. Compared with that under the control D25 treatment, the average storage root number per plant reached the peak value under the D20 treatment at a significant difference level ($p < 0.05$). The lowest CV and L/D ratio were also observed in the D20 treatment.

Table 1. Effect of planting density on storage root yield, yield components, and appearance quality.

Years	Cultivar	Treatment †	SR Diameter (mm)	L/D Ratio	Average SR Weight (g)	CV (%)	Average SR Number Per Plant	Yield (kg·hm^{-2})	Yield Increment (%) ‡
2021 (Haikou)	YS-25	D15	36.95 b	2.7	104.56 d	5.64	3.33 c	29,017.72 b	5.45
		D20	38.75 a	2.5	111.13 c	1.77	4.67 a	33,030.38 a	19.73
		D25	39.39 a	2.6	127.26 b	11.87	4.33 ab	27,565.56 c	-
		D30	39.15 a	3.0	133.97 ab	5.49	4.33 ab	24,154.95 e	−14.11
	PS-32	D15	32.00 c	2.6	103.94 d	11.37	3.00 d	25,968.37 d	1.56
		D20	34.09 bc	2.5	113.57 c	5.74	4.00 b	28,401.60 bc	10.93
		D25	39.09 a	2.7	127.82 b	6.76	4.00 b	25,576.78 d	-
		D30	38.56 a	2.9	141.53 a	7.47	3.67 bc	21,628.44 f	−18.51
	ANOVA								
	C		7.22 *	-	2.46 *	-	2.78 ^{ns}	10.10 **	-
	T		5.21 **	-	3.43 *	-	3.07 *	11.39 ***	-
	C × T		1.66 ^{ns}	-	0.93 ^{ns}	-	0.41 ^{ns}	0.47 ^{ns}	-
2022 (Sanya)	YS-25	D15	38.04 c	2.6	88.50 d	7.69	5.26 bc	38,795.60 b	3.46
		D20	39.81 b	2.3	110.81c	2.90	6.19 a	42,883.33 a	14.13
		D25	39.72 b	2.4	131.66 b	6.71	5.71b	37,494.90 bc	-
		D30	42.71 a	2.7	170.61a	7.64	5.01 c	35,592.04 c	−5.34
	PS-32	D15	36.68 c	2.4	91.91 d	5.00	4.67 d	35,569.17 c	9.87
		D20	38.14 b	2.3	99.40 cd	2.06	5.76 b	35,795.45 c	10.18
		D25	38.97 b	2.6	131.42 b	4.40	4.93 c	32,437.80 d	-
		D30	40.47 b	3.1	160.09 a	6.61	4.83 cd	32,197.00 d	−0.62
ANOVA									
C			3.95 ^{ns}	-	1.65 ^{ns}	-	9.27 **	10.63 **	-
T			3.14 *	-	86.25 ***	-	10.11 **	12.36 **	-
C × T			8.36 ^{ns}	-	1.12 ^{ns}	-	0.61 ^{ns}	0.56 *	-

Note: SR, storage root; Y, year; C, cultivar; T, treatment; YS-25, Yanshu25; PS-32, Pushu32. Two-way ANOVA, LSD. Values followed by different letters present significant differences among planting density treatments ($p < 0.05$). * $p < 0.05$; ** $p < 0.01$; *** $p < 0.001$; ^{ns}, no significance. † D15, 15 cm plant distance and 83,280 plants ha^{-1}; D20, 20 cm plant distance and 62,520 plants ha^{-1}; D25, 25 cm plant distance and 50,025 plants ha^{-1}; D30, 30 cm plant distance and 41,640 plants ha^{-1}. The row distance was set at 80 cm for all. ‡ Compared with the D25 control treatment.

2.2. *Commercial Storage Root Characteristics*

The results of the two-year field experiment depicted that the commercial storage root weight and large storage roots significantly increased with the reduction in planting density ($p < 0.05$). However, compared with the control D25 treatment, the D20 treatment significantly increased the commercial storage root number, primary medium storage root number, and commercial storage root yield ($p < 0.05$). In addition, the commercial storage root yields increased by 23.35–66.42% and 19.02–33.46% in 2021 and 2022, respectively (Table 2).

Table 2. Effect of planting density on storage root commercial characteristics at harvest period.

Years	Cultivars	Treatment †	Commercial SR Weight Per Plant (g)	Commercial SR Number Per Plant	Large SR Number	Medium SR Number	Small SR Number	Commercial SR Yield ($kg{\cdot}hm^{-2}$)	Increment (%) ‡
2021 (Haikou)	YS-25	D15	122.61 b	2.33 d	0.33 c	1.67 c	0.33 a	19,558.23 b	−0.51
		D20	129.91 b	3.00 b	1.00 b	2.00 b	-	24,367.17 a	23.35
		D25	158.45 a	2.33 d	1.33 a	1.00 e	-	19,697.51 b	-
		D30	162.41 a	2.67 c	1.33 a	1.33 d	-	16,998.00 c	−16.16
	PS-32	D15	115.20 c	2.00 e	1.00 b	1.00 e	-	19,202.65 b	43.28
		D20	119.26 c	3.33 a	1.00 b	2.33 a	-	22,368.40 a	66.42
		D25	124.90 b	2.33 d	1.00 b	1.33 d	-	13,377.52 d	-
		D30	133.31 b	2.33 d	1.33 a	0.67 f	0.33 a	12,952.40 d	−3.24
ANOVA									
C			13.67 **	0.17 ns	0.00 ns	0.80 ns	0.25 ns	4.61 ns	-
T			8.33 **	4.61 *	0.67 ns	2.93 ns	2.92 ns	5.35 *	-
C × T			5.21 ns	0.61 ns	1.33 ns	7.20 **	1.58 ns	0.40 ns	-
2022 (Sanya)	YS-25	D15	98.89 bc	4.00 c	-	2.00 c	2.00 a	33,272.02 a	29.60
		D20	117.89 bc	4.67 b	0.67 b	3.00 a	1.00 c	34,342.08 a	33.46
		D25	136.87 b	4.00 c	0.33 c	2.00 c	1.67 b	25,676.50 de	-
		D30	201.50 a	3.00 e	1.00 a	1.33 d	0.67 d	25,198.06 e	−1.95
	PS-32	D15	92.96 c	3.67 cd	-	1.67 c	2.00 a	28,412.07 c	5.97
		D20	102.05 bc	5.00 a	0.33 c	3.00 a	1.67 b	31,899.79 b	19.02
		D25	135.20 b	4.00 c	0.33 c	2.67 b	1.00 c	26,851.42 d	-
		D30	191.99 a	3.33 d	1.00 a	2.00 c	0.33 e	26,533.47 d	−1.13
ANOVA									
C			0.88 ns	0.40 ns	0.25 ns	3.00 ns	0.25 ns	0.13 ns	-
T			26.02 ***	7.07 **	6.25 **	17.2 ***	13.58 ***	7.23 **	-
C × T			0.12 ns	0.13 ns	0.25 ns	3.00 ns	2.92 ns	0.41 ns	-

SR, storage root; Y, year; C, cultivar; T, treatment; YS-25, Yanshu25; PS-32, Pushu32. Two-way ANOVA, LSD was used. Values followed by different letters present significant differences among planting density treatments ($p < 0.05$). * $p < 0.05$; ** $p < 0.01$; *** $p < 0.001$; ns, no significance. † D15, 15 cm plant distance and 83,280 plants ha^{-1}; D20, 20 cm plant distance and 62,520 plants ha^{-1}; D25, 25 cm plant distance and 50,025 plants ha^{-1}; D30, 30 cm plant distance and 41,640 plants ha^{-1}. The row distance was set at 80 cm for all. ‡ Compared with the D25 control treatment.

2.3. Storage Root Traits at Closure Period

During the canopy closure period, the 2-year field experiment showed a similar pattern of planting density on the storage root traits (Table 3). The average storage root weight, diameter, and mature storage root number significantly increased with the reduction in planting density ($p < 0.05$). Meanwhile, the storage root number per plant and the developing storage root number did not significantly increase at a higher planting density (D15 and D20) compared with those under the control D25 treatment ($p < 0.05$). Moreover, the developing storage root number was highest in the D20 treatment ($p < 0.05$).

Table 3. Effect of planting density on storage root traits at canopy closure period.

Years	Cultivar	Treatment †	SR Weight (g)	SR Number Per Plant	Young SR Number ($2 < \Phi < 5$ mm)	Developing SR Number ($5 < \Phi < 20$ mm)	Mature SR Number ($\Phi > 20$ mm)	SR Diameter (mm)
2021 (Haikou)	YS-25	D15	1.35 f	4.67 b	2.33 a	2.33 c	-	5.14 d
		D20	2.60 c	5.67 a	1.67 b	4.00 a	-	7.84 bc
		D25	4.87 a	4.33 c	0.67 e	3.67 ab	-	8.07 bc
		D30	4.70 a	3.67 d	0.67 e	3.00 b	-	10.62 a
	PS-32	D15	1.97 e	3.67 d	2.00 b	1.67 e	-	4.91 d
		D20	2.23 d	4.33 c	1.33 c	2.33 c	0.67a	7.13 c
		D25	3.15 bc	4.00 bc	2.00 b	2.00 d	-	8.71 b
		D30	3.47 b	3.33 e	1.00 d	2.33 c	-	8.35 bc
ANOVA								
C			3.30 ns	8.33 **	1.80 ns	39.20 ***	4.00 ns	0.79 ns
T			10.38 ***	4.56 **	8.73 **	6.93 **	4.00 *	7.02 **
C × T			1.88 ns	8.00 ns	4.47 *	4.00 *	4.00 *	0.73 ns

Table 3. *Cont.*

Years	Cultivar	Treatment †	SR Weight (g)	SR Number Per Plant	Young SR Number (2 < Φ < 5 mm)	Developing SR Number (5 < Φ < 20 mm)	Mature SR Number (Φ > 20 mm)	SR Diameter (mm)
2022 (Sanya)	YS-25	D15	14.27 e	5.00 c	0.33 c	3.67 c	1.00 f	13.00 e
		D20	16.98 d	7.00 a	0.67 b	4.33 b	2.00 c	15.10 de
		D25	24.60 b	5.00 cd	-	2.67 e	2.33b	23.47 b
		D30	34.08 a	4.00 e	-	1.00 f	3.33 a	27.54 a
	PS-32	D15	7.40 f	5.67 b	2.00 a	2.67 e	1.00 f	13.38 e
		D20	13.35 e	6.67a	-	5.33 a	1.33 d	16.45 d
		D25	19.34 c	4.67 d	-	3.00 d	1.67 cd	16.65 d
		D30	20.53 c	3.67 f	-	1.33 f	2.33 b	19.28 c
ANOVA								
C			4.56 **	10.36 *	1.33 ***	1.50 ^{ns}	12.80 **	1.23 ^{ns}
T			3.33 *	8.33 **	1.20 ***	49.50 ***	17.33 ***	2.05 **
C × T			2.69 ^{ns}	6.23 ^{ns}	1.33 ***	4.61 *	1.60 ^{ns}	2.45 ^{ns}

SR, storage root; Y, year; C, cultivar; T, treatment; YS-25, Yanshu25; PS-32, Pushu32. Two-way ANOVA, LSD was used. Values followed by different letters present significant differences among planting density treatments ($p < 0.05$). * $p < 0.05$; ** $p < 0.01$; *** $p < 0.001$; ^{ns}, no significance. † D15, 15 cm plant distance and 83,280 plants ha^{-1}; D20, 20 cm plant distance and 62,520 plants ha^{-1}; D25, 25 cm plant distance and 50,025 plants ha^{-1}; D30, 30 cm plant distance and 41,640 plants ha^{-1}. The row distance was set at 80 cm for all.

2.4. Effect on Root Development System and Potential Storage Root Traits

The planting density significantly affected sweet potato's adventitious root formation and potential storage root development (Table 4). Compared with the control D25 treatment, the D20 treatment significantly improved the adventitious root number, potential storage root diameter, weight, and ratio of potential storage root weight during the storage root initiation stage (during 0–25 days after planting; $p < 0.05$). By contrast, these parameters slightly decreased or were similar to those at the lower planting density 35 days after planting. However, the total root fresh weight significantly increased with the decrease in planting density during the storage root formation period ($p < 0.05$).

Table 4. Effect of planting density on root traits during storage root formation (2022 Sanya).

DAP(d)	Cultivar	Treatment †	Adventitious Root Number	Total Root Fresh Weight (g)	Potential SR Diameter (mm)	Potential SR Weight (g)	Potential SR Weight Ratio (%)
15 d	YS-25	D15	13.60 b	2.84 e	1.03 c	1.75 b	61.62 a
		D20	15.20 b	3.44 c	1.16 b	1.96 ab	56.97 ab
		D25	15.20 b	3.69 b	1.12 b	1.89 ab	51.22 b
		D30	13.80 b	4.29 a	1.05 c	2.07 a	48.25 b
	PS-32	D15	17.60 a	2.95 de	1.29 ab	1.64 c	55.59 a
		D20	19.00 a	3.10 d	1.41 a	1.84 ab	59.35 a
		D25	18.20 a	3.64 bc	1.25 ab	1.75 b	48.08 b
		D30	18.00 a	3.80 b	1.26 ab	1.76 b	46.31 b
ANOVA							
C			66.57 ***	12.59 ***	35.50 ***	20.55 ***	6.30 *
T			2.28 ^{ns}	81.94 ***	2.86 *	7.38 **	14.15 ***
C × T			0.33 ^{ns}	6.21 **	0.67 ^{ns}	1.62 ^{ns}	1.13 ^{ns}
25 d	YS-25	D15	14.00 bc	6.11 b	1.64 c	3.96 a	64.32 a
		D20	15.00 b	6.59 a	2.39 a	3.84 ab	58.27 b
		D25	13.20 c	6.74 a	2.00 b	3.21 c	47.62 c
		D30	14.00 bc	6.94 a	1.96 b	2.75 d	39.62 d
	PS-32	D15	16.60 ab	5.19 c	1.59 c	3.40 b	65.51 a
		D20	17.40 a	5.37 c	2.20 a	3.47 b	64.61 a
		D25	15.40 b	5.41c	2.19 a	3.44 b	63.58 a
		D30	15.20 b	6.18b	2.19 a	3.76 ab	60.84 a

Table 4. *Cont.*

DAP(d)	Cultivar	Treatment †	Adventitious Root Number	Total Root Fresh Weight (g)	Potential SR Diameter (mm)	Potential SR Weight (g)	Potential SR Weight Ratio (%)
ANOVA							
C			55.98 ***	112.96 ***	1.86 ns	0.61 ns	0.92 ns
T			5.14 **	14.29 ***	16.12 ***	4.61 **	8.51 ***
C × T			0.38 ns	12.04 ***	3.94 *	1.81 ns	5.60 ***
35 d	YS-25	D15	-	28.90 cd	8.77 c	4.32 c	14.94 c
		D20	-	36.94 b	8.81 c	8.21 a	22.22 a
		D25	-	39.83 ab	10.75 b	8.69 a	21.81 a
		D30	-	39.79 ab	9.66 b	7.38 ab	18.54 b
	PS-32	D15	-	21.19 d	9.27 bc	4.39 c	20.72 a
		D20	-	31.53 c	10.67 b	6.77 b	21.47 a
		D25	-	35.80 b	11.13 ab	7.83 ab	21.87 a
		D30	-	41.26 a	13.43 a	7.70 ab	18.66 b
ANOVA							
C			-	2.12 ns	14.91 ***	2.25 ns	0.64 ns
T			-	12.43 ***	6.92 ***	49.01 ***	20.86 ***
C × T			-	2.96 **	2.67 ns	1.64 ns	6.21 **

DAP, days after planting; SR, storage root; Y, year; C, cultivar; T, treatment; YS-25, Yanshu25; PS-32, Pushu32. Two-way ANOVA, LSD were used. Values followed by different letters present significant differences among planting density treatments ($p < 0.05$). * $p < 0.05$; ** $p < 0.01$; *** $p < 0.001$; ns, no significance. † D15, 15 cm plant distance and 83,280 plants ha^{-1}; D20, 20 cm plant distance and 62,520 plants ha^{-1}; D25, 25 cm plant distance and 50,025 plants ha^{-1}; D30, 30 cm plant distance and 41,640 plants ha^{-1}. The row distance was set at 80 cm for all.

2.5. Effect on Dry Biomass Accumulation and Allocation

A high accumulation and allocation ratio of dry weight in the root system benefits storage root formation. The results indicated that the total plant dry weight and the root and shoot dry weight significantly increased with the increment in the plant distance ($p < 0.05$, Table 5). However, the allocation ratio of root dry weight and the root/shoot ratio dramatically increased at a higher planting density (D15 and D20) and reached the peak values under the D20 treatment ($p < 0.05$). The allocation ratio of shoot dry weight had no significant difference among treatments at 15–25 days after planting ($p > 0.05$), but it significantly decreased under the D20 treatment at 35 days after planting ($p < 0.05$).

Table 5. Effect of planting density on dry matter accumulation and allocation during storage root formation (2022 Sanya).

DAP (d)	Cultivar	Treatment †	Total Plant Dry Weight (g)	Root Dry Weight (g)	Shoot Dry Weight (g)	Root Dry Weight Allocation (%)	Shoot Dry Weight Allocation (%)	Root/Shoot Ratio
15 d	YS-25	D15	1.98 c	0.21 d	1.77 d	10.61bc	89.39 a	0.12 b
		D20	2.40 b	0.32 a	2.08 bc	13.33 a	86.67 a	0.15 a
		D25	2.55 a	0.24 c	2.31 a	9.41 cd	90.59 a	0.10 c
		D30	2.56 a	0.28 b	2.28 a	9.94 c	89.06 a	0.11 c
	PS-32	D15	2.04 c	0.23 c	1.86 cd	8.83 d	91.17 a	0.10 c
		D20	2.29 bc	0.31 a	1.98 c	13.54 a	86.46 a	0.16 a
		D25	2.42 b	0.27 b	2.15 b	11.16 b	88.84 a	0.13 b
		D30	2.25 bc	0.28 b	2.05 bc	8.89 d	91.11 a	0.10 c
ANOVA								
C			4.87 *	2.15 ns	6.59 *	10.39 **	8.75 **	11.37 **
T			3.43 *	27.47 ***	2.02 ns	21.20 ***	22.02 ***	19.72 ***
C × T			19.17 ***	1.73 ns	20.93 **	10.25 ***	10.49 ***	8.52 ***
25 d	YS-25	D15	4.23 c	0.62 b	3.61 c	14.80 ab	85.20 a	0.17 b
		D20	4.49 c	0.71 a	3.78 c	15.82 a	84.12 a	0.19 a
		D25	5.06 b	0.67 a	4.39 b	13.00 c	87.00 a	0.15 c
		D30	5.67 b	0.63 b	5.04 a	11.20 d	88.20 a	0.12 d
	PS-32	D15	4.14 c	0.58 c	3.56 c	14.00 b	86.00 a	0.16 bc
		D20	4.27 c	0.61 b	3.66 c	14.20 b	85.80 a	0.16 bc
		D25	4.51 c	0.62 b	3.89 c	11.00 d	89.00 a	0.16 bc
		D30	6.20 a	0.69 a	5.51 a	11.20 d	88.80 a	0.12 d

Table 5. *Cont.*

DAP (d)	Cultivar	Treatment †	Total Plant Dry Weight (g)	Root Dry Weight (g)	Shoot Dry Weight (g)	Root Dry Weight Allocation (%)	Shoot Dry Weight Allocation (%)	Root/Shoot Ratio
ANOVA								
C			0.52 ^{ns}	5.95 *	0.20 ^{ns}	0.64 ^{ns}	0.64 ^{ns}	1.29 ^{ns}
T			47.21 **	4.64 **	46.85 ***	20.81 ***	20.81 ***	21.56 ***
C × T			3.96 *	6.61 ***	3.17 *	2.29 ^{ns}	2.29 ^{ns}	1.56 ^{ns}
35 d	YS-25	D15	13.44 e	3.58 d	9.88 d	26.60 c	73.40 b	0.36 c
		D20	18.23 c	4.91 c	13.32 c	27.00 c	73.00 b	0.37 c
		D25	23.01 b	5.04 c	17.96 b	22.00 d	78.00 a	0.28 d
		D30	25.30a	5.34 bc	19.95 a	21.20 d	78.80 a	0.27 d
	PS-32	D15	15.34 d	4.76 c	10.08 d	31.00 b	69.00 c	0.47 b
		D20	15.69 d	5.95 b	9.74 d	37.60 a	62.40 d	0.61 a
		D25	23.62 b	7.08 a	16.53 b	30.00 b	70.00 bc	0.43 bc
		D30	24.34 ab	6.84 a	17.49 b	28.20 c	71.80 bc	0.39 c
ANOVA								
C			0.51 ^{ns}	72.60 ***	24.68 ***	83.33 ***	83.33 ***	28.50 ***
T			208.66 ***	28.38 ***	150.35 ***	16.80 ***	16.80 ***	5.09 **
C × T			7.75 **	1.69 ^{ns}	7.11 ***	2.44 ^{ns}	2.44 ^{ns}	1.67 ^{ns}

DAP, days after planting; SR, storage root; Y, year; C, cultivar; T, treatment; YS-25, Yanshu25; PS-32, Pushu32. Two-way ANOVA, LSD was used. Values followed by different letters present significant differences among planting density treatments ($p < 0.05$). * $p < 0.05$; ** $p < 0.01$; *** $p < 0.001$; ^{ns}, no significance. † D15, 15 cm plant distance and 83,280 plants ha^{-1}; D20, 20 cm plant distance and 62,520 plants ha^{-1}; D25, 25 cm plant distance and 50,025 plants ha^{-1}; D30, 30 cm plant distance and 41,640 plants ha^{-1}. The row distance was set at 80 cm for all.

2.6. Influence on Expression of Genes Associated with Carbohydrate and Lignin Metabolism

The expression of genes' regulating sucrose enzymolysis and starch synthesis had a similar pattern during the storage root formation period. The gene expression levels of *SuSy*, *AGPase*, *SPS*, *SSS*, *SBE1*, and *GBSS* in the D20 treatment were significantly enhanced compared with those in the control D25 treatment. Furthermore, a significant decrease in the expression of these genes was observed as the planting density decreased ($p < 0.05$; Figure 1).

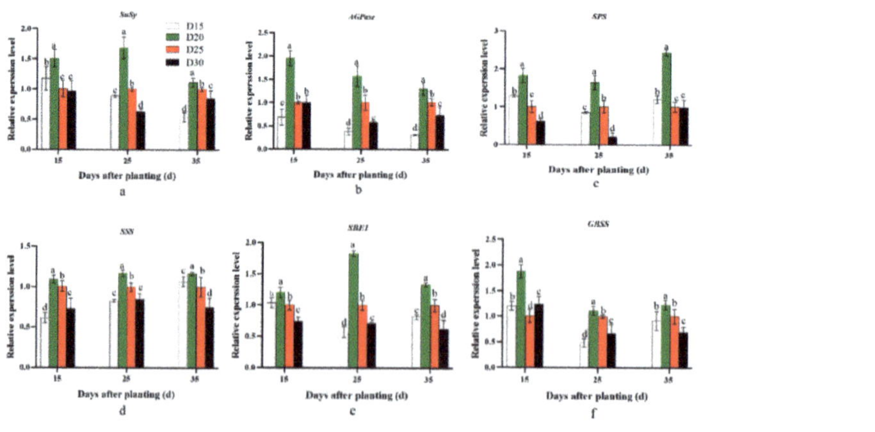

Figure 1. Effect of planting density on the expression of carbohydrate metabolism genes, namely *SuSy* (**a**), *AGPase* (**b**), *SPS* (**c**), *SSS* (**d**), *SBE1* (**e**), and *GBSS* (**f**), in YS-25 during storage root formation period (2022 Sanya). D15, 15 cm plant distance and 83,280 plants ha^{-1}; D20, 20 cm plant distance and 62,520 plants ha^{-1}; D25, 25 cm plant distance and 50,025 plants ha^{-1}; D30, 30 cm plant distance and 41,640 plants ha^{-1}. The row distance was set at 80 cm for all. 15 d, 15 days after planting; 25 d, 25 days after planting; 35 d, 35 days after planting. Error bars represent 1 SD (n = 3) within the same column, and different letters indicate a significant difference between treatments ($p < 0.05$).

Differences were observed in the expression of *PAL*, *C4H*, *CCoAOMT*, *HCT*, *4CL*, and *CAD*. The expression of these genes (referring to lignin biosynthesis) in the D20 treatment significantly decreased compared with that in the control D25 treatment ($p < 0.05$). However, a significant increase was observed with a further decrease in the planting density ($p < 0.05$; Figure 2).

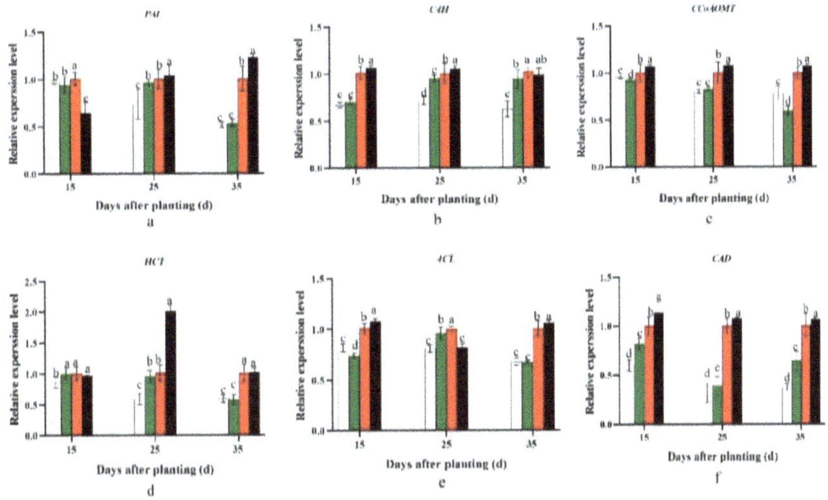

Figure 2. Effect of planting density on expression levels of lignin metabolism genes, namely *PAL* (**a**), *C4H* (**b**), *CCoAOMT* (**c**), *HCT* (**d**), *4CL* (**e**), and *CAD* (**f**), in YS-25 during storage root formation period (2022 Sanya). D15, 15 cm plant distance and 83,280 plants ha^{-1}; D20, 20 cm plant distance and 62,520 plants ha^{-1}; D25, 25 cm plant distance and 50,025 plants ha^{-1}; D30, 30 cm plant distance and 41,640 plants ha^{-1}. The row distance was set at 80 cm for all. 15 d, 15 days after planting; 25 d, 25 days after planting; 35 d, 35 days after planting. Error bars represent 1 SD (n = 3) within the same column, and different letters indicate a significant difference between treatments ($p < 0.05$).

2.7. Effect on Carbohydrate and Lignin Content

The starch and sucrose content in the potential storage root had the same effect. Compared with the control, the starch and sucrose content in the potential storage root significantly increased in the D20 treatment ($p < 0.05$) and decreased with a reduction in planting density. Furthermore, the increment in starch and sucrose content at 35 days after planting was more dramatic than at 15 and 25 days after planting (Figure 3a,b).

The lignin content in the potential storage root significantly increased with the reduction in planting density ($p < 0.05$), but the content in each treatment gradually decreased with the prolongation of the planting period ($p < 0.05$; Figure 3c).

The ratios of starch to sucrose and starch to lignin showed a similar trend. The value was significantly higher in D20 than in the control at 25 and 35 days after planting ($p < 0.05$). Meanwhile, 15 days after planting, the ratio of starch to lignin was steadily reduced as the planting density decreased ($p < 0.05$), whereas the ratio of starch to sucrose showed the reverse trend ($p < 0.05$; Figure 3d,e).

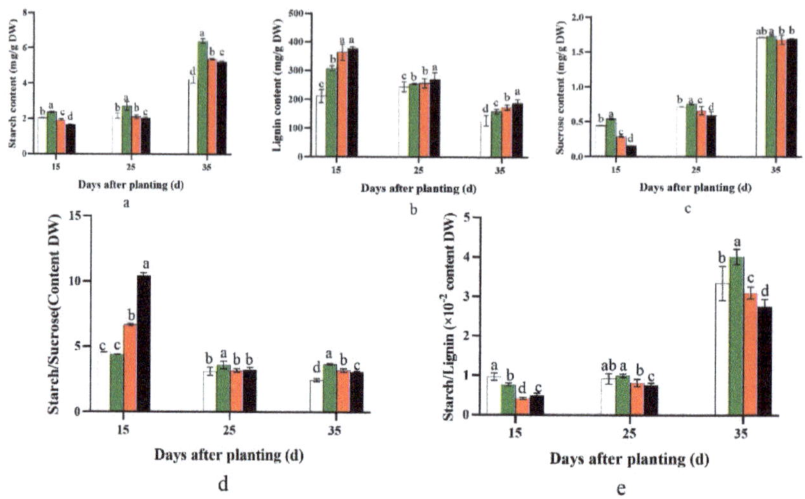

Figure 3. Effect of planting density on starch (**a**), sucrose (**b**), and lignin content (**c**), and ratios of starch/sucrose (**d**) and starch/lignin (**e**), in YS-25 during storage root formation period (2022 Sanya). D15, 15 cm plant distance and 83,280 plants ha^{-1}; D20, 20 cm plant distance and 62,520 plants ha^{-1}; D25, 25 cm plant distance and 50,025 plants ha^{-1}; D30, 30 cm plant distance and 41,640 plants ha^{-1}. The row distance was set at 80 cm for all. 15 d, 15 days after planting; 25 d, 25 days after planting; 35 d, 35 days after planting. Error bars represent 1 SD (n = 3) within the same column, and different letters indicate a significant difference between treatments ($p < 0.05$).

2.8. *Influence on Genes Involved in Cambium Development and Potential Storage Root Anatomy*

Ibkn1, *Ibkn2*, and *Ibkn3*, which regulate potential storage root cambium development, were significantly higher in the D20 treatment than in the control D25 treatment at 0–25 days after planting. However, the expression of *Ibkn2* and *Ibkn3* significantly decreased 25–35 days after planting, and *Ibkn1* was similar to the control D25 treatment. Furthermore, the gene expression levels were significantly reduced when the planting density was further decreased (Figure 4a–c). Consequently, the D20 treatment significantly enhanced the number of protoxylems and secondary xylem, the diameter of the stele and potential root, and the cross-sectional area of the stele and potential storage root 15 days after planting compared with the control D25 treatment (Figures 4d–f and 5).

2.9. *Correlation Analysis of Storage Root Number Per Plant, Fresh Weight, and Yield with Relative Gene Expression*

The correlation analysis demonstrated that the storage root number per plant was positively correlated with *SuSy*, *SPS* ($p < 0.05$), *SSS*, *AGPase*, *GBSS* ($p < 0.01$), and *SBE*, which are involved in carbohydrate metabolism. Furthermore, no significant positive correlation was found in *PAL*, *4CL*, *C4H*, *CCoAOMT*, *CAD*, and *HCT*, which are involved in lignin biosynthesis, whereas *C4H* and *HCT* were found to be negatively correlated. *Ibkn1*, *Ibkn2*, and *Ibkn3*, which regulate cambium development, were also positively correlated with the storage root number per plant but with no significant difference (Table 6). The storage root fresh weight was insignificantly negatively correlated with all genes ($p > 0.05$; Table 6), and the yield was positively correlated with the genes involved in carbohydrate metabolism and cambium development and almost negatively correlated with the genes involved in lignin biosynthesis (Table 6).

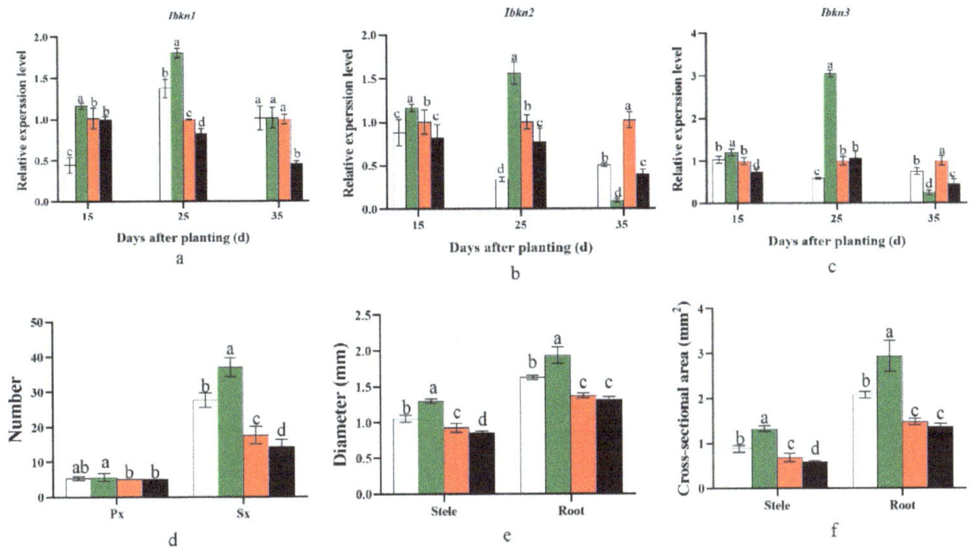

Figure 4. Effect of planting density on gene expression levels of *Ibkn1* (**a**), *Ibkn2* (**b**), and *Ibkn3* (**c**) during storage root formation period. Analysis of the potential storage root section anatomy, including the number of Px and Sx (**d**), the diameter of the stele and root (**e**), and the cross-sectional area of the stele and root (**f**) at 15 days after planting. D15, 15 cm plant distance and 83,280 plants ha^{-1}; D20, 20 cm plant distance and 62,520 plants ha^{-1}; D25, 25 cm plant distance and 50,025 plants ha^{-1}; D30, 30 cm plant distance and 41,640 plants ha^{-1}. The row distance was set at 80 cm for all. 15 d, 15 days after planting; 25 d, 25 days after planting; 35 d, 35 days after planting. Error bars represent SD (n = 3) within the same column, and different letters indicate a significant difference between treatments ($p < 0.05$).

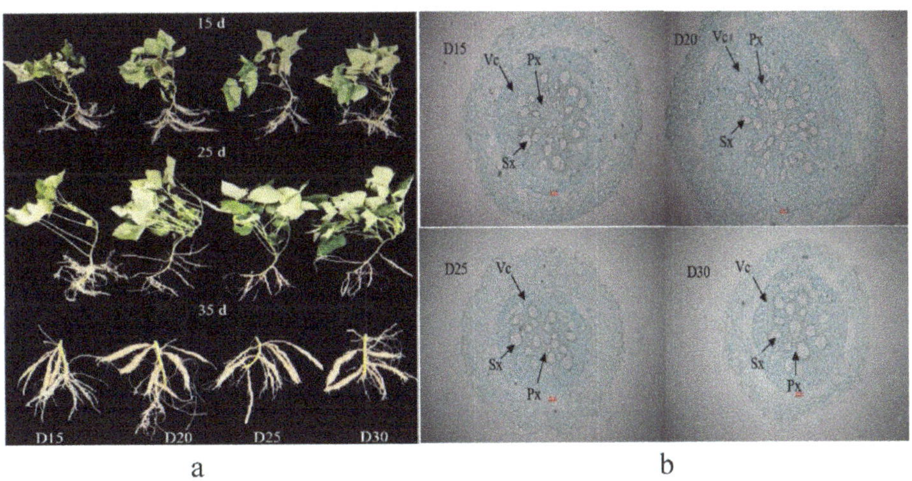

Figure 5. Effect of planting density on sweet potato YS-25 root growth and development during storage root formation period (**a**) and cambium development at 15 days after planting (**b**). D15, 15 cm plant distance and 83,280 plants ha^{-1}; D20, 20 cm plant distance and 62,520 plants ha^{-1}; D25, 25 cm plant distance and 50,025 plants ha^{-1}; D30, 30 cm plant distance and 41,640 plants ha^{-1}. The row distance was set at 80 cm for all. 15 d, 15 days after planting; 25 d, 25 days after planting; 35 d, 35 days after planting. Px, protoxylem; Sx, secondary xylem; Vc, vascular cambium. Scale bar = 100 µm.

Table 6. Correlation analysis of storage root number, weight, and yield with gene expression.

	SuSy	*SPS*	*SSS*	*AGPase*	*GBSS*	*SBE*	*PAL*	*4CL*
SRN †	0.44	0.61 *‡	0.165	0.31	0.72 **	0.30	0.27	0.33
SRFW	−0.39	−0.39	−0.32	−0.23	−0.09	−0.38	−0.26	−0.05
Yield	0.44	0.62 *	0.43	0.29	0.33	0.47	−0.11	0.00
	C4H	*CCoAOMT*	*CAD*	*HCT*	*Ibkn1*	*Ibkn2*	*Ibkn3*	-
SRN	−0.20	0.62	0.17	−0.14	0.41	0.39	0.25	-
SRFW	−0.21	−0.41	−0.04	−0.03	−0.35	−0.10	−0.27	-
Yield	−0.24	0.16	−0.43	−0.25	0.51	0.08	0.32	-

† SRN, storage root number per plant; SRFW, storage root fresh weight. ‡, * and ** indicate significant differences at $p < 0.05$ and $p < 0.01$, respectively.

3. Discussion

3.1. Effect of Planting Density on Carbohydrate and Lignin Metabolism and Cambium Cell Development

The upregulation of starch biosynthesis and downregulation of lignin biosynthesis in potential storage roots are the main events during the storage root formation period [8,15,22], probably because the activity of carbon flow towards carbohydrate metabolism is stronger than the delivery into phenylpropanoid biosynthesis during the storage root formation period [14,30]. Previous findings have shown that low sucrose and high hexose content in the root system could facilitate storage root formation [14,31]. This process could be explained by the fact that sucrose, the primary type of carbohydrate, is delivered from the source leaves to the root through the phloem, where it is subsequently broken down into hexoses to supply energy and the carbon skeletons needed for cell growth and root swelling. Sucrose is decomposed into hexoses by the SuSy and sucrose invertase pathways, and the former is the main source for fructose and UDP-glucose biosynthesis, which provides the substrate for starch biosynthesis [14,31]. Crop root development is adversely correlated with lignin deposition, and lignin accumulation has a negative effect on root development [8,32].

Furthermore, lignin and cellulose content have a negative correlation with starch accumulation in cassava tuber root formation [33,34]. Lignin content is a common physiological indicator of sweet potato lignification, which strongly inhibits sweet potato storage root formation [8,33,34]. In the present study, compared with the control D25 treatment, the D20 treatment significantly promoted sucrose and starch biosynthesis by significantly increasing the expression levels of *SuSy*, *SPS*, *SSS*, *AGPase*, *SBE1*, and *GBSS* (Figure 1). By contrast, it inhibited lignin biosynthesis by dramatically decreasing the *PAL*, *C4H*, *CCoAOMT*, *HCT*, *4CL*, and *CAD* expression levels (Figure 2). In addition, the ratios of starch to sucrose and starch to lignin were significantly higher in D20 than in the control D25 treatment at 15–35 days after planting, and significantly higher or lower at 0–15 days after planting, respectively (Figure 3). Previous studies revealed that sweet potato storage root formation was determined by stele cell lignification and cambium proliferation [5,6]. Furthermore, only the appearance of anomalous cambium could prevent stele lignification and favor storage root formation [33,34]. The KNOX1 protein stimulates cell differentiation by negatively regulating lignin biosynthesis [35]. *Ibkn1*, *Ibkn2*, and *Ibkn3* belong to the KNOX gene family, and their expression in storage root was two-fold higher than that in fiber roots [36]. In the present study, the D20 treatment promoted primary cambium cell development by upregulating *Ibkn1*, *Ibkn2*, and *Ibkn3* expression 0–25 days after planting (Figure 4). Therefore, 15 days after planting, the D20 treatment significantly increased protoxylems and secondary xylem, the stele and potential storage root diameter, and the stele and potential storage root cross-sectional area compared with the control D25 treatment. The correlation analysis found that the genes involved in carbohydrate metabolism were positively correlated with the storage root number per plant, and *SPS* and *GBSS* reached a significant level ($p < 0.05$ and $p < 0.01$, respectively). However, no significant positive correlation was found with the lignin biosynthesis genes, and *C4H* and *HCT* were negatively correlated (Table 6). Furthermore, the storage root fresh weight was negatively correlated with all genes' expression, and the yield was positively correlated with the genes involved in carbohydrate metabolism and cambium development and almost negatively correlated with the genes involved in lignin biosynthesis. The correlation analysis strongly improved the theoretical understanding of these processes; specifically, upregulated carbohydrate and downregulated lignin biosynthesis could improve the yield by promoting storage root formation in sweet potatoes.

3.2. Effect of Planting Density on Plant Dry Matter Dynamic and Root Development

Coordinating the shoot and root relationship is a considerable factor in promoting storage root formation. The formation of sweet potato storage roots comprises four main events: the initiation of cambial cells, cell division, cell expansion and growth, and carbohydrate storage [36,37]. Major developmental activities strongly rely on dry matter accumulation in roots. The current study indicated that the total plant dry matter significantly increased with the decrease in planting density. Furthermore, compared with the control D25 treatment, the D20 treatment significantly increased the root dry weight, root dry weight ratio, and R/T ratio. By contrast, the shoot dry matter weight significantly decreased during the storage root formation period. However, the shoot dry matter ratio was similar 0–25 days after planting and it dramatically decreased 35 days after planting (Table 5). Root development is primarily measured by the number, weight, diameter, and length of the roots, among other parameters, and these roots morphology could indicate environmental changes, such as stress, the degree of nutrition and water uptake, and, more crucially, the strength of root differentiation in sweet potatoes [38]. The number of adventitious roots controls the quantity of storage roots during the early growth stage; the more adventitious roots, the more storage roots that form [15,38]. The root diameter is an important parameter to indicate the degree of root thickening caused by root cambium cell division and differentiation. An adventitious root with a length of more than 20 cm could facilitate storage root formation [39]. The results of the present study indicated that an appropriate planting density under the D20 treatment significantly increased the adventitious root number but decreased the total root fresh weight. The potential storage root parameters, namely the potential storage root diameter and potential storage root weight, significantly increased in the D20 treatment at 0–25 days after planting and decreased at 25–35 days after planting compared with those in other treatments. Furthermore, the potential storage root weight ratio was always significantly higher in the D20 treatment than in other treatments (Table 4).

3.3. Effect of Planting Density on Storage Root Yield, Components, and Commercial Characteristics

Altering the planting density is an effective practice in cereal crop production, balancing the relationship between the kernel number per spike and the average weight per hundred kernels [40]. A higher planting density seems inclined to support kernel formation while reducing the average weight per hundred kernels [41]. The maximum output from cereal crops could only be reached by balancing the number of kernels per spike and the average weight of a hundred kernels. Sweet potato storage root yields were determined by the average storage root number per plant and average storage root weight, and the storage root per plant was the most significant contributing factor to the yield [15,38]. An increased storage root number could lead to a good appearance and improve the commercial traits of this crop [38]. The canopy closure period is a key factor in determining the storage root number of sweet potatoes, and the storage root number is known at this point [15,38]. Hence, the maximum storage root number at the canopy closure period could result in a high sweet potato yield. The current study showed that an appropriate planting density (D20 treatment) could significantly increase the storage root number (mainly the developing storage root number) compared with the control D25 treatment.

By contrast, the storage root weight, mature storage root number, and storage root diameter remarkably increased during the canopy closure period as the planting density decreased (Table 3). Therefore, compared with the control D25 treatment, the appropriate planting density under the D20 treatment significantly increased the yield by dramatically increasing the storage root number per plant, instead of the storage root diameter and average storage root weight. As a result, it improved the storage roots' commercial characteristics by significantly increasing the number of commercial and middle-sized storage roots and reducing the storage root weight CV and L/D ratio to create a good-quality appearance (Tables 1 and 2).

4. Materials and Methods

4.1. Materials

Orange-fleshed and widely cultivated sweet potato cultivars in China, YS-25 and PS-32, with approximately four storage roots, were selected for this experiment. Vegetative terminal cuttings with the following characteristics were used: the length was approximately 25 cm, the excess buds and leaves were removed, and the top three fully unfolded leaves were retained. The cutting base was soaked with 30 mg kg^{-1} carbendazim for 5 min. In this experiment, the fertilizers were urea (46% N, Sinopec, Co., Ltd., Dongfang, China), calcium superphosphate (16% P_2O_5, SDIC Xinjiang Lop Nur Potassium Salt Co., Ltd., Hami, China), and potassium sulfate (52% K_2O, Guangdong Zhanhua Group Co., Ltd., Zhanjiang, China).

4.2. Methods

4.2.1. Experimental Design

The two-year and two-site field experiments were conducted from 2020 to 2022. The first round of field experiments was arranged on 1 November 2020 and plants were harvested on 1 March 2021. The second round was conducted on 5 November 2021 and plants were harvested on 5 March 2022. The China Meteorological Data Service Center provided the two growth seasons' climate data, as shown in Table S1.

The first field experiment was carried out at the agricultural base of Hainan University, Haikou, China (20°06′ N, 110°33′ E), and the second was conducted at the research base of the Institute of Nanfan & Seed Industry, Guangdong Academic of Sciences at Yazhou District, Sanya, China (18°21′30″ N, 109°9′54″ E). The soil type of the two fields was sandy loam. All the soil physical and chemical properties at the 0–30 cm soil tillage layer are presented in Table S2.

The field experiments were carried out as a two-factor split-plot design with five replicates in a randomized block arrangement. The two sweet potato cultivars, YS-25 and PS-32, were assigned to the main plot with four plant distances, namely D15, 15 cm plant distance and 83,280 plants ha^{-1}; D20, 20 cm plant distance and 62,520 plants ha^{-1}; D25, 25 cm plant distance and 50,025 plants ha^{-1}; and D30, 30 cm plant distance and 41,640 plants ha^{-1}. The row distance was set at 80 cm for all. Each sub-plot had five ridges covering 12.60, 16.80, 21.00, and 25.20 m^2. Before planting, each treatment was initiated with 120 kg ha^{-1} N, 112 kg ha^{-1} P_2O_5, and 240 kg ha^{-1} K_2O as a base fertilizer.

Stem cuttings were planted with four nodes introduced into the soil by the oblique planting method. During the experimental period, the field management, including pest, disease, and weed control, consisted of local high-yield field practices.

4.2.2. Sampling and Measurements

Ten representative plants from each treatment were divided into two groups evenly at 15, 25, and 35 days after planting. In one group, the six thickest roots from five plants were selected as potential storage roots [15,38], and the potential storage roots were washed with distilled water. Then, they were dried with a tissue, immersed in liquid nitrogen, and stored at −80 °C. These samples were obtained to analyze the gene expression. The root and shoot systems of five plants from the other group were separated. Each plant's root system was examined to count the number of adventitious roots, weigh the total roots and potential storage root by using an electric weighing balance, and measure the potential storage root's thickest part's diameter by using a Vernier caliper. The thickest part, with approximately a 1.00 cm length of the potential storage root, was immersed in 70% FAA fixative solution (Scientific Phygene) to observe cambium development. The shoots and roots of each plant were blanched at 105 °C for 30 min and 80 °C to a constant weight by oven drying. The root dry matter was preserved to quantify the lignin and carbohydrate content.

Meanwhile, several parameters were obtained as follows:
Root/shoot ratio = root dry weight/shoot dry weight × 100%;
Ratio of root dry biomass allocation = root dry weight/total plant dry biomass × 100%;
Ratio of shoot dry biomass allocation = shoot dry biomass/total plant dry biomass × 100%;
Potential SR weight ratio = potential storage fresh weight/total root fresh weight × 100%.

At 45 days after planting, the division method of Wang [23] was used to calculate the potential storage roots weighed and the number per plant ($\Phi > 2.00$ mm). Furthermore, the number of young storage roots ($2.00 < \Phi \leq 5.00$ mm), developing storage roots ($5 < \Phi \leq 20$ mm), and mature storage roots ($\Phi > 20$ mm) was calculated. All store roots were harvested 120 days after planting, and the total weight of the fresh storage roots in each plot was recorded. Then, the average storage root numbers per plant, weight, and yield were calculated. Moreover, five representative plants per plot were selected to measure the diameter and length of the storage root, and then the length/diameter ratio (L/D ratio) was calculated [42]. The L/D ratio has been used to describe the shape in agricultural products (if an object has L/D ratio = 1, it is considered circular). The uniformity of the storage root weight is expressed by the coefficient of variation (CV), where CV = standard deviation/average. The smaller the CV, the better the uniformity of the storage root weight. In accordance with the grading standard of fresh sweet potato introduced by Si [38] and the actual production in China in recent years, the storage roots ($\Phi > 1.00$ cm, FW > 50 g) were considered commercial storage roots. They were divided into large commercial storage roots (FW of 250–500 g), medium-sized commercial storage roots (FW of 100–250 g), and small commercial storage roots (FW 50~100 g). The number of commercial storage roots of each grade was investigated manually.

4.2.3. Carbohydrate Content Determination

Sucrose and starch content was analyzed by anthrone colorimetry [43]. Approximately 200 mg of dry root sample was crushed into a powder and placed in a 10 mL centrifugal tube with 5 mL distilled water. Then, the sample was centrifuged at 3000 rpm for 5 min. The supernatants were collected in a 50 mL volumetric flask, and then distilled water was added to the scale mark. This sample was noted as solution A. The residue was dissolved by 10 mL of 3 mol mL^{-1} HCl and placed in a 50 mL volumetric flask. Then, it was placed in a water bath for 40 min, cooled down to room temperature, and combined with 10 mL of 3 mol mL^{-1} NaOH. Afterwards, distilled water was added to the scale mark, and 2 mL of this sample was transferred to a new 50 mL volumetric flask. Distilled water was added to the scale mark, and this sample was noted as solution B. Solution A was used for sucrose determination, and solution B was used for starch determination.

Approximately 2 mL of 2 mol L^{-1} KOH solution was transferred to 10 mL of solution A, boiled for 10 min, cooled down to room temperature, and diluted with distilled water to 50 mL. Afterwards, 2 mL of this sample was transferred to a new tube to react with the anthrone reagent. The sucrose content was measured spectrophotometrically at 640 nm. A sucrose solution (0.1%) was used as a standard solution.

Approximately 2 mL of solution B was transferred to the anthrone reagent and placed in a boiling water bath for 5 min. Then, it was cooled to room temperature, and the starch content was determined at 640 nm. Furthermore, a glucose solution (0.1%) was used to generate a standard curve.

4.2.4. Lignin Content Determination

A lignin content assay kit was used to measure the lignin content in the potential storage root (Suzhou Comin Biotechnology Co., Ltd., Suzhou, China). Dried roots (15 mg) were crushed and sieved (0.25 mm). Each sample was transferred to a 10 mL stoppered glass test tube, and each tube was immersed with 1000 μL of reagent 1 and 40 μL of perchloric acid and then placed at 80 °C for 40 min. Then, 5 mg quartz sand per tube for three repetitions was used as the control. Each tube was shaken for 10 min. Then, 1000 μL of reagent 2 was added and mixed well. The mixture was centrifuged at 5000 rpm for 2 min, and then 40 μL of the supernatant was removed and mixed with 1960 μL of reagent 3. Subsequently, the mixture was evaluated at 280 nm. Moreover, the lignin content was calculated as follows:

$$\text{Lignin content (mg/g dry weight)} = (\Delta A - 0.0068) \div 0.0694 \times Vt \times 10 - 3 \div W \times T,$$

where ΔA = Asample − Actrl; Vt = total reaction system volume; W = sample weight; and T = dilution ratio.

4.2.5. Root Anatomical Observation and Histochemical Analysis

Root samples were soaked in a 70% FAA fixation solution (FTY. Phygene Life Sciences Co., Ltd., Fuzhou, China) and dehydrated using an ethanol dilution series before being embedded in paraffin wax. Three samples from each treatment were cut into 15-μm-thick sections with a microtome (Campden Instruments Ltd., London, UK). They were deparaffinized in a histoclear solution and rehydrated with an ethanol dilution series to further prepare root sections for histochemical staining and autofluorescence imaging. Safranin–fast green staining was used to investigate the root vascular system. The deparaffinized samples were stained in safranin-O (1%) for 2 h, the excess dye was washed off with distilled water, and then the samples were stained in fast green (0.5%) for 10 s. Images were observed and captured by a Nikon DS-Fi1 digital camera. The number of protoxylems and secondary xylem elements, the stele diameter, and the stele cross-sectional area were counted or measured.

4.2.6. Real-Time Quantitative PCR Performance

The total RNA of potential storage roots was extracted in accordance with the Plant Total RNA Isolation Kit Plus (Foregene, RE05024, Chengdu, China). The quality of isolated RNA was described by the RNA concentration and strip integrity measured by a micro-spectrophotometer (UV–Vis spectrophotometer Q5000, Quawell Technology, Inc., Sunnyvale, CA, USA). The first-strain cDNA was generated by the MonScriptTM RTIII ALL-in-One with dsDNase (One-Step) (Monad Biotech Co., Ltd., Wuhan, China). Real-time quantitative PCR was performed in a 20 μL reaction volume containing 1× MonAMPTM ChemoHS qPCR Mix (Monad Biotech Co., Ltd., Wuhan, China). The procedures of real-time quantitative PCR were performed in two steps as follows: initiated with predenaturation at 95 °C for 10 min, followed by 40 cycles at 95 °C for 15 s and 60 °C for 30 s, and default settings were used to collect the melt curves. Quantitative analysis was conducted using the ABI QuanStudioTM 5 System with standard mode. qRT-PCR detection was performed in three biological replicates. The relative expression levels were estimated using the $2^{-\Delta\Delta Ct}$ method. *β-actin* was used as the internal control. The primers used in qRT-PCR are listed in Table S3.

4.2.7. Statistical Analysis

Two-way ANOVA was applied to determine the statistical significance, with a significance level at $p < 0.05$, which was tested by LSD. The statistical analysis was performed using the SPSS software (Version 19), and the figures were designed using the GraphPad Prism software (Version 8.4.2 for Windows).

5. Conclusions

In this study, the response mechanism of the storage root number, yield, and storage root commercial traits to the planting density was explained from the aspects of carbohydrate and lignin metabolism, such as carbohydrate content, lignin content, and the related regulatory gene expression in potential storage roots, as well as the potential storage root histochemical analysis, plant dry matter dynamics, and young root traits. The results proved that an appropriate planting density (D20 treatment in this study) could promote carbohydrate accumulation and inhibit lignin biosynthesis in potential storage roots. This finding could facilitate storage root formation to increase the storage root number by stimulating cambium cell division and inhibiting stele cell lignification. The planting density is an important factor in regulating carbohydrate and lignin metabolism in potential storage roots, affecting the number and yield and the commercial traits of sweet potato storage roots. However, storage root formation and storage root bulking are two important events that determine the final yield of this crop. Therefore, the findings were confirmed by taking into account the storage root bulking mechanism and photosynthetic features in response to planting density during the storage root bulking period (50–90 days after planting, data not given in this study).

Supplementary Materials: The following supporting information can be downloaded at: https://www.mdpi.com/article/10.3390/plants12102039/s1, Table S1: Climatic conditions; Table S2: Experimental soil physical and chemical properties; Table S3: Gene primer sequences used in qRT-PCR.

Author Contributions: Q.L., G.Z. and Y.C. (Yanli Chen) conceived the study. Q.L. and H.C. (Hongrong Chen) designed the experimental procedures. Q.L., H.C. (Hailong Chang), H.C. (Hongrong Chen), Y.L. (Yi Liu) and Y.C. (Yue Chen) carried out the experiment. Q.L. wrote the manuscript. Q.W., S.K., Y.L. (Yonghua Liu) and J.W. modified the language and reviewed the manuscript. All authors have read and agreed to the published version of the manuscript.

Funding: This work was financially supported by the earmarked fund for CARS-10-Sweetpotato, the key research and development program of Hainan Province (grant number ZDYF2020226).

Institutional Review Board Statement: Not applicable.

Informed Consent Statement: Not applicable.

Data Availability Statement: All data are included in the present study.

Conflicts of Interest: The authors declare no conflict of interest.

References

1. Dong, T.; Zhu, M.; Yu, J.; Han, R.; Tang, C.; Xu, T.; Liu, J.; Li, Z. RNA-Seq and iTRAQ reveal multiple pathways involved in storage root formation and development in sweet potato (*Ipomoea batatas* L.). *BMC Plant Biol.* **2019**, *19*, 136. [CrossRef] [PubMed]
2. Karan, Y.B.; Şanli, Ö. The assessment of yield and quality traits of sweet potato (*Ipomoea batatas* L.) genotypes in middle Black Sea region, Turkey. *PLoS ONE* **2021**, *16*, e0257703. [CrossRef] [PubMed]
3. Rukundo, P.; Shimelis, H.; Laing, M.; Gahakwa, D. Combining ability; maternal effects, and hertability of drought tolerance, yield and yield components in sweetpotato. *Front. Plant Sci.* **2017**, *7*, 1981. [CrossRef] [PubMed]
4. Truong, V.D.; Avula, R.Y.; Pecota, K.V.; Yencho, G.C. Sweetpotato production; processing; nutritional quality. *Handb. Veg. Veg. Process.* **2018**, *2*, 811–838.
5. Singh, V.; Sergeeva, L.; Ligterink, W.; Aloni, R.; Zemach, H.; Doron-Faigenboim, A.; Yang, J.; Zhang, P.; Shabtai, S.; Firon, N. Gibberellin promotes sweetpotato root vascular lignification and reduces storage-root formation. *Front. Plant Sci.* **2019**, *10*, 1320. [CrossRef]
6. Dong, H.T.; Li, Y.; Henderson, C.; Brown, P.; Xu, C.Y. Optimum Nitrogen Application Promotes Sweetpotato Storage Root Initiation. *Horticulturae* **2022**, *8*, 710. [CrossRef]
7. Meng, Y.Y.; Wang, N.; Si, C.C. Glutamate promotes sweet potato storage root swelling by enhancing starch accumulation. *Acta Physiol. Plant.* **2020**, *42*, 58.
8. Meng, Y.Y.; Wang, N.; Si, C.C. The Application of Nitrogen Source in Regulating Lignin Biosynthesis, Storage Root Development and Yield of Sweet Potato. *Agronomy* **2022**, *12*, 2317. [CrossRef]
9. Liu, Q.; Li, J.; Han, R.; Wang, H.; Liu, W. Comparative analysis of anatomical structure, assimilate accumulation and gene expression in lignin and carbohydrate metabolism pathway during taproot thickening of taicai and pak choi. *Sci. Hortic.* **2022**, *301*, 111046. [CrossRef]
10. Bilska-Kos, A.; Mytych, J.; Suski, S.; Magoń, J.; Ochodzki, P.; Zebrowski, J. Sucrose phosphate synthase (SPS), sucrose synthase (SUS) and their products in the leaves of *Miscanthus* × *giganteus* and *Zea mays* at low temperature. *Planta* **2020**, *252*, 23. [CrossRef]
11. Stein, O.; Granot, D. An overview of sucrose synthases in plants. *Front. Plant Sci.* **2019**, *10*, 95. [CrossRef]

12. Yu, H.; Wang, T. Proteomic dissection of endosperm starch granule associated proteins reveals a network coordinating starch biosynthesis and amino acid metabolism and glycolysis in rice endosperms. *Front. Plant Sci.* **2016**, *7*, 707. [CrossRef] [PubMed]
13. Li, H.; Gidley, M.J.; Dhital, S. High-amylose starches to bridge the "Fiber Gap" development, structure, and nutritional functionality. *Compr. Rev. Food Sci. Food Saf.* **2019**, *18*, 362–379. [CrossRef]
14. Du, F.; Liu, H.; Yin, X.; Zhao, Q.; Shi, C. Potassium-mediated regulation of sucrose metabolism and storage root formation in sweet potato. *Arch. Agron. Soil Sci.* **2021**, *67*, 703–713. [CrossRef]
15. Si, C.; Shi, C.; Liu, H.; Zhan, X.; Liu, Y.; Wang, D.; Meng, D.; Tang, L. Influence of two nitrogen forms on hormone metabolism in potential storage roots and storage root number of sweetpotato. *Crop Sci.* **2018**, *58*, 2558–2568. [CrossRef]
16. Wu, Y.; Ren, Z.; Gao, C.; Sun, M.; Li, S.; Min, R.; Wu, J.; Li, D.; Wang, X.; Wei, Y.; et al. Change in sucrose cleavage pattern and rapid starch accumulation govern lily shoot-to-bulblet transition in vitro. *Front. Plant Sci.* **2021**, *11*, 564713. [CrossRef]
17. Mottiar, Y.; Vanholme, R.; Boerjan, W.; Ralph, J.; Mansfield, S.D. Designer lignins harnessing the plasticity of lignification. *Curr. Opin. Biotechnol.* **2016**, *37*, 190–200. [CrossRef]
18. Escamilla-Treviño, L.L.; Shen, H.; Hernandez, T.; Yin, Y.; Xu, Y.; Dixon, R.A. Early lignin pathway enzymes and routes to chlorogenic acid in switchgrass (*Panicum virgatum* L.). *Plant Mol. Biol.* **2014**, *84*, 565–576. [CrossRef]
19. Xu, R.X.; Ni, R.; Gao, S.; Fu, J.; Xiong, R.L.; Zhu, T.T.; Liu, H.X.; Cheng, A.X. Molecular cloning and characterization of two distinct caffeoyl CoA O-methyltransferases (CCoAOMTs) from the liverwort Marchantia paleacea. *Plant Sci.* **2022**, *314*, 111102. [CrossRef]
20. Chao, N.; Huang, S.; Kang, X.; Yidilisi, K.; Dai, M.; Liu, L. Systematic functional characterization of cinnam yl alcohol dehydrogenase family members revealed their functional divergence in lignin biosynthesis and stress responses in mulberry. *Plant Physiol. Biochem.* **2022**, *186*, 145–156. [CrossRef]
21. Li, L.; Yang, K.; Wang, S.; Lou, Y.; Zhu, C.; Gao, Z. Genome-wide analysis of laccase genes in moso bamboo highlights PeLAC10 involved in lignin biosynthesis and in response to abiotic stresses. *Plant Cell Rep.* **2020**, *39*, 751–763. [CrossRef] [PubMed]
22. Cai, Z.; Cai, Z.; Huang, J.; Wang, A.; Ntambiyukuri, A.; Chen, B.; Zheng, G.; Li, H.; Huang, Y.; Zhan, J.; et al. Transcriptomic analysis of tuberous root in two sweet potato varieties reveals the important genes and regulatory pathways in tuberous root development. *BMC Genom.* **2022**, *23*, 473. [CrossRef] [PubMed]
23. Wang, H.; Yang, J.; Zhang, M.; Fan, W.; Firon, N.; Pattanaik, S.; Yuan, L.; Zhang, P. Altered phenylpropanoid metabolism in the maize Lc-expressed sweet potato (*Ipomoea batatas*) affects storage root development. *Sci. Rep.* **2016**, *6*, 18645. [CrossRef] [PubMed]
24. Shao, H.; Xia, T.; Wu, D.; Chen, F.; Mi, G. Root growth and root system architecture of field-grown maize in response to high planting density. *Plant Soil* **2018**, *430*, 395–411. [CrossRef]
25. Wang, Y.; Qin, Y.; Chai, Q.; Feng, F.; Zhao, C.; Yu, A. Interspecies interactions in relation to root distribution across the rooting profile in wheat-maize intercropping under different plant densities. *Front. Plant Sci.* **2018**, *9*, 483. [CrossRef]
26. Zhou, J.; Zhang, Z.; Xin, Y.; Chen, G.; Wu, Q.; Liang, X.; Zhai, Y. Effects of Planting Density on Root Spatial and Temporal Distribution and Yield of Winter Wheat. *Agronomy* **2022**, *12*, 3014. [CrossRef]
27. Muñoz-Parra, E.; Pelagio-Flores, R.; Raya-González, J.; Salmerón-Barrera, G.; Ruiz-Herrera, L.F.; Valencia-Cantero, E.; López-Bucio, J. Plant–plant interactions influence developmental phase transitions, grain productivity and root system architecture in *Arabidopsis* via auxin and PFT1/MED25 signalling. *Plant Cell Environ.* **2017**, *40*, 1887–1899. [CrossRef]
28. Sukumaran, S.; Reynolds, M.P.; Lopes, M.S.; Crossa, J. Genome-wide association study for adaptation to agronomic plant density: A component of high yield potential in spring wheat. *Crop Sci.* **2015**, *55*, 2609–2619. [CrossRef]
29. Zhang, H.; Xie, B.; Wang, B.; Wang, Q.; Dong, S.; Li, A.; Hou, F.; Duan, W.; Zhang, L. Effects of Planting Density on Yield and Source-sink Characteristics of Sweet Potato [*Lpomoea batatas* (L.) Lam]. *Agric. Sci. Technol.* **2015**, *16*, 1628–1642.
30. Firon, N.; LaBonte, D.; Villordon, A.; Kfir, Y.; Solis, J.; Lapis, E.; Perlman, T.S.; Doron-Faigenboim, A.; Hetzroni, A.; Althan, L.; et al. Transcriptional profiling of sweetpotato (*Ipomoea batatas*) roots indicates down-regulation of lignin biosynthesis and up-regulation of starch biosynthesis at an early stage of storage root formation. *BMC Genom.* **2013**, *14*, 460. [CrossRef]
31. Si, C.; Shi, C.; Liu, H.; Zhan, X.; Liu, Y. Effects of nitrogen forms on carbohydrate metabolism and storage-root formation of sweet potato. *J. Plant Nutr. Soil Sci.* **2018**, *181*, 419–428. [CrossRef]
32. Singh, V.; Zemach, H.; Shabtai, S.; Aloni, R.; Yang, J.; Zhang, P.; Sergeeva, L.; Ligterink, W.; Firon, N. Proximal and distal parts of sweetpotato adventitious roots display differences in root architecture, lignin, and starch metabolism and their developmental fates. *Front. Plant Sci.* **2021**, *11*, 609923. [CrossRef] [PubMed]
33. Sun, J.; Hui, K.; Guo, Z.; Li, Y.; Fan, X. Cellulose and lignin contents are negatively correlated with starch accumulation, and their correlation characteristics vary across cassava varieties. *J. Plant Growth Regul.* **2023**, *42*, 658–669. [CrossRef]
34. Wang, C.; Wu, J.; Tang, Y.; Min, Y.; Wang, D.; Ma, X.; Li, H.; Li, J.; Chen, Y.; Chen, S.; et al. Understanding the changes of phenylpropanoid metabolism and lignin accumulation in wounded cassava root during postharvest storage. *Sci. Hortic.* **2023**, *310*, 111765. [CrossRef]
35. Townsley, B.T.; Sinha, N.R.; Kang, J. KNOX1 genes regulate lignin deposition and composition in monocots and dicots. *Front. Plant Sci.* **2013**, *4*, 121. [CrossRef]
36. Tanaka, M. Recent progress in molecular studies on storage root formation in sweetpotato (*Ipomoea batatas*). *Jpn. Agric. Res. Q.* **2016**, *50*, 293–299. [CrossRef]
37. Ravi, V.; Chakrabarti, S.K.; Makeshkumar, T.; Saravanan, R. Molecular regulation of storage root formation and development in sweet potato. In *Horticultural Reviews: Volume 42*; Wiley Online Library: Hoboken, NJ, USA, 2014; pp. 157–208.

38. Si, C.C.; Liang, Q.G.; Liu, H.J.; Wang, N.; Kumar, S.; Chen, Y.L.; Zhu, G.P. Response Mechanism of Endogenous Hormones of Potential Storage Root to Phosphorus and Its Relationship with Yield and Appearance Quality of Sweetpotato. *Front. Plant Sci.* **2022**, *13*, 872422. [CrossRef]
39. Villordon, A.Q.; Clark, C.A. Variation in virus symptom development and root architecture attributes at the onset of storage root initiation in 'Beauregard'sweetpotato plants grown with or without nitrogen. *PLoS ONE* **2014**, *9*, e107384. [CrossRef]
40. Liu, Y.; Liao, Y.; Liu, W. High nitrogen application rate and planting density reduce wheat grain yield by reducing filling rate of inferior grain in middle spikelets. *Crop J.* **2021**, *9*, 412–426. [CrossRef]
41. Severini, A.D.; Borrás, L.; Westgate, M.E.; Cirilo, A.G. Kernel number and kernel weight determination in dent and popcorn maize. *Field Crops Res.* **2011**, *120*, 360–369. [CrossRef]
42. Villordon, A.; LaBonte, D.; Firon, N.; Carey, E. Variation in nitrogen rate and local availability alter root architecture attributes at the onset of storage root initiation in 'Beauregard'sweetpotato. *HortScience* **2013**, *48*, 808–815. [CrossRef]
43. Laurentin, A.; Edwards, C.A. A microtiter modification of the anthrone-sulfuric acid colorimetric assay for glucose-based carbohydrates. *Anal. Biochem.* **2003**, *315*, 143–145. [CrossRef] [PubMed]

plants

MDPI

Article

Variation in the Root System Architecture of Peach × (Peach × Almond) Backcrosses

Ricardo A. Lesmes-Vesga [1], Liliana M. Cano [2], Mark A. Ritenour [1], Ali Sarkhosh [3], Josè X. Chaparro [3] and Lorenzo Rossi [1,*]

1 Indian River Research and Education Center, Horticultural Sciences Department, Institute of Food and Agricultural Sciences, University of Florida, Fort Pierce, FL 34945, USA; ricardolesmes@ufl.edu (R.A.L.-V.)
2 Indian River Research and Education Center, Plant Pathology Department, Institute of Food and Agricultural Sciences, University of Florida, Fort Pierce, FL 34945, USA; lmcano@ufl.edu
3 Horticultural Sciences Department, Institute of Food and Agricultural Sciences, University of Florida, Gainesville, FL 32603, USA
* Correspondence: l.rossi@ufl.edu; Tel.: +1-(772)-577-7341

Abstract: The spatial arrangement and growth pattern of root systems, defined by the root system architecture (RSA), influences plant productivity and adaptation to soil environments, playing an important role in sustainable horticulture. Florida's peach production area covers contrasting soil types, making it necessary to identify rootstocks that exhibit soil-type-specific advantageous root traits. In this sense, the wide genetic diversity of the *Prunus* genus allows the breeding of rootstock genotypes with contrasting root traits. The evaluation of root traits expressed in young seedlings and plantlets facilitates the early selection of desirable phenotypes in rootstock breeding. Plantlets from three peach × (peach × almond) backcross populations were vegetatively propagated and grown in rhizoboxes. These backcross populations were identified as BC1251, BC1256, and BC1260 and studied in a completely randomized design. Scanned images of the entire root systems of the plantlets were analyzed for total root length distribution by diameter classes, root dry weight by depth horizons, root morphological components, structural root parameters, and root spreading angles. The BC1260 progeny presented a shallower root system and lower root growth. Backcross BC1251 progeny exhibited a more vigorous and deeper root system at narrower root angles, potentially allowing it to explore and exploit water and nutrients in deep sandy entisols from the Florida central ridge.

Keywords: *Prunus*; stone fruit; rootstock breeding; stem cutting; root system architecture; rhizotron

Citation: Lesmes-Vesga, R.A.; Cano, L.M.; Ritenour, M.A.; Sarkhosh, A.; Chaparro, J.X.; Rossi, L. Variation in the Root System Architecture of Peach × (Peach × Almond) Backcrosses. *Plants* **2023**, *12*, 1874. https://doi.org/10.3390/plants12091874

Academic Editor: Erica Lumini

Received: 30 March 2023
Revised: 1 May 2023
Accepted: 2 May 2023
Published: 3 May 2023

1. Introduction

Given the continuing increase in food demand and the environmental impacts of agricultural production worldwide, the improvement of water and nutrient use efficiency is essential for sustainable horticulture. Plant genotype determines root morphology, which influences the plants' efficiency in nutrients and water uptake [1]. There is widespread evidence for genotypic variation in root traits of many species [2] and the genus *Prunus*, to which peach [*Prunus persica* (L.) Batsch] belongs, is no exception. Peach is the third most produced temperate tree fruit species behind apple and pear [3].

Root architectural traits determine the temporal and spatial distribution of the root systems in soil, playing a fundamental role in the ability of plants to uptake soil resources. According to Manschadi et al. [2], root system architecture (RSA) is defined as the in situ space-filling properties or the spatial distribution of the root system within the rooting volume. Lynch [4] defines RSA as the spatial arrangement and growth pattern of roots, which influences the water and nutrient uptake ability of plants and their exploration capacity in the growing media in response to resource distribution. Thus, RSA determines the productivity and adaptation of horticultural crops to suboptimal soil environments [2,5]. Rootstocks with longer roots and numerous lateral branches and root hairs explore the soil more efficiently as they uptake water and nutrients at different depths and soil textures [6]. RSA traits play a fundamental role in achieving this goal, influencing the ability of root systems to take up water and nutrients [7]. This was confirmed by Fitter et al. [8], who demonstrated that nutrient and water uptake efficiency is a function of root system architecture. The development and configuration of the root system affect plants' soil exploration and resource exploitation in the niche occupied by plants [9]. A herringbone root architecture (branches from a single primary root axis) is typically developed by plants adapted to nutrient- or water-poor soils [10].

In general, the main challenges for peach production are associated with drought, waterlogging, alkalinity tolerance, and soil-borne diseases, especially nematodes [11,12]. Since the *Prunus* genus encompasses a wide number of species (over 230), exploiting this genetic diversity can significantly enhance the development of rootstocks with improved resource use efficiency and field performance [13]. The range of rootstocks available for peach production has increased dramatically in the last few decades [3]. Despite the fundamental importance of studying root systems, this "hidden half" has not been studied as detailed as the aerial part of the plants, particularly in perennial fruit trees such as peach. Studying the RSA and diversity of root traits among peach genotypes can be important for rootstocks breeding given the horticultural applications for water and nutrient management of orchards [9,14]. The relative tolerance of rootstocks to water stress is influenced by RSA traits, such as rooting depth, root density, specific root length, and root/shoot ratio [15].

Extensive research has reported the potential for RSA improvement since traits such as root depth appear to be controlled by multiple genes in crops such as wheat [7]. Nevertheless, root growth and architecture are also influenced by the local soil environment, depending upon the plasticity of the genotype [16]. Soil is often heterogeneous and a complex medium with high spatial and temporal environmental variability (i.e., soil texture, structure, nutrient, and water content) [17,18]. There are contrasting RSAs between plant materials and soil types, as found in the peach production area in Florida. Root architectural traits that increase the acquisition efficiency for one soil resource may not be appropriate to capture other soils' resources. Optimizing root architecture to improve the acquisition efficiency for one soil resource may incur trade-offs for acquiring other resources [2]. For instance, shallower RSAs are more efficient in acquiring immobile nutrients such as phosphorus (P). In contrast, deeper RSAs optimize water uptake and increase the acquisition of mobile nutrients such as nitrate (NO_3-N) [5]. As pointed out by Manschadi et al. [2], the advantages of architectural root traits must be interpreted in the context of the type of environment in which the crops are grown. Therefore, it is highly relevant to identify the appropriate rootstock root system suitable to the specific limitations, and that can offer advantages that the in situ-soil type presents.

One of the particularities of the peach industry in Florida (USA) is that most of the main production areas cover contrasting soil types represented by psamments (sandy entisols) in the central ridge and aquods (wet spodosols) in the Flatwoods [19]. Since choosing the best rootstock for a particular soil is one of the most important decisions for successful peach orchard establishment [20], the study of rootstock RSA is crucial for understanding their adaptability to given edaphic conditions [21].

For the reasons given above, it is very important to develop studies that provide information about the potential horticultural performance of rootstocks in light of their RSA analysis. However, the study of the below-ground part of perennial fruit trees, such as stone fruits, under field conditions is highly challenging. The development of methods and equipment such as rhizoboxes has helped to overcome these issues by allowing direct and repeated observations of the roots within the rhizosphere [22,23]. Rhizoboxes also permit the imaging and monitoring of root growth dynamics without disturbance [1,24].

The total root length of a system is the most important measure of its growth [2,25–27]. According to Manschadi et al. [28], the analysis of early expressed root traits, such as root angle, represents potential selection criteria for breeding. Many studies of RSA analysis are carried out in young seedlings under laboratory conditions since seedling roots can be observed and measured rapidly and relatively easily. These laboratory conditions allow for higher detail, replication and standardization, allowing the comparison between species and genotypes [14].

'Flordaguard' is a seed propagated rootstock released by the University of Florida in 1991, with complex parentage that includes *P. persica* and *P. davidiana* in its pedigree. It is currently the only rootstock recommended for commercial peach production in Florida mainly because of its lower chill requirement and resistance to root-knot nematode, including *Meloidogyne floridensis* found in Florida soils [29]. Among other advantages that 'Flordaguard' exhibits, this rootstock is compatible with all peach cultivars, propagates easily by seed, has quicker readiness for grafting [30,31], and its red leaves facilitates the suckers detection and removal. However, 'Flordaguard' is susceptible to iron deficiency chlorosis under alkaline conditions [32]. The backcrossing with almond for peach production confers better adaptation to low-chill areas [33,34], and more tolerance to drought and alkalinity avoiding iron chlorosis (known as lime-induced chlorosis) [20,35,36]. In this study, 'Flordaguard' was used in one of the peach × (peach × almond) backcross families studied in this experiment. The peach × almond parental selections used in this study, being 'Flordaguard' the female parental line, are highly resistant to *Botryosphaeria dothidea*. These materials segregate rootstock populations for peach production on contrasting soils in Florida (USA): Sandy Entisols of the Central Ridge and Wet Spodosols of the Flatwoods [19]. The main objective of this study was to compare the root system architecture of vegetatively propagated peach × (peach × almond) backcrosses.

2. Results

2.1. Root Growth Parameters

The average total root length of Backcross BC1260 was 243 cm, which was significantly lower than that of BC1251 (806 cm) and BC1256 (591 cm) (Table 1). Similar differences were obtained for total root surface area between the backcrosses BC1251 (82.1 cm^2), BC1256 (71.9 cm^2), and BC1260 (29.4 cm^2) (Table 1). There were no significant differences between the average root diameters (0.39–0.44 cm) of all the backcrosses (Table 1). The total root volume of BC1260 (0.29 cm^3) was significantly lower than BC1251 (0.69 cm^3) and BC1256 (0.72 cm^3), which were not significantly different from each other (Table 1). BC1251 generated significantly more root tips (3608) compared with the backcrosses BC1256 (1920) and BC1260 (1156), which were also significantly different from each other (Table 1). These relative differences were also evident in the number of root forks, where BC1251 (3207) had a significantly higher number of root tips than backcrosses BC1256 (1803) and BC1260 (1179), which did not report a significantly different number of root forks from each other (Table 1).

2.2. Root Structural Parameters

The root specific length of BC1251 (6500 cm/g) was significantly higher than backcrosses BC1256 (5195 cm/g) and BC1260 (5534 cm/g), which show non-significant differences between each other. Similarly, the root fineness of BC1251 (1201 cm/cm^3) was significantly higher than BC1256 (832 cm/cm^3) and BC1260 (860 cm/cm^3) (Table 1). Finally, the root tissue density of backcross BC1256 (0.16 g/cm^3) was not significantly different from BC1251 and BC1260 (0.19 and 0.15 g/cm^3, respectively). However, BC1251 was significantly higher than BC1260 for this structural root parameter (Table 1).

2.3. Root Dry Weight and Morphological Components

The root dry weight of BC1260 (0.04 g) was significantly lower than BC1251 (0.14 g) and BC1256 (0.12 g), which were not significantly different from each other (Table 1).

BC1251 had a root mass ratio (0.13 g/g) significantly higher than BC1260 (0.06 g/g) (Table 1). The root mass ratio of BC1256 (0.11 g/g) was intermediate and not significantly different from BC1251. These relative differences were mirrored in the root length ratio for BC1251 (661 cm/g), BC1256 (512 cm/g), and BC1260 (334 cm/g) (Table 1).

Table 1. Statistical results of the comparisons between the backcrosses BC1251, BC1256, and BC1260 for their root growth parameters, and structural parameters, dry weight, and morphological components. Mean values ($n = 4$) not connected by the same letter are significantly different according to Tukey's studentized range (HSD) test ($p \leq 0.05$).

Root Growth Parameters							
Backcross	**Response**	**SE**	**Group**	**Backcross**	**Response**	**SE**	**Group**
	Total root length (cm)				Total Root Volume (cm^3)		
BC1251	806	91.5	a	BC1251	0.69	0.10	a
BC1256	591	148.5	a	BC1256	0.72	0.16	a
BC1260	243	21.3	b	BC1260	0.29	0.02	b
	Total Root Surface Area (cm^2)				Number of Root Tips		
BC1251	82.1	10.6	a	BC1251	3608	41.85	a
BC1256	71.9	16.5	a	BC1256	1920	529.09	b
BC1260	29.4	0.5	b	BC1260	1156	164.60	c
	Average Root Diameter (mm)				Number of Root Forks		
BC1251	0.41	0.06	ns	BC1251	3207	649.73	a
BC1256	0.44	0.04	ns	BC1256	1803	572.20	b
BC1260	0.39	0.03	ns	BC1260	1179	238.19	b
	Root Structural Parameters				**Root Dry Weight and Morphological Components**		
Backcross	Response	SE	Group	Backcross	Response	SE	Group
	Root Specific Length (cm/g)				Root Dry Weight (g)		
BC1251	6500	650.35	a	BC1251	0.14	0.03	a
BC1256	5195	377.91	b	BC1256	0.12	0.03	a
BC1260	5534	341.82	b	BC1260	0.04	0.00	b
	Root Tissue Density (g/cm^3)				Root Mass Ratio (g/g)		
BC1251	0.19	0.01	a	BC1251	0.13	0.03	a
BC1256	0.16	0.02	ab	BC1256	0.11	0.02	a
BC1260	0.15	0.02	b	BC1260	0.06	0.01	b
	Root Fineness (cm/cm^3)				Root Length Ratio (cm/g)		
BC1251	1201	47.15	a	BC1251	661	70.71	a
BC1256	832	133.90	b	BC1256	512	90.18	a
BC1260	860	123.63	b	BC1260	334	63.16	b

2.4. Total Root Length Distribution by Diameter Classes

The total length of very fine roots (≤0.5 mm) of BC1251 (673.41 cm) was significantly higher than BC1256 (442.04 cm), which was significantly higher than BC1260 (189.59 cm) (Figure 1a). However, the total length of fine (>0.5–≤1.0 mm) and thin (>1.0 mm) roots from BC1251 (118.59 cm and 13.60 cm, respectively) were not significantly different from BC1256 (129.84 cm and 19.23 cm). Conversely, BC1260 had significantly lower values in fine and thin class roots (47.29 cm and 6.62 cm) compared with the other two backcrosses.

2.5. Root Spreading Angle

The total root length of all the backcrosses were not significantly different within the shallower (0–25°) and shallow (25–45°) spreading angles (Figure 1b). However, the root length within the deep spreading angle (45–65°) in the backcross BC1260 (85.68 cm) was significantly lower than the other two backcrosses. Finally, within the deeper spreading angle (65–90°), BC1251 showed a significantly higher root length (352.10 cm) than the other two backcrosses, which were significantly different to each other, where BC1256 showed a higher root length (248.62 cm) than BC1260 (85.68 cm) (Figure 1b).

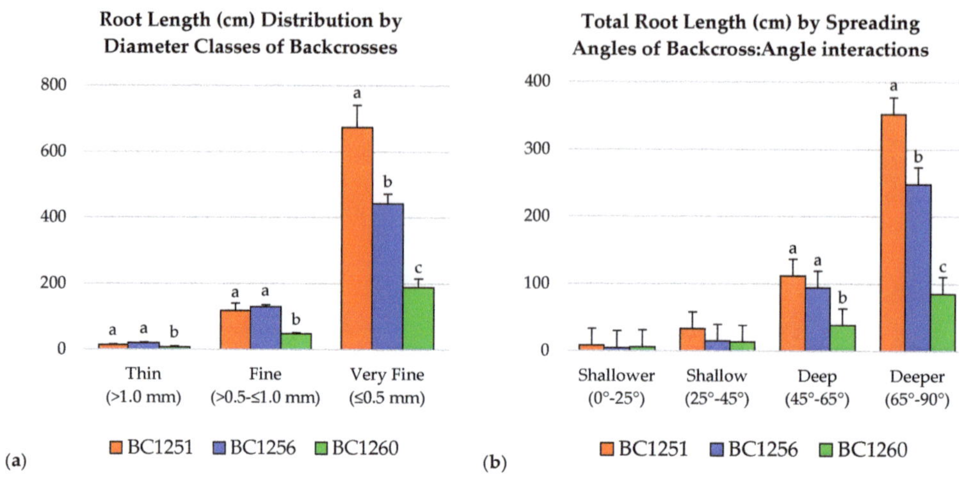

Figure 1. (**a**) Total root length comparisons grouped in three root diameter classes (very fine, fine, and thin) for families BC1251, BC1256, and BC1260, showing the interaction backcross/diameter class. (**b**) Root Spreading Angles (RSG) values estimated in terms of the total root length distributed between angular sections: shallower (0–25°), shallow (25–45°), deep (45–65°), and deeper (65–90°). Bars represent the standard error from the mean (n = 4), and different letters represent significant differences at $p \leq 0.05$.

2.6. Root Depth Pattern and Root Depth Index

There were no significant differences between BC1251 (Figure 2a) and BC1256 (Figure 2b) within the percentage of total root length between their root horizons (A, B, and C). Conversely, in BC1260 (Figure 2c), the percentage of the total root length from horizon C (8.78%) was significantly lower than horizon A (51.49%) and horizon B (41.92%), which were not significantly different from each other. Regarding the root depth index (RDI), the backcross BC1260 (10.5) had significantly lower values than BC1251 (15.0) and BC1256 (14.4), which were not significantly different from each other (Figure 2d).

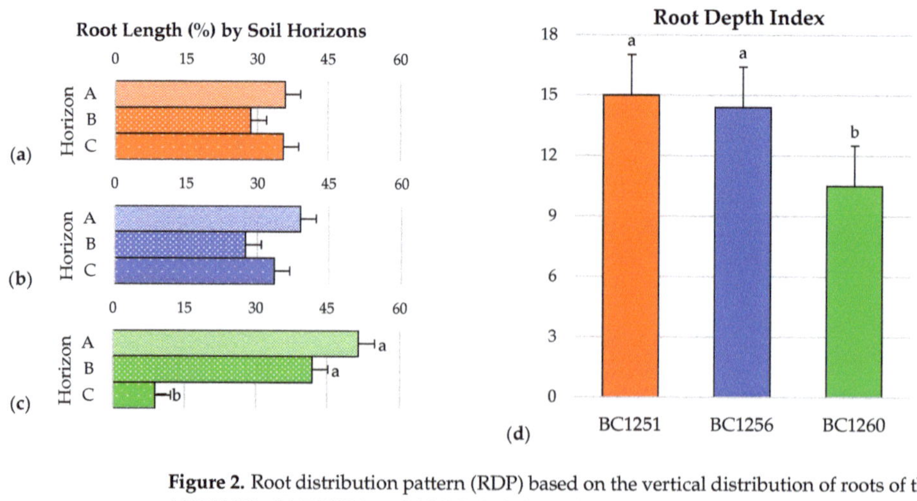

Figure 2. Root distribution pattern (RDP) based on the vertical distribution of roots of the backcrosses: (**a**) BC1251, (**b**) BC1256, and (**c**) BC1260 along the root system horizons (A, B, and C). (**d**) Root depth index (RDI) of the three backcrosses. Bars represent the standard error from the mean (n = 4), and different letters represent significant differences at $p \leq 0.05$.

3. Discussion

Most of the root growth parameters of backcross BC1260 exhibited a smaller average size. A possible explanation for the smaller root system in BC1260 may be due to the vigor reduction provoked by inbreeding depression, where the reduction in vigor traits occurs in offspring of related parents [37,38]. BC1260 was the only backcross that has 'Flordaguard' in the pedigrees of the male (peach × almond male parent) and female parents ('R95654.16'). The selection 'R95654.16' originated from a USDA selection × 'Flordaguard' F_2 population. It is possible that backcross BC1260 is potentially a dwarfing rootstock. In peach, dwarf trees exhibit smaller root systems, and compact trees tend to exhibit high root branching [9].

On the other hand, the backcross BC1251 exhibited superior values in most root growth parameters resulting in a more robust root system in general. This is a potential advantage for BC1251 and BC1256 over BC1260 in the psamments (sandy entisols) of the Florida central ridge. The root system architecture is critically important for soil exploration and nutrient acquisition (Lynch, 2007), and trees with smaller root systems may be more sensitive to soil resource limitations typical of sandy entisols [9].

The backcrosses BC1251 and BC1256 exhibited an interesting superiority in root diameter classes. Fine roots (<1 mm in diameter) are believed to play an important role in water and nutrient uptake [39]. According to Solari et al. [40], the efficiency in water transport of peach rootstocks may be influenced by this trait, where most of the water uptake is presumed to occur in roots with <1 mm of diameter, with a direct effect on the radial hydraulic conductance. Moreover, BC1251 exhibited a higher number of root tips with a diameter <1 mm at the deepest horizon (C) compared with the other horizons and compared with the other backcrosses in this horizon (data not shown). This agreed with Basile et al. [41], who found in 'K119-50', a rootstock with almond in its genetic background, the highest number of fine roots in the deepest soil layer (below 69 cm) compared with 'Nemaguard' and 'Hiawatha' rootstocks.

Root dry matter and root morphological components of backcross BC1251 were higher in general, and significantly different from BC1260. This is especially advantageous for BC1251, since the root mass ratio can affect tree–soil–water relations [39]. Additionally, a higher root length ratio may suggest a more efficient soil exploration and higher hydraulic conductance, leading to superior water uptake efficiency [40]. Similar studies in citrus rootstocks have reported this correlation between a higher root length ratio and high root hydraulic conductance [42].

The root length per unit root biomass (root specific length) of BC1251 was also higher than the other backcrosses with no significant differences. Higher values of this trait suggest an efficient soil exploration at lower carbohydrate costs, which is beneficial under limited water supply, as reported by Eissenstat [43] in citrus rootstocks in sandy soil (quartzipsamment) in Florida. It is noteworthy that this soil type is similar to the psamments (sandy entisols) from the Florida central ridge, where a good part of Florida's peach production is located. These soil types have a low water-holding capacity and essentially no horizon development or soil structure [19].

The root depth pattern of backcrosses BC1251 and BC1256 was similar, showing a deeper and more evenly distributed average root length between the substrate profiles. Conversely, the backcross BC1260 exhibited a significantly higher percentage of its total root length distributed in the shallowest horizon (A). This shallower root system would be more appropriate for flood-prone 'Flatwood' soils, where root asphyxiation can be a problem. Conversely, the root systems of BC1251 and BC1256 have a higher proportion of the roots in the deepest profile. This morphology suggests that BC1251 and BC1256 be evaluated in the deep, well-drained soils of the Florida central ridge. Such higher root-length distribution at depth allows potential access to water at greater depth, potentially making the rootstocks more drought-tolerant [28,44,45]. These features are also convenient in response to soil drying at the surface layers [2], as they tend to occur in psamments because of percolation [19]. Moreover, Glenn and Welker [46] found circumstantial evidence, indicating that the development of deep roots is important to maintain the root system when the soil's top layers are dry in peach. In this sense, deeper root distribution patterns and higher root depth indexes are key for capturing leach-prone nutrients in such soils as well. A deeper and more vigorous root system enables access to leached nitrates (NO^{3-}) and enhances drought adaptation by improving access to water stored in subsoil [45].

The higher values of root length distributed within the deeper spreading angles (65–90°) in the backcrosses BC1251 and BC1256 were consistent with the root distribution pattern and root depth index, confirming the feasible higher adaptability of BC1251 and BC1256 to deep sandy soils. According to Lynch [5], root gravitropism is an architectural trait under genetic control, and genotypes that express a narrow growth angle may be suitable for environments where plants rely largely on subsoil water [47]. Moreover, this type of root spreading may enhance the energy efficiency of these backcrosses for this purpose by inhibiting the elongation of lateral roots while maintaining primary root growth downwards [16,48].

Based on the principle of reiteration found in the entire root system of plum trees [49], we inferred the performance of adult rootstocks from the studied rootstock plantlets. Breeding for root system architecture traits may be a more efficient method to select for drought tolerance compared to breeding for physiological tolerance [36].

One of the goals of this study was to identify promising rootstock materials for the Florida central ridge and/or the Florida flatwoods. The root architecture of BC1251 and BC1256 is different to that observed in BC1260. Our results suggest that the three backcross populations should be evaluated under field conditions in locations with contrasting flatwood and central ridge-type soils. Such an experiment would demand the use of a technique different from rhizoboxes to study the RSA in situ with a wider time span. The results from such an experiment would indicate if the architectural data obtained in the rhizoboxes is predictive of root architecture in the field and, therefore, tree performance under varying soil conditions. Low-chill rootstocks that include plum (*Prunus* subg. *prunus*) in their genetic background are likely to be tolerant to flooding stress, whereas those crossed with almonds, such as the backcrosses BC1251, BC1256, and BC1260, are likely to be highly susceptible [50]. Finally, no anatomical studies were carried out on the rootstock materials used in this project. Therefore, we recommend for future studies to consider, in addition to the root architectural traits analyzed in this study, to include the additional factors such as the anatomical features of these backcrosses that may potentially influence the rootstock hydraulic conductivity in peach [40].

4. Materials and Methods

4.1. Plant Material

Seedlings of peach backcrosses: peach [*Prunus persica* (L.) Batsch] × (peach × almond [*Prunus dulcis* (Mill.) D.A. Webb]) were obtained. The mother trees were the peach selection 'R95654.16', located at the Southeastern Fruit and Tree Nut Research Laboratory of the USDA-ARS in Byron, GA, USA. The pollen parents were three different peach × almond hybrid selections: 1251, 1256, and 1260, located at the Fruit Tree Breeding and Genetics Laboratory of the Horticultural Sciences Department at the University of Florida in Gainesville, FL, USA. The parentages of the backcross families (BC) pedigrees are described in Table 2. The Peach × 'Tardy Nonpareil' almond hybrids used in this study have been identified as resistant to peach gummosis caused by *Botryosphaeria dothidea* [42].

4.2. Production of Backcross Populations

During the bloom season, "popcorn" stage (still closed and with expanded petals) were collected daily from peach × almond F_1 hybrids for pollen extraction. The collected flowers were stored at 9 °C in sealed plastic bags, and the pollen was extracted by removing and drying the anthers at ambient temperature. The extracted pollen was stored in sealed plastic bags at 1 °C.

For controlled crosses, all open flowers were removed from the mother trees prior to pollination. Pollinations were performed daily on "popcorn" stage flowers. Flowers were emasculated by removing the sepals, petals, and stamens, exposing the pistil. Pollen was immediately applied to the stigma using a pencil eraser. Pollinations were performed over a 5-week period.

Table 2. List of interspecific peach × (peach × almond) backcross populations from which the leafy softwood cuttings used in this study were obtained.

Backcross ID	Female ♀(peach)	Male ♂(peach × almond)
BC1251	'R95654.16'	'Fla. 97-47c' × 'Tardy-Nonpareil'
BC1256	'R95654.16'	'Fla. 97-42c' × 'Tardy-Nonpareil'
BC1260	'R95654.16'	'Flordaguard' × 'Tardy-Nonpareil'

4.3. Seeds Stratification and Germination

The mature fruits were harvested from the mother trees, discarding the extracted seed that floated in the water. Non-floating seeds were hydrated by immersing them in water for 96 h, renewing the water every 24 h. After this period, the seeds were submerged in Captan fungicide (Drexel Chemical, Memphis, TN, USA) (0.15% *w*/*v*) for 24 h. Seeds were then stratified in a plastic bag containing perlite moistened with Captan (0.15% *w*/*v*) for six weeks at 4–8 °C, until germination. Germinating seeds were sown in plastic trays (720700C SureRoots®; T.O. Plastics, Inc.; Clearwater, MN, USA) containing a 1:1 blend of potting mix sphagnum (Jolly Gar-dener® Pro-Line C/20 Growing Mix; Jolly Gardener Products, Inc.; Poland Spring, ME, USA) and coarse perlite (Specialty Vermiculite Corp.; Pompano Beach, FL, USA). Before filling up the germination trays, these were disinfected using a sodium hypochlorite solution (1.5% *v*/*v*) for 30 min. The potting blend was autoclaved at 121 °C for 90 min prior to use. The germination trays were covered with shade cloth at 70% shading in the greenhouse and watered manually. In the case of early fruit drops or pest/pathogen damage, the embryo rescue protocol developed by Chaparro and Sherman [43] was used: fruits were incubated in a sodium hypochlorite solution (1.88% *v*/*v*) for 20 min. Seeds were extracted under a laminar flow hood and cultured in test tubes containing sterile K_2 tissue culture media with 30 mg of sucrose for peach ovule culture. Tubes were appropriately labeled, sealed with cellophane film (regenerated cellulose) and stored in a dark room at 4–8 °C until germination (radicle tip emergence).

4.4. Seedling Development

After germination, the seedlings were transplanted into plastic pots (10 cm × 10 cm × 35.5 cm) containing a 3:1 mixture of potting mix sphagnum and coarse perlite from the same manufacturers used for the germination trays. The plastic pots were previously disinfected using a sodium hypochlorite solution (1.5% *v*/*v*) for 30 min. The seedlings were grown within a plastic-covered greenhouse located in Fort Pierce, FL, USA, 27°25′34.2″ N–80°24′34.0″ W. Leaf color was used to rogue the seedling populations. Red leaved seedlings were hybrids and progeny from self-pollination green-leaved.

4.5. Softwood Cuttings Obtaining

Leafy softwood cuttings obtained from five selected 1-year-old plants of each backcross were treated with K-IBA at 0.2% (*w*/*v*) and rooted aeroponically, following the protocol described by Lesmes-Vesga et al. [51] (Table 2).

4.6. Plantlets Growing Conditions

After 28 days, three selected rooted cuttings (plantlets) from each BC plant were established in rhizoboxes (40 cm × 40 cm × 2.5 cm) to serve as biological replications. The rhizoboxes consisted of a three-sided plastic frame with two clear glass panes (40 cm × 40 cm) attached to both faces (front and back) (Figure 3a). The bottom of the rhizoboxes frames were perforated for drainage. The rhizoboxes were disinfected using a sodium hypochlorite solution (1.5% *v*/*v*) for 30 min prior to filling with a sphagnum/coarse perlite (3:1) potting mix (same manufacturers as germination trays substrate) containing 0.5% (*v*/*v*) Osmocote®

Plus (15-9-12) (A.M. Leonard, Inc., Piqua, OH, USA), a 3-month controlled release fertilizer. Additionally, a preventive control of pathogens was conducted prior to transplanting by drenching the substrate in each rhizobox and soaking the roots of the plantlets for 10 min in the systemic fungicide Luna® Experience (Bayer CropScience, Research Triangle Park, NC, USA) [Fluopyram (17.6%) and Tebuconazole (17.6%)] diluted in distilled water at 0.2% (v/v).

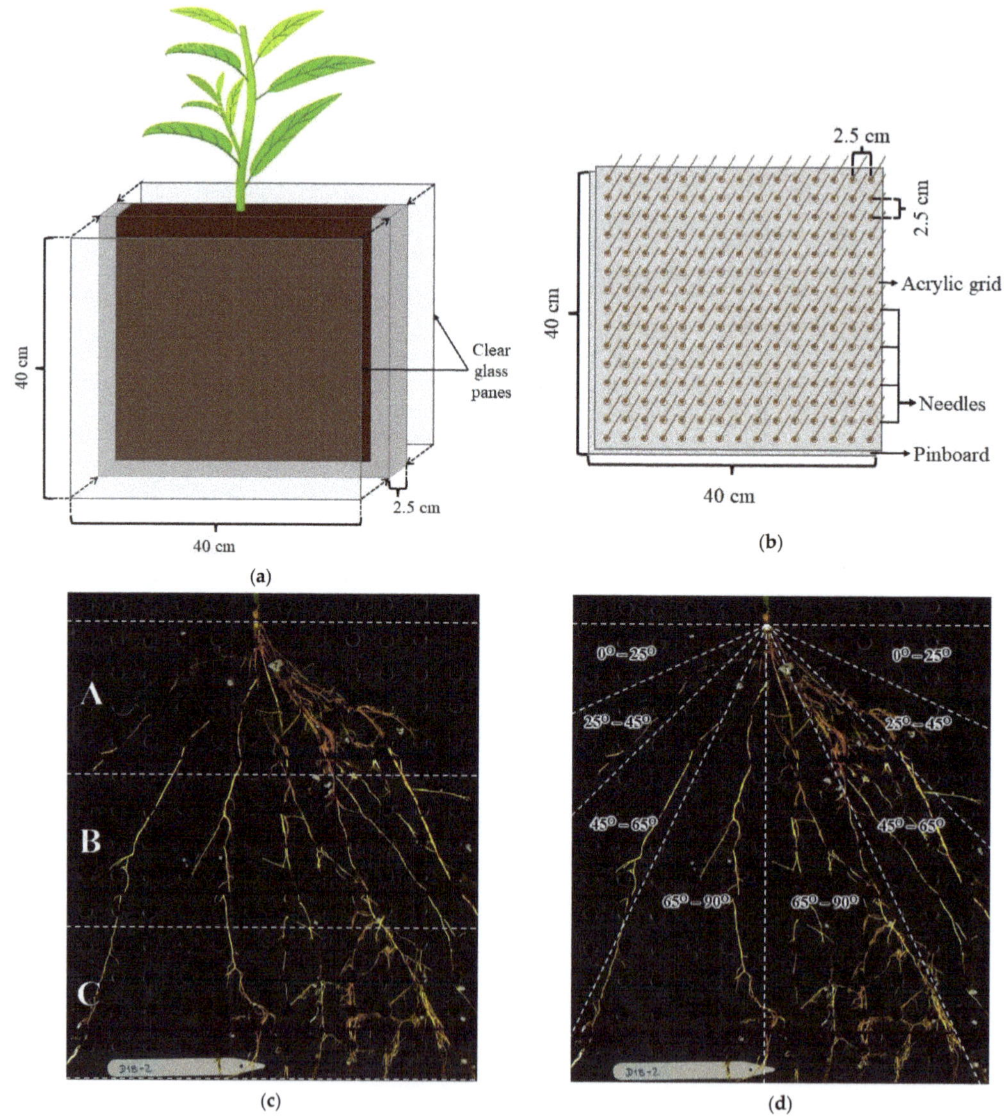

Figure 3. (**a**) Diagram of the rhizoboxes with their respective dimensions. (**b**) Diagram of the pinboard and the fitted acrylic grid to extract the root systems from the rhizoboxes. (**c**) Root system image of one of the root systems divided by horizons. The letters represent the three horizons (A, B, and C) in which the root systems were split for scanning. Each horizon had a depth of 10 cm. (**d**) Root system image of the root system, throughout the 3 horizons, divided by spreading angles to estimate the Root Spreading Angle (RSG). RSG was estimated by measuring the root length (cm) within each angular section: shallower (0–25°), shallow (25–45°), deep (45–65°), and deeper (65–90°). Photos courtesy of Ricardo A. Lesmes-Vesga.

The plantlets were transplanted in the center of the rhizoboxes to ensure their successful growth by avoiding air gaps around their base that generate root dehydration (Figure 3a). The glass panes of each rhizobox were covered with aluminum foil to keep the roots in the substrate under darkness. The rhizoboxes with plantlets were set standing in black plastic containers with bottom draining holes. The experiment was maintained under laboratory conditions (23 °C and 65% RH), where supplementary light from LED (light-emitting diode) bulbs and HPS (high-pressure sodium) lamps were installed to keep the photoperiod at 16–8 h (day-night). The plantlets were treated twice with a Luna® Experience drench for pathogen control for the duration of the experiment. The plantlets were watered manually daily throughout the study.

4.7. Scanning of the Root Systems

The root systems of the plantlets were extracted from each rhizobox after 70 days for root analysis. The root system was accessed by removing one of the glass panes and inserting a pinboard into the roots-substrate mass to keep the spatial distribution of the roots. The pinboard consisted of a grid of acrylic panels with holes inserted on another panel with 2.5 cm-long nails separated 2.5 cm × 2.5 cm (Figure 3b). The pinboard-acrylic grid was inserted into the exposed side of the rhizoboxes, and the potting substrate was gently washed off the root systems inserted in the pinboard. The whole root system of each plantlet was scanned using a flatbed scanner EPSON Expression 10000XL (EPSON America, Inc.). Full-color images were captured in TIFF (Tagged Image Format File) at a resolution of 400 dpi (dots per inch). The aerial portion of each plantlet was removed, and the root systems were split into three horizons (A, B, and C) (Figure 3c). The roots from each horizon were placed in a Plexiglas tray (20 cm × 30 cm × 1.5 cm) containing water prior to scanning with an EPSON Perfection V800/V850 (EPSON America, Inc.) flatbed scanner. The Plexiglas tray with water was used to untangle the roots and minimize root overlap prior to scanning. The images were captured in TIFF at 600 dpi resolution.

4.8. Image Analysis Software and Measurements

The images obtained with the EPSON Perfection V800/V850 scanner were analyzed using the root image analysis software WinRHIZO™ Pro (Regent Instruments Inc., Quebec City, Quebec, Canada). The root growth parameters measured were as follows: total root length (cm), total root surface area (cm^2), average root diameter (cm), total root volume (cm^3), number of root tips, and number of root forks. Additionally, total root length was distributed into three root diameter classes, following the criteria applied by Caruso et al. [52]: very fine ($\leq$0.5 mm), fine (>0.5–$\leq$1.0 mm), and large (>1.0 mm). The dry weight (g) of the aerial portion of the plant and the roots from each horizon was estimated by placing the samples in paper bags at 70 °C for 7 days, following the Ryser and Lambers [53] protocol.

Based on the root growth measurements and the dry weight of the aerial portion, the morphological components' root mass ratio (RMR) (g/g) and root length ratio (RLR) (cm/g) were estimated. The RMR, which indicates the relative biomass allocated to the roots [53], was estimated by dividing the root's dry weight by the whole plant's dry weight. The RLR was estimated by dividing the total root length by the whole plant's dry weight. This parameter expresses the root's potential for the acquisition of below-ground resources. The structural root parameters were estimated as follows: root specific length (RSL) (cm/g), root fineness (RF) (cm/cm^3), and root tissue density (RTD) (g/cm^3). RSL was estimated by dividing the total root length by the root dry weight [54], the RF was estimated by dividing the total root length by the total root volume, and RTD by dividing the root dry weight by the total root volume.

The whole root system images, obtained with the flatbed scanner EPSON Expression 10000XL, were used to measure the root spreading angle (RSG) based on the protocol applied by Ramalingam et al. [55]. For this purpose, the root length (cm) was measured in four angular sections from the root systems images, being the cutting base the central point, with four categories as follows: shallower (0–25°), shallow (25–45°), deep (45–65°), and deeper (65–90°) (Figure 3d).

4.9. Experimental Design and Statistical Analysis

The experiment was arranged in a completely randomized design with five replications ($n = 5$) per backcross, using three plantlets per replicate. The collected data were analyzed by Analysis of Variance (ANOVA) using RStudio software (R, 2019), with a significance level of 0.05, and the means separation conducted using the Tukey HSD (Honestly Significance Difference) test.

5. Conclusions

The large genotypic diversity in *Prunus* is a valuable resource for breeding peach rootstock cultivars that overcome production challenges. Significant phenotypic differences in root system architecture of rootstocks that belong to the same genus may be found, with potential applications in their horticulture. This study demonstrates that similar populations can have different root architectures that may represent peach backcross. BC1251 represents a promising rootstock material, combining resistance to root-knot nematodes and peach gummosis (given its genetic background) with variable root architectures. As such, further field studies in these environments are warranted in addition to anatomical studies.

Author Contributions: Conceptualization: J.X.C., L.R. and R.A.L.-V.; methodology: J.X.C., L.R. and R.A.L.-V.; validation, J.X.C., L.R., L.M.C., M.A.R. and A.S.; resources, L.R. and J.X.C.; writing—original draft preparation, R.A.L.-V.; writing—review and editing, J.X.C., L.R., L.M.C., M.A.R. and A.S.; project administration, L.R. All authors have read and agreed to the published version of the manuscript.

Funding: This research was partially supported by the U.S. Department of Agriculture National Institute of Food and Agriculture, Hatch project #FLA-429 IRC-005743.

Institutional Review Board Statement: Not applicable.

Informed Consent Statement: Not applicable.

Data Availability Statement: The datasets for this study can be provided upon request.

Acknowledgments: The authors want to thank the valuable advice from James Colee (statistical data analysis), and the help and support from Laura Muschweck.

Conflicts of Interest: The authors declare no conflict of interest. The funders had no role in the design of the study; in the collection, analyses, or interpretation of data; in the writing of the manuscript; or in the decision to publish the results.

References

1. Downie, H.F.; Adu, M.O.; Schmidt, S.; Otten, W.; Dupuy, L.X.; White, P.J.; Valentine, T.A. Challenges and Opportunities for Quantifying Roots and Rhizosphere Interactions through Imaging and Image Analysis. *Plant Cell Environ.* **2015**, *38*, 1213–1232. [CrossRef] [PubMed]
2. Manschadi, A.M.; Manske, G.; Vlek, P. Root Architecture and Resource Acquisition: Wheat as a Model Plant. In *Plant Roots: The Hidden Half*; Eshel, A., Beeckman, T., Eds.; CRC Press: Boca Raton, FL, USA, 2013; pp. 319–336. ISBN 978-1-4398-4649-0.
3. Byrne, D.; Raseira, M.; Bassi, D.; Piagnani, M.C.; Gasic, K.; Reighard, G.; Moreno, M.A.; Pérez, S. Peach. In *Fruit Breeding*; Badenes, M.L., Byrne, D., Eds.; Springer: New York, NY, USA, 2012.
4. Lynch, J.P. Root Architecture and Plant Productivity. *Plant Physiol.* **1995**, *109*, 7–13. [CrossRef]
5. Lynch, J. Roots of the Second Green Revolution. *Aust. J. Bot.* **2007**, *55*, 493–512. [CrossRef]
6. Vahdati, K.; Sarikhani, S.; Arab, M.M.; Leslie, C.A.; Dandekar, A.M.; Aletà, N.; Bielsa, B.; Gradziel, T.M.; Montesinos, Á.; Rubio-Cabetas, M.J.; et al. Advances in Rootstock Breeding of Nut Trees: Objectives and Strategies. *Plants* **2021**, *10*, 2234. [CrossRef]
7. Danjon, F.; Stokes, A.; Bakker, M.R. Root Systems of Woody Plants. In *Plant Roots: The Hidden Half*; Eshel, A., Beeckman, T., Eds.; CRC Press: Boca Raton, FL, USA, 2013; pp. 435–460. ISBN 978-1-4398-4649-0.
8. Fitter, A.H.; Stickland, T.R.; Harvey, M.L.; Wilson, G.W. Architectural Analysis of Plant Root Systems 1. Architectural Correlates of Exploitation Efficiency. *New Phytol.* **1991**, *118*, 375–382. [CrossRef]
9. Tworkoski, T.; Scorza, R. Root and Shoot Characteristics of Peach Trees with Different Growth Habits. *J. Am. Soc. Hortic. Sci.* **2001**, *126*, 785–790. [CrossRef]

10. Fitter, A.H.; Stickland, T.R. Architectural Analysis of Plant Root Systems 2. Influence of Nutrient Supply on Architecture in Contrasting Plant Species. *New Phytol.* **1991**, *118*, 383–389. [CrossRef]
11. Pinochet, J.; Calvet, C.; Hernández-Dorrego, A.; Bonet, A.; Felipe, A.; Moreno, M. Resistance of Peach and Plum Rootstocks from Spain, France, and Italy to Root-Knot Nematode Meloidogyne Javanica. *HortScience* **1999**, *34*, 1259–1262. [CrossRef]
12. Anthony, B.M.; Minas, I.S. Optimizing Peach Tree Canopy Architecture for Efficient Light Use, Increased Productivity and Improved Fruit Quality. *Agronomy* **2021**, *11*, 1961. [CrossRef]
13. Lesmes-Vesga, R.A.; Cano, L.M.; Ritenour, M.A.; Sarkhosh, A.; Chaparro, J.X.; Rossi, L. Rootstocks for Commercial Peach Production in the Southeastern United States: Current Research, Challenges, and Opportunities. *Horticulturae* **2022**, *8*, 602. [CrossRef]
14. Gregory, P. Measuring Root System Architecture: Opportunities and Challenges. In Proceedings of the International Symposium "Root Research and Applications" Root RAP, Vienna, Austria, 2–4 September 2009.
15. Psarras, G.; Merwin, I.A. Water Stress Affects Rhizosphere Respiration Rates and Root Morphology of Young 'Mutsu' Apple Trees on M.9 and MM.111 Rootstocks. *J. Am. Soc. Hortic. Sci.* **2000**, *125*, 588–595. [CrossRef]
16. Malamy, J.E. Intrinsic and Environmental Response Pathways That Regulate Root System Architecture. *Plant Cell Environ.* **2005**, *28*, 67–77. [CrossRef] [PubMed]
17. Puhe, J. Growth and Development of the Root System of Norway Spruce (*Picea abies*) in Forest Stands—A Review. *For. Ecol. Manag.* **2003**, *175*, 253–273. [CrossRef]
18. Moore, J.; Tombleson, J.; Turner, J.; van der Colff, M. Wind Effects on Juvenile Trees: A Review with Special Reference to Toppling of Radiata Pine Growing in New Zealand. *For. Int. J. For. Res.* **2008**, *81*, 377–387. [CrossRef]
19. Obreza, T.A.; Collins, M.E. *Common Soils Used for Citrus Production in Florida*; IFAS Extension, University of Florida: Gainesville, FL, USA, 2008.
20. Reighard, G.L.; Loreti, F. Rootstock Development. In *The Peach: Botany, Production and Uses*; Layne, D., Bassi, D., Eds.; CAB International: Cambridge, MA, USA, 2008; pp. 193–215.
21. Pandey, R.; Chinnusamy, V.; Rathod, G.; Paul, V.; Jain, N. Evaluation of Root Growth and Architecture. In Proceedings of the Manual of ICAR Sponsored Training Programme on Physiological Techniques to Analyze the Impact of Climate Change on Crop Plants, New Delhi, India, 16–25 January 2017; ICAR-Indian Agricultural Research Institute (IARI): New Delhi, India, 2017; pp. 16–25.
22. Lesmes-Vesga, R.A.; Cano, L.M.; Ritenour, M.A.; Sarkhosh, A.; Chaparro, J.X.; Rossi, L. Rhizoboxes as Rapid Tools for the Study of Root Systems of Prunus Seedlings. *Plants* **2022**, *11*, 2081. [CrossRef]
23. Rewald, B.; Ephrath, J.E. Minirhizotron Techniques. In *Plant Roots: The Hidden Half*; Eshel, A., Beeckman, T., Eds.; CRC Press: Boca Raton, FL, USA, 2013; pp. 728–743. ISBN 978-1-4398-4649-0.
24. McMichael, B.L.; Zak, J.C. The Role of Rhizotrons and Minirhizotrons in Evaluating the Dynamics of Rhizoplane-Rhizosphere Microflora. In *Microbial Activity in the Rhizoshere*; Mukerji, K.G., Manoharachary, C., Singh, J., Eds.; Springer: Berlin/Heidelberg, Germany, 2006; pp. 71–87. ISBN 978-3-540-29420-7.
25. Singh, V.; van Oosterom, E.J.; Jordan, D.R.; Messina, C.D.; Cooper, M.; Hammer, G.L. Morphological and Architectural Development of Root Systems in Sorghum and Maize. *Plant Soil* **2010**, *333*, 287–299. [CrossRef]
26. Tian, H.; Smet, I.D.; Ding, Z. Shaping a Root System: Regulating Lateral versus Primary Root Growth. *Trends Plant Sci.* **2014**, *19*, 426–431. [CrossRef]
27. Zobel, R.W.; Waisel, Y. A Plant Root System Architectural Taxonomy: A Framework for Root Nomenclature. *Plant Biosyst.* **2010**, *144*, 507–512. [CrossRef]
28. Manschadi, A.M.; Christopher, J.; deVoil, P.; Hammer, G. The Role of Root Architectural Traits in Adaptation of Wheat to Water-Limited Environments. *Funct. Plant Biol.* **2006**, *33*, 823–837. [CrossRef]
29. Sarkhosh, A.; Olmstead, M.A.; Chaparro, J.X.; Beckman, T.G. *Rootstocks for Florida Stone Fruit*; Horticultural Sciences Department, University of Florida IFAS Extension: Gainesville, FL, USA, 2018.
30. Kaur, S. Evaluation of Different Doses of Indole-3-Butyric Acid (IBA) on the Rooting, Survival and Vegetative Growth Performance of Hardwood Cuttings of Flordaguard Peach (*Prunus persica* L. Batch). *J. Appl. Nat. Sci.* **2017**, *9*, 173–180. [CrossRef]
31. Menegatti, R.D.; Souza, A.d.G.; Bianchi, V.J. Growth and Nutrient Accumulation in Three Peach Rootstocks until the Grafting Stage. *Comun. Sci.* **2019**, *10*, 467–476. [CrossRef]
32. Egilla, J.N.; Byrne, D. The Search for Peach Rootstocks Tolerant to Alkalinity. *Fruit Var. J.* **1989**, *43*, 7–11.
33. Reighard, G.L. Peach Rootstocks for the United States: Are Foreign Rootstocks the Answer? *HortTechnology* **2000**, *10*, 714–718. [CrossRef]
34. Felipe, A.J. 'Felinem', 'Garnem', and 'Monegro' Almond x Peach Hybrid Rootstocks. *HortScience* **2009**, *44*, 196–197. [CrossRef]
35. Başar, H. Factors Affecting Iron Chlorosis Observed in Peach Trees in the Bursa Region. *Turk. J. Agric. For.* **2000**, *24*, 237–245.
36. Beckman, T.G.; Lang, G.A. Rootstock Breeding for Stone Fruits. In Proceedings of the Acta Horticulturae; International Society for Horticultural Science (ISHS), Leuven, Belgium, 31 August 2003; pp. 531–551.
37. Charlesworth, D.; Willis, J.H. The Genetics of Inbreeding Depression. *Nat. Rev. Genet.* **2009**, *10*, 783–796. [CrossRef]
38. García-Dorado, A. A Simple Method to Account for Natural Selection When Predicting Inbreeding Depression. *Genetics* **2008**, *180*, 1559–1566. [CrossRef]
39. Eissenstat, D.M. Costs and Benefits of Constructing Roots of Small Diameter. *J. Plant Nutr.* **1992**, *15*, 763–782. [CrossRef]

40. Solari, L.I.; Pernice, F.; DeJong, T.M. The Relationship of Hydraulic Conductance to Root System Characteristics of Peach (*Prunus persica*) Rootstocks. *Physiol. Plant.* **2006**, *128*, 324–333. [CrossRef]
41. Basile, B.; Bryla, D.R.; Salsman, M.L.; Marsal, J.; Cirillo, C.; Johnson, R.S.; DeJong, T.M. Growth Patterns and Morphology of Fine Roots of Size-Controlling and Invigorating Peach Rootstocks. *Tree Physiol.* **2007**, *27*, 231–241. [CrossRef]
42. Syvertsen, J.P.; Graham, J.H. Hydraulic Conductivity of Roots, Mineral Nutrition, and Leaf Gas Exchange of Citrus Rootstocks. *J. Am. Soc. Hortic. Sci.* **1985**, *110*, 865–869. [CrossRef]
43. Eissenstat, D.M. On the Relationship between Specific Root Length and the Rate of Root Proliferation: A Field Study Using Citrus Rootstocks. *New Phytol.* **1991**, *118*, 63–68. [CrossRef]
44. Hammer, G.L.; Dong, Z.; McLean, G.; Doherty, A.; Messina, C.; Schussler, J.; Zinselmeier, C.; Paszkiewicz, S.; Cooper, M. Can Changes in Canopy and/or Root System Architecture Explain Historical Maize Yield Trends in the U.S. Corn Belt? *Crop Sci.* **2009**, *49*, 299–312. [CrossRef]
45. Dunbabin, V.; Diggle, A.; Rengel, Z. Is There an Optimal Root Architecture for Nitrate Capture in Leaching Environments? *Plant Cell Environ.* **2003**, *26*, 835–844. [CrossRef] [PubMed]
46. Glenn, D.M.; Welker, W.V. Root Development Patterns in Field Grown Peach Trees. *J. Am. Soc. Hortic. Sci.* **1993**, *118*, 362–365. [CrossRef]
47. Watt, M.; Magee, L.J.; McCully, M.E. Types, Structure and Potential for Axial Water Flow in the Deepest Roots of Field-Grown Cereals. *New Phytol.* **2008**, *178*, 135–146. [CrossRef]
48. Deak, K.I.; Malamy, J. Osmotic Regulation of Root System Architecture. *Plant J.* **2005**, *43*, 17–28. [CrossRef]
49. Vercambre, G.; Pagès, L.; Doussan, C.; Habib, R. Architectural Analysis and Synthesis of the Plum Tree Root System in an Orchard Using a Quantitative Modelling Approach. *Plant Soil* **2003**, *251*, 1–11. [CrossRef]
50. Ziegler, V.H.; Ploschuk, E.L.; Weibel, A.M.; Insausti, P. Short-Term Responses to Flooding Stress of Three Prunus Rootstocks. *Sci. Hortic.* **2017**, *224*, 135–141. [CrossRef]
51. Lesmes-Vesga, R.A.; Chaparro, J.X.; Sarkhosh, A.; Ritenour, M.A.; Cano, L.M.; Rossi, L. Effect of Propagation Systems and Indole-3-Butyric Acid Potassium Salt (K-IBA) Concentrations on the Propagation of Peach Rootstocks by Stem Cuttings. *Plants* **2021**, *10*, 1151. [CrossRef]
52. Caruso, T.; Mafrica, R.; Bruno, M.; Vescio, R.; Sorgonà, A. Root Architectural Traits of Rooted Cuttings of Two Fig Cultivars: Treatments with Arbuscular Mycorrhizal Fungi Formulation. *Sci. Hortic.* **2021**, *283*, 110083. [CrossRef]
53. Ryser, P.; Lambers, H. Root and Leaf Attributes Accounting for the Performance of Fast- and Slow-Growing Grasses at Different Nutrient Supply. *Plant Soil* **1995**, *170*, 251–265. [CrossRef]
54. Sorgonà, A.; Abenavoli, M.R.; Gringeri, P.G.; Cacco, G. Comparing Morphological Plasticity of Root Orders in Slow- and Fast-Growing Citrus Rootstocks Supplied with Different Nitrate Levels. *Ann. Bot.* **2007**, *100*, 1287–1296. [CrossRef] [PubMed]
55. Ramalingam, P.; Kamoshita, A.; Deshmukh, V.; Yaginuma, S.; Uga, Y. Association between Root Growth Angle and Root Length Density of a Near-Isogenic Line of IR64 Rice with DEEPER ROOTING 1 under Different Levels of Soil Compaction. *Plant Prod. Sci.* **2017**, *20*, 162–175. [CrossRef]

plants

MDPI

Article

Interactive Effect of Cultivars, Crop Years and Rootstocks on the Biochemical Traits of *Prunus persica* (L.) Batsch Fruits

Ivan I. Morales-Manzo [1], Ana M. Ribes-Moya [1], Claudia Pallotti [1], Ana Jimenez-Belenguer [2,*], Clara Pérez Moro [1], María Dolores Raigón [1], Adrián Rodríguez-Burruezo [1] and Ana Fita [1,*]

1 Instituto de Conservación y Mejora de la Agrodiversidad Valenciana, Edificio 8E Escalera J, CPI, Universitat Politècnica de València, 46022 Valencia, Spain; ivmoman@posgrado.upv.es (I.I.M.-M.)
2 Centro Avanzado de Microbiología Aplicada, Universitat Politècnica de València, Camino de Vera s/n, 46022 Valencia, Spain
* Correspondence: anjibe@upvnet.upv.es (A.J.-B.); anfifer@btc.upv.es (A.F.); Tel.: +34-9-6387-9418 (A.F.)

Abstract: Modern agriculture has boosted the production of food based on the use of pesticides and fertilizers and improved plant varieties. However, the impact of some such technologies is high and not sustainable in the long term. Although the importance of rhizospheres in final plant performance, nutrient cycling, and ecosystems is well recognized, there is still a lack of information on the interactions of their main players. In this paper, four accessions of pepper are studied at the rhizosphere and root level under two farming systems: organic and conventional. Variations in soil traits, such as induced respiration, enzymatic activities, microbial counts, and metabolism of nitrogen at the rhizosphere and bulk soil, as well as measures of root morphology and plant production, are presented. The results showed differences for the evaluated traits between organic and conventional management, both at the rhizosphere and bulk soil levels. Organic farming showed higher microbial counts, enzymatic activities, and nitrogen mobilization. Our results also showed how some genotypes, such as Serrano or Piquillo, modified the properties of the rhizospheres in a very genotype-dependent way. This specificity of the soil–plant interaction should be considered for future breeding programs for soil-tailored agriculture.

Keywords: rhizosphere; soil respiration; enzymatic activity; N cycle; *Capsicum annuum*; soil microorganism

Citation: Morales-Manzo, I.I.; Ribes-Moya, A.M.; Pallotti, C.; Jimenez-Belenguer, A.; Moro, C.P.; Raigón, M.D.; Rodríguez-Burruezo, A.; Fita, A. Root–Soil Interactions for Pepper Accessions Grown under Organic and Conventional Farming. *Plants* **2023**, *12*, 1873. https://doi.org/10.3390/plants12091873

Academic Editors: Lorenzo Rossi and Gianluca Caruso

Received: 16 March 2023
Revised: 12 April 2023
Accepted: 28 April 2023
Published: 3 May 2023

1. Introduction

Capsicum peppers are one of the most relevant vegetables (and spices) in the world. They are an economically important crop, particularly appreciated for their nutritional properties and antioxidant content [1]. In Spain, which is the main pepper producer within the EU [2], peppers are cultivated mainly as an intensive high input crop in the Andalucía and Murcia region [3]. However, in recent years, organic farming has increased in importance [4]. In contrast with conventional farming, where production is based on creating an "ideal" environment for plant development by limiting abiotic or biotic restrictions by any means, organic agriculture is based on maintaining an equilibrated ecosystem (especially in the soil, as no other substrates are allowed) compatible with agricultural production. Therefore, their management and consequences for the environment can be very different.

It is well known that plant characteristics, development, yield, and quality are affected by the environment [5–7]. The majority of research focuses on temperature or hydrological conditions, but little is known about the soil environment, which can be profoundly different from one field to another, especially if they are under distinct management systems. Different soil-environment conditions affect nutrition [8] and the health condition of the soil. For example, the effect of plant growth-promoting rhizobacteria (PGPR) in eliciting so-called "induced systemic tolerance (IST)" in plants under different abiotic stresses is well known [9–11]. Most of these interactions take place at the rhizosphere level.

The rhizosphere is the biologically active zone of soil where plant roots and soil interact and is of great importance for plant performance as well as for nutrient cycling and ecosystem functioning [12]. Rhizosphere processes are poorly understood and in situ agricultural soils are largely uncharacterized [13]. The rhizosphere dynamics involve complex interactions among roots, root exudates, the physical and chemical properties of the soil, and soil microorganisms, among others. All these factors change according to the others in the soil system. First, root architecture and root exudates depend mainly on plant species, genotype, and farming techniques, especially fertilization regimes [14,15]. It has been demonstrated how root exudates can shape the microbial community in the rhizosphere [16]. Second, different farming systems also modify the rhizosphere's microbial communities [17]. For example, chemical fertilizers used in conventional farming supply nutrients, mainly N, P and K, whereas organic fertilizers also supply different amounts of C with macro and micronutrients, thus selecting microbial communities with different nutritional requirements [18,19]. Therefore, the different techniques applied (tilling, inorganic fertilization, etc.) modify the soil properties and microbiome [20]. Finally, soil microbes can play an important role in growth and nutrient uptake by plants as well as modifying their tolerance to biotic and abiotic stresses. Genotypes, farming systems, and microorganisms shape the performance of a crop in the field; thus, they are important aspects to consider in plant breeding.

Microbial community diversity and soil functional diversity are effective measures to illustrate the effects of natural and anthropological actions on the soil [21,22]. The characterization of soil microbiological community diversity and functional diversity can be assessed by different means, such as determining the microbial profiles [23], analyzing microbial catabolic potential [24], analyzing soil enzymatic activities [25], and by substrate-induced respiration measures [26]. Assessments of microbial activity with the MicroResp™ system need small volumes of soil, making the method suitable for rhizosphere studies [27]. These traits can be used as appropriate indicators of soil heterogeneity and performance, which have been used extensively to check the health status of soil under different agricultural systems [28].

To sum up, to change from intensive agricultural systems to less harmful agriculture, increased knowledge of the rhizosphere's processes is needed. In this paper, four accessions of pepper, selected for having diverse root systems and phosphorous uptake efficiency [14], are studied at the rhizosphere and root level under two farming systems (organic and conventional) to better understand root–soil interactions for more sustainable agriculture.

2. Results

All data obtained were subjected to exploratory MANOVA to identify the possible effects of the accessions and the farming system on the rhizosphere's characteristics and plant performance. This analysis was carried out by separating the two sampling times, which correspond with two phenological stages of the plant: at the vegetative stage (T1, summer) and at the late fruit stage (T2, autumn). The MANOVA demonstrated that the farming system had a significant effect on the rhizosphere's studied traits at T1 (Figure 1a). At this time, 9 out of 17 evaluated rhizosphere traits showed significant differences due to the farming system, in contrast with just 1 trait that showed significant differences according to the accession. Interestingly, accession $\times$ farming system interaction was observed in five studied traits. This situation changed at T2 (Figure 1b), where although the farming system effect was the highest, being significant for 12 out of 17 traits, the accession effect increased the number of meaningful traits, being significant for 7 out of 17 traits. In this phenological stage, the interaction effect was significant for five traits again, but only one was the same as T1. In the case of root and biomass traits, influence at T1 of the accession was as important as the farming system and interaction effects, being significant in 9, 9, and 10 out of 14 traits, respectively (Figure 1a). At T2, the accession and farming system effects increased (15 and 11, respectively, Figure 1b). In the following sections, the values of the different rhizosphere and plant traits in each farming system and sampling time are described.

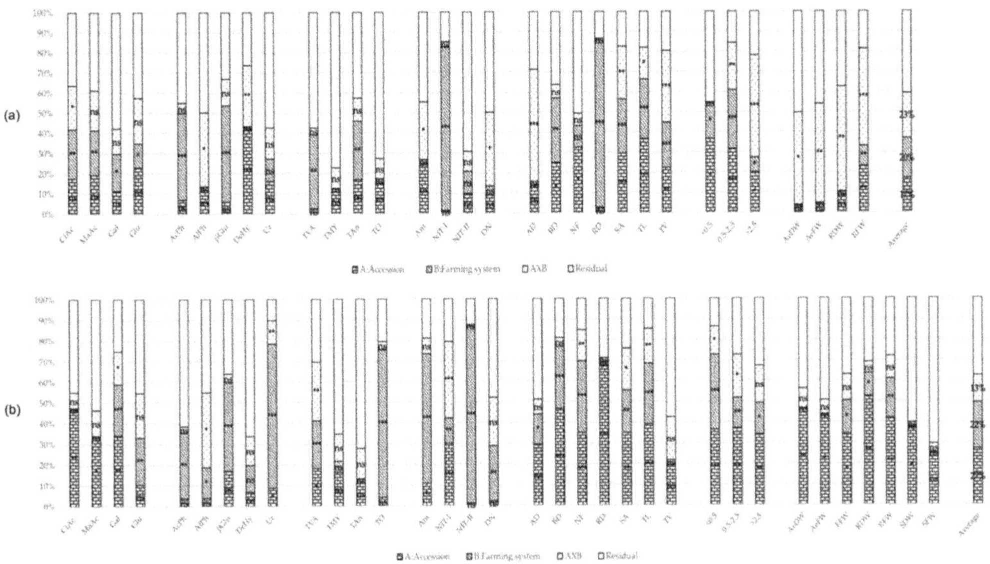

Figure 1. Percentage of MANOVA effects' sum of squares (SS) contribution for trait; (**a**) At the vegetative stage (T1); (**b**) At the late fruit stage (T2). Percentage contribution of variance of each effect: Accessions and bulk (when available) (A, bricks pattern), Farming system (B, lines pattern), A × B interaction (dots pattern) and Residual (plain white). Substrate induced respiration: CiAc, MaAc, Gal and Glu; Enzymatic activity: AcPh, AlPh, BGlu, DeHy, and Ur; Microbial count: TVA, TMY, TAn, and TO; N. cycle: Am, NIT-I, NIT-II, and DN; Root traits: AD, BD, NF, RD, SA, TL, TV, and diameter classes: ≤0.5, 0.5–2.5, and >2.5; Biomass: AeDW, AeFW, FFW, RDW, RFW, SDW, and SFW; Average percentage MANOVA effects' SS; *–***: significant at p-values of 0.05–0.001, ns non-significant.

2.1. *Rhizosphere's Traits*

2.1.1. Substrate Induced Respiration

At T1, induced respiration was significantly different between farming systems regardless of the substrate (Figure 1a and Table 1). Induction with citric acid (CiAc) and malic acid (MaAc) produced lower respiration under the organic farming system compared to the conventional farming system: 17.70 μg CO_2 g soil^{-1} h^{-1} at organic vs. 53.02 μg CO_2 g soil^{-1} h^{-1} at conventional, 21.47 μg CO_2 g soil^{-1} h^{-1} at organic vs. 47.02 μg CO_2 g soil^{-1} h^{-1} at conventional, respectively (Table 1). CiAc also showed accession × farming system interaction, probably due to the different performance of *Bola* accession, which has higher respiration in organic, 45.82 μg CO_2 g soil^{-1} h^{-1}, than conventional soil, 38.19 μg CO_2 g soil^{-1} h^{-1}.

Respiration with sugar substrates such as Gal and Glu was also lower in the organic field than the conventional (1.58 μg CO_2 g soil^{-1} h^{-1} vs. 2.96 μg CO_2 g soil^{-1} h^{-1} and 4.02 μg CO_2 g soil^{-1} h^{-1} vs. 13.80 μg CO_2 g soil^{-1} h^{-1}, respectively). None of the accessions showed significantly different respiration rates from the ones shown in the bulk soil.

At T2, the trend for all the substrates in the organic system was to show a slight increase with respect to T1 (Table 1), whereas in the conventional system, the induced respiration was lower at T2 than at T1, especially for Gal and Glu. Therefore, the induced respiration for CiAc and MaAc at T2 was not significantly different between farming systems, but there were significant differences for Gal: 1.83 μg CO_2 g soil^{-1} h^{-1} at organic vs. 1.15 μg CO_2 g soil^{-1} h^{-1} at conventional, and for Glu: 4.73 μg CO_2 g soil^{-1} h^{-1} at organic vs. 2.61 μg CO_2 g soil^{-1} h^{-1} at conventional (Figure 1b, Table 1). Interestingly, at

T2, the accession factor was the major contributor to the variance for CiAc, MaAc, and Gal (Figure 1b). For these sources of carbon, the rhizospheres of *Serrano* and *Bola* produced significantly higher amounts of CO_2 than the rest of the accessions, both in conventional and organic farming systems (Table 1) and were stable between phenological stages.

Table 1. Mean values (n = 3) of substrate induced respiration (μg CO_2 g soil^{-1} h^{-1}) from rhizospheres and bulk soil samples using different substrates: citric acid (CiAc), malic acid (MaAc), galactose (Gal), glucose (Glu), in two farming systems (organic; org and conventional; con) at two sampling times (T1: summer, T2: autumn).

Sample	Substrate	Farming System	*BOL-58*	*Serrano*	*Bola*	*Piquillo*	Bulk	Average
T1	CiAc	Org	08.35 a	08.09 a	45.82 b	15.58 a	10.65 a	17.70
		Con	60.92 ns	06.17 ns	38.19 ns	82.02 ns	77.80 ns	53.02 *
	MaAc	Org	08.86 a	08.88 a	46.58 b	18.66 a	24.35 a	21.47
		Con	70.90 ns	13.51 ns	43.07 ns	61.08 ns	46.54 ns	47.02 *
	Gal	Org	01.69 ns	01.66 ns	01.89 ns	01.36 ns	01.29 ns	01.58
		Con	02.75 ns	01.07 ns	04.02 ns	02.91 ns	04.06 ns	02.96 *
	Glu	Org	03.68 ns	04.33 ns	04.00 ns	03.92 ns	04.16 ns	04.02
		Con.	06.14 ns	02.46 ns	12.58 ns	07.67 ns	40.16 ns	13.80 *
T2	CiAc	Org.	25.75 ab	58.76 b	37.33 ab	08.13 a	12.24 a	28.44 NS
		Con.	22.62 ns	51.84 ns	42.64 ns	35.99 ns	06.87 ns	31.99 NS
	MaAc	Org.	29.80 ns	43.62 ns	50.53 ns	08.06 ns	05.51 ns	27.50 NS
		Con.	27.98 ns	30.28 ns	43.57 ns	32.39 ns	22.37 ns	31.32 NS
	Gal	Org.	01.81 ns	01.80 ns	02.22 ns	01.64 ns	01.66 ns	01.83 *
		Con.	01.13 ab	02.08 b	01.66 ab	00.41 a	00.47 a	01.15
	Glu	Org.	04.01 ns	05.87 ns	06.56 ns	03.41 ns	03.82 ns	04.73 *
		Con.	01.69 ns	02.39 ns	01.77 ns	02.12 ns	05.10 ns	02.61

* Mean values with different lower-case letters within rows indicate significant differences among rhizospheres including bulk soil based on the Student–Newman–Keuls multiple range test at *p*-value < 0.05, ns indicates no significant differences. Asterisk in last column (average), indicates significant differences between farming systems for pairs of data from the same substrate at *p*-value < 0.05, NS in the last column indicates no significant differences between farming systems for pairs of data from the same substrate.

2.1.2. Soil Enzymatic Activity

Acid phosphatase activity (AcPh) was significantly different among farming systems at both phenological stages (Figure 1a,b), with mean values significantly higher at organic (91.09 μmol g^{-1} h^{-1} at T1 and 89.88 μmol p-nitrophenol soil g^{-1} h^{-1} at T2) than at conventional (73.89 μmol g^{-1} h^{-1} and 74.51 μmol g^{-1} h^{-1} at T1 and T2, respectively) (Table 2). No significant differences were observed between the rhizospheres or the bulk soil for this enzymatic activity. The case of alkaline phosphatase (AlPh) was different. At T1, the only significant effect was the accession × farming system interaction (Figure 1a), which was due mainly to the higher AlPh activity registered for *Serrano*'s rhizosphere in comparison to the rest of the accessions or the bulk soil at organic (563.92 μmol p-nitrophenol soil g^{-1} h^{-1}) but not at conventional (402.87 μmol g^{-1} h^{-1}). This trend was conserved at T2, with significant differences in AlPh activity in the rhizosphere of *Serrano* (569.41 μmol g^{-1} h^{-1} at organic). In addition, at T2, there were significant differences between farming systems for AlPh (485.33 μmol g^{-1} h^{-1} at organic and 427 μmol g^{-1} h^{-1} at conventional, Table 2).

β-glucosidase (βGlu) activity was similar at both sampling times; the activity of this enzyme was significantly higher in organic 0.72 μmol p-nitrophenol soil g^{-1} h^{-1} at T1 and 0.62 μmol g^{-1} h^{-1} at T2 than for conventional cultivation, 0.42 μmol g^{-1} h^{-1} at T1 and 0.45 μmol g^{-1} h^{-1} at T2. For this enzyme activity, there were no differences regarding the accession rhizospheres. Dehydrogenase (DeHy) activity in the soil was around 1 μmol INTF soil g^{-1} h^{-1}. At T1, there were accession and accession x farming system interaction effects, mainly due to DeHy activity in the *BOL-58* rhizosphere, 1.18 μmol g^{-1} h^{-1}, in the organic field, which was higher than other genotypes and bulk soil, and by the low activity of DeHy in the rhizosphere of *Serrano* in the conventional field, 0.78 μmol g^{-1} h^{-1}

(Table 2). At T2, there were no significant effects. The activity of the urease (Ur) was around 0.165 μmol $N\text{-}NH_4^+$ soil g^{-1} h^{-1} T1 and there were no differences due to the farming system or accession. However, at T2, although the activity was similar, an increase was observed for all accessions and bulk, as well as significant differences among the two farming systems (0.36 μmol g^{-1} h^{-1} at organic and 0.20 μmol g^{-1} h^{-1} at conventional). For the accession, Ur activity from *Bola*'s rhizosphere was higher than the rest in organic cultivation, at 0.47 μmol g^{-1} h^{-1} (Table 2).

Table 2. Mean values (n = 3) of enzymatic activity of the rhizospheres and the bulk soil for acid phosphatase (AcPh, μmol p-nitrophenol soil g^{-1} h^{-1}), alkaline phosphatase (AlPh, μmol p-nitrophenol soil g^{-1} h^{-1}), β-glucosidase (βGlu, μmol p-nitrophenol soil g^{-1} h^{-1}), dehydrogenase (DeHy, μmol INTF soil g^{-1} h^{-1}), and urease (Ur, μmol $N\text{-}NH_4^+$ soil g^{-1} h^{-1}) in two farming systems (organic: org or conventional: con) at two sampling times (T1, summer and T2, autumn).

Sampling	Enzyme	Farming System	*BOL-58*	*Serrano*	*Bola*	*Piquillo*	Bulk	Average
T1	AcPh	Org.	90.16 ^{ns}	89.34 ^{ns}	98.49 ^{ns}	87.59 ^{ns}	89.85 ^{ns}	91.09 *
		Con.	76.15 ^{ns}	65.68 ^{ns}	76.81 ^{ns}	80.23 ^{ns}	70.56 ^{ns}	73.89
	AlPh	Org.	455.69 ^{a}	563.92 ^{b}	450.40 ^{a}	429.01 ^{a}	467.92 ^{a}	473.39 ^{NS}
		Con.	460.04 ^{ns}	402.87 ^{ns}	435.10 ^{ns}	459.57 ^{ns}	526.41 ^{ns}	456.80 ^{NS}
	βGlu	Org.	00.72 ^{ns}	00.61 ^{ns}	00.80 ^{ns}	00.68 ^{ns}	00.79 ^{ns}	00.72 *
		Con.	00.40 ^{ns}	00.60 ^{ns}	00.55 ^{ns}	00.44 ^{ns}	00.39 ^{ns}	00.48
	DeHy	Org.	01.18 ^{b}	00.95 ^{a}	00.91 ^{a}	00.99 ^{a}	00.94 ^{a}	00.99 ^{NS}
		Con.	01.08 ^{b}	00.78 ^{a}	01.14 ^{b}	01.07 ^{b}	01.02 ^{b}	01.02 ^{NS}
	Ur	Org.	00.16 ^{ns}	00.17 ^{ns}	00.17 ^{ns}	00.18 ^{ns}	00.20 ^{ns}	00.18 ^{NS}
		Con.	00.14 ^{ns}	00.13 ^{ns}	00.15 ^{ns}	00.20 ^{ns}	00.14 ^{ns}	00.15 ^{NS}
T2	AcPh	Org.	93.99 ^{ns}	85.00 ^{ns}	95.04 ^{ns}	87.75 ^{ns}	87.60 ^{ns}	89.88 *
		Con.	76.11 ^{ns}	69.38 ^{ns}	72.79 ^{ns}	76.03 ^{ns}	78.22 ^{ns}	74.51
	AlPh	Org.	477.13 ^{a}	569.41 ^{b}	472.26 ^{a}	449.11 ^{a}	458.76 ^{a}	485.33 *
		Con.	406.98 ^{ns}	350.58 ^{ns}	420.99 ^{ns}	447.58 ^{ns}	510.30 ^{ns}	427.29
	βGlu	Org.	00.62 ^{ns}	00.70 ^{ns}	00.67 ^{ns}	00.60 ^{ns}	00.52 ^{ns}	00.62 *
		Con.	00.53 ^{ns}	00.43 ^{ns}	00.50 ^{ns}	00.44 ^{ns}	00.34 ^{ns}	00.45
	DeHy	Org.	01.04 ^{ns}	00.94 ^{ns}	00.96 ^{ns}	00.99 ^{ns}	00.98 ^{ns}	00.98 ^{NS}
		Con.	01.02 ^{ns}	01.03 ^{ns}	01.14 ^{ns}	00.99 ^{ns}	01.07 ^{ns}	01.05 ^{NS}
	Ur	Org.	00.31 ^{a}	00.36 ^{a}	00.47 ^{b}	00.35 ^{a}	00.30 ^{a}	00.36 *
		Con.	00.23 ^{ns}	00.19 ^{ns}	00.19 ^{ns}	00.20 ^{ns}	00.18 ^{ns}	00.20

* Mean values with different lower-case letters within rows indicate significant differences among rhizospheres including bulk soil based on the Student–Newman–Keuls multiple range test at p-value < 0.05, whereas, ns indicates no significant differences. Asterisk in last column (average), indicates significant differences between farming systems for pairs of data from the same substrate at p-value < 0.05, NS in the last column indicates no significant differences between farming systems for pairs of data from the same substrate.

2.1.3. Microbial Counts

At T1, the counts of total viable aerobic mesophilic bacteria (TVA) and the counts of total anaerobic bacteria (TAn) were significantly different depending on the farming system (Figure 1a). The higher means were observed for the organic system, with 7.40 log CFU g soil^{-1} for TVA and 6.45 log CFU g soil^{-1} for TAn (Table 3). For total molds and yeast counts (TMY) and for total oligotrophic bacteria counts (TO), no significant effects or differences were observed among systems, nor between the rhizosphere's accessions or the bulk soil, with the average count for molds and yeast being 5.56 log CFU g soil^{-1} and the average count for oligotrophic bacteria being 7.27 log CFU g soil^{-1}. At T2, the difference among systems for the TVA was maintained as in T1 (Figure 1b, Table 3). However, during this period, the accession and accession x system interactions also had significant effects. The rhizosphere of *Serrano* in the conventional system (7.79 log CFU g soil^{-1}) had higher counts of viable bacteria than *Piquillo*'s rhizosphere (6.76 log CFU g soil^{-1}) (Table 3). No differences were observed due to the farming systems, rhizospheres, or bulk soil at T2 for TMY (5.21 log CFU g soil^{-1}) and for TAn (5.92 log CFU g soil^{-1}). It is interesting to note that the TAn was lower at T2 in organic farming but not in conventional farming. Finally, TO showed significant differences among systems (Figure 1b, Table 3) at T2. In this case, the conventional system showed lower counts (6.66 log CFU g soil^{-1}) in comparison with the organic system, or the values registered at T1.

Table 3. Mean values of microbial counts (log CFU g soil^{-1}) from rhizospheres and bulk soil samples using different growing media: total viable aerobic mesophilic bacterial count (TVA), total molds and yeast count (TMY), total anaerobic bacteria (TAn), total oligotrophic bacterial count (TO) in two farming systems (organic; org and conventional; con) at two sampling times (T1: summer, T2: autumn).

Sampling	Microbial Count	Farming System	*BOL-58*	*Serrano*	*Bola*	*Piquillo*	Bulk	Average
T1	TVA	Org	7.34 ^{ns}	7.51 ^{ns}	7.53 ^{ns}	7.37 ^{ns}	7.27 ^{ns}	7.40 *
		Con	6.88 ^{ns}	7.00 ^{ns}	6.78 ^{ns}	7.04 ^{ns}	6.89 ^{ns}	6.92
	TMY	Org	8.30 ^{ns}	4.67 ^{ns}	5.57 ^{ns}	5.12 ^{ns}	4.45 ^{ns}	5.62 ^{NS}
		Con	5.64 ^{ns}	5.78 ^{ns}	5.59 ^{ns}	5.06 ^{ns}	5.40 ^{ns}	5.50 ^{NS}
	TAn	Org	7.29 ^{ns}	5.98 ^{ns}	6.67 ^{ns}	6.57 ^{ns}	5.74 ^{ns}	6.45 *
		Con	5.64 ^{ns}	5.69 ^{ns}	5.53 ^{ns}	5.90 ^{ns}	5.39 ^{ns}	5.63
	TO	Org	7.18 ^{ns}	7.33 ^{ns}	7.55 ^{ns}	7.33 ^{ns}	7.23 ^{ns}	7.33 ^{NS}
		Con.	7.18 ^{ns}	7.49 ^{ns}	7.38 ^{ns}	6.76 ^{ns}	7.24 ^{ns}	7.21 ^{NS}
T2	TVA	Org	7.62 ^{ns}	7.40 ^{ns}	7.43 ^{ns}	7.46 ^{ns}	7.36 ^{ns}	7.45 *
		Con	6.19 ^{a}	7.79 ^{b}	7.05 ^{ab}	6.42 ^{a}	7.01 ^{ab}	6.89
	TMY	Org	4.76 ^{ns}	5.47 ^{ns}	5.43 ^{ns}	5.33 ^{ns}	5.44 ^{ns}	5.329 ^{NS}
		Con	5.10 ^{ns}	5.32 ^{ns}	5.11 ^{ns}	5.05 ^{ns}	4.97 ^{ns}	5.11 ^{NS}
	TAn	Org	5.81 ^{ns}	5.66 ^{ns}	6.18 ^{ns}	5.82 ^{ns}	5.68 ^{ns}	5.83 ^{NS}
		Con	5.42 ^{ns}	6.86 ^{ns}	6.13 ^{ns}	5.68 ^{ns}	5.94 ^{ns}	6.01 ^{NS}
	TO	Org	7.41 ^{ns}	7.29 ^{ns}	7.19 ^{ns}	7.31 ^{ns}	7.22 ^{ns}	7.28 *
		Con.	6.73 ^{ns}	6.58 ^{ns}	6.85 ^{ns}	6.64 ^{ns}	6.48 ^{ns}	6.66

* Mean values with different lower-case letters within rows indicate significant differences among rhizospheres including bulk soil based on the Student–Newman–Keuls multiple range test at p-value < 0.05, ns indicates no significant differences. Asterisk in last column (average), indicates significant differences between farming systems for pairs of data from the same substrate at p-value < 0.05, NS in the last column indicates no significant differences between farming systems for pairs of data from the same substrate.

2.1.4. Nitrogen Catabolism Potential (N-Cycle)

At T1, for ammonification (Am), the only significant effect was the accession × farming system interaction (Figure 1a), mainly due to the rhizosphere of *Piquillo* in the conventional field (310.4 mg NH_4^+-N g^{-1}), which showed a higher rate of ammonification in comparison with the others, contravening the general tendency to have higher ammonification potential in the organic field (Table 4). At T1, the nitrification I (NIT-I) potential was significantly higher in the organic (20.27 mg NO_2-N g^{-1}) than the conventional (2.56 mg NO_2-N g^{-1}). No differences were found among samples for nitrification II (NIT-II), with an average of 20.34 mg NO_3-N g^{-1}. For potential denitrification (DN), there were no significant differences among samples, which were, on average, 2.8.

At T2, all N-cycle parameters showed differences among farming systems with higher activity in the organic field, except for DN, which had higher levels in the conventional field (Figure 1b, Table 3). Interestingly, again, the rhizosphere of *Piquillo* in conventional farming showed the highest Am among accessions and bulk (31.04 mg NH_{4^+}-N g^{-1}). For NIT-I, the general values were very low, ranging from 0.15 to 0.76 mg NO_{2^-}-N g^{-1}, exceeding the values of the rhizosphere of *Bola* with values of 2.53 mg NO_{2^-}-N g^{-1}, although this value is lower than in T1. Lastly, at T2, there were significant differences in DN levels, which were higher in the conventional field (Table 4).

Table 4. Mean values of nitrogen catabolism products of the rhizospheres and the bulk soil for ammonification potential (Am, mg NH_4^+-N g^{-1}), nitrification I potential (NIT-I, mg NO_2^+-N g^{-1}), nitrification II potential (NIT-II, mg NO_3^+-N g^{-1}), denitrification potential (DN, 0–4 scale) in two farming systems (organic: org or conventional: con) at two sampling times (T1, summer and T2, autumn).

Sampling	N-Cycle Stage	Farming System	*BOL-58*	*Serrano*	*Bola*	*Piquillo*	Bulk	Average
T1	Am	Org	232.8 ns	119 ns	206.9 ns	144.8 ns	75.0 ns	155.7 NS
		Con	100.9 a	119 a	31.0 a	310.4 b	51.7 a	122.6 NS
	NIT-I	Org	24.32 ns	16.21 ns	20.27 ns	20.27 ns	20.27 ns	20.27 *
		Con	2.28 ns	3.24 ns	1.72 ns	2.03 ns	3.55 ns	2.56
	NIT-II	Org	22.60 ns	15.07 ns	18.83 ns	18.83 ns	18.83 ns	18.83 NS
		Con	22.60 ns	22.60 ns	18.83 ns	22.60 ns	22.60 ns	21.85 NS
	DN	Org	2.50 ns	3.33 ns	1.00 ns	2.33 ns	3.67 ns	2.57 NS
		Con.	3.50 ns	2.67 ns	3.33 ns	3.33 ns	2.33 ns	3.03 NS
T2	Am	Org.	77.60 ns	38.80 ns	77.60 ns	67.25 ns	103.5 ns	72.94 *
		Con.	6.73 a	3.10 a	4.91 a	31.04 b	15.52 a	12.26
	NIT-I	Org.	0.20 a	0.76 a	2.53 b	0.20 a	0.20 a	0.78 *
		Con.	0.15 ns	0.22 ns	0.15 ns	0.30 ns	0.35 ns	0.24
	NIT-II	Org.	18.83 ns	16.95 ns	18.83 ns	18.83 ns	22.60 ns	19.21 *
		Con.	1.51 ns	1.88 ns	2.26 ns	2.26 ns	2.26 ns	2.03
	DN	Org.	1.33 ns	2.50 ns	1.67 ns	1.00 ns	2.33 ns	1.77
		Con.	3.00 ns	2.00 ns	3.67 ns	3.67 ns	2.67 ns	3.00 *

* Mean values with different lower-case letters within rows indicate significant differences among rhizospheres including bulk soil based on the Student–Newman–Keuls multiple range test at p-value < 0.05, ns indicates no significant differences. Asterisk in last column (average) indicates significant differences between farming systems for pairs of data from the same substrate at p-value < 0.05, NS in the last column indicates no significant differences between farming systems for pairs of data from the same substrate.

2.2. Plant's Traits

2.2.1. Biomass and Yield

At T1, there were significant accession × system interaction effects for all biomass traits (Figure 1a). For organic fields, *BOL-58* and *Serrano* always had heavier shoots (AeDW for dry and AeFW for fresh matter) and roots (RDW for dry and RFW for fresh matter); for the conventional field, *Bola* and *Piquillo* were the heaviest and significantly different (Table 5).

Table 5. Mean values (n = 3) of biomass and yield traits (g) from accessions: Aerial dry weight (AeDW), aerial fresh weight (AeFW), root dry weight (RDW), root fresh weight (RFW), fruit fresh weight (FFW), stump dry weight (SDW), stump fresh weight (SFW) in two farming systems (organic; org and conventional; con) at two sampling times (T1: summer, T2: autumn).

Sample	Biomass Traits	Farming System	*BOL-58*	*Serrano*	*Bola*	*Piquillo*	Average
T1	AeDW	Org	15.40 ns	14.74 ns	9.73 ns	8.69 ns	12.14 NS
		Con	5.97 ns	7.95 ns	17.27 ns	15.91 ns	11.78 NS
	AeFW	Org	106.15 ns	84.86 ns	56.36 ns	49.56 ns	74.23 NS
		Con	44.85 ns	48.75 ns	118.65 ns	116.83 ns	82.27 NS
	RDW	Org	1.79 ns	1.98 ns	1.35 ns	1.22 ns	1.58 NS
		Org	0.83 a	0.69 a	2.24 b	1.85 b	1.40 NS
	RFW	Org	7.98 ns	10.72 ns	6.94 ns	5.48 ns	7.78
		Con	4.90 a	4.84 a	21.49 b	15.63 b	11.72 *

Table 5. *Cont.*

Sample	Biomass Traits	Farming System	*BOL-58*	*Serrano*	*Bola*	*Piquillo*	Average
T2	AeDW	Org	239.67 ns	329.00 ns	117.67 ns	175.00 ns	215.3 NS
		Con	369.33 b	262.00 ab	112.33 a	122.67 a	216.6 NS
	AeFW	Org	1103.7 ns	1173.0 ns	407.00 ns	723.33 ns	851.7 NS
		Con	1533.00 b	883.00 ab	471.18 a	574.57 a	865.4 NS
	RDW	Org	7.60 ns	15.06 ns	3.45 ns	12.13 ns	9.56 *
		Con	6.30 b	10.66 c	2.62 a	5.68 b	6.32
	RFW	Org	19.91 ab	33.69 ab	13.43 a	43.68 b	27.68 *
		Con	14.83 ab	23.90 b	9.64 a	18.86 ab	16.80
	FFW	Org	557.00 ns	1849.0 ns	419.33 ns	714.67 ns	885.0
		Con	1047.4 ns	1629.5 ns	1261.0 ns	1657.3 ns	1398.8 *
	SDW	Org	17.61 ns	25.10 ns	10.89 ns	19.45 ns	18.26 NS
		Con	16.48 ns	26.41 ns	15.24 ns	19.84 ns	19.49 NS
	SFW	Org	47.94 ns	69.60 ns	35.82 ns	58.58 ns	52.99 NS
		Con	44.72 ns	65.79 ns	52.14 ns	61.64 ns	56.08 NS

* Mean values with different lower-case letters within rows indicate significant differences among rhizospheres including bulk soil based on the Student–Newman–Keuls multiple range test at p-value < 0.05, ns indicates no significant differences. Asterisk in last column (average) indicates significant differences between farming systems for pairs of data from the same substrate at p-value < 0.05, NS in the last column indicates no significant differences between farming systems for pairs of data from the same substrate.

At T2, there were significant differences among accessions for almost all biomass traits (Figure 1b, Table 6). *BOL-58* stood out for its great vegetative growth in the aerial part (AeFW of 1103 g at organic and 1533 g at conventional), whereas *Serrano* had the heavier root system (RFW of 33.69 g at organic and 23.9 g at conventional). There were no significant differences in the development of the aerial parts of the accessions depending on the farming system (Figure 1b, Table 5); however, the roots of all genotypes were heavier in the organic field than in the conventional field (Table 5). Contrarily, the total yield (FFW for fresh matter) was higher in conventional than in organic for the accessions tested, except for *Serrano*. There were no significant differences in the stump weight (SDW for dry and SFW for fresh matter) in any accession or farming system (Figure 1b, Table 5), but *Serrano* had the highest values, as seen with root weight.

2.2.2. Root Parameters

At T1, there were significant differences in the roots among accessions, but in most cases, there were also differences due to the farming system and, in some cases, interactions (Figure 1a). The general trend according to the farming system was for the roots to be more branched (0.95 forks cm^{-1}) and dense (0.33 g cm^{-3}) in the organic field than in the conventional field (Table 6), but to be shorter with less volume. However, the roots of the plants growing in the organic system were longer, with a diameter higher than 2.5 mm (>2.5 class).

Regarding the performance of the genotypes at T1, *Bola* and *Piquillo* showed significantly longer (TL, 2356 cm, and 1650 cm, respectively) and more voluminous roots with a higher number of forks than the other genotypes, especially in conventional conditions (NF, Table 6) in the conventional field. *Serrano* stood out for its great root diameter (AD: 1.06 mm) and longer thicker roots (>2.5: 25.11 cm) in the organic field.

At T2, the situation was different, and in this case, higher values were found not only for branching parameters (BD, NF) but also length (TL, >0.5, 0.5–2.5, and >2.5), root volume (TV, cm^3), and root area (SA, cm^2) in organic farming (Table 6). There were also differences among accessions and accession x farming system interactions. At this time, *Piquillo* performed the best in terms of root length (TL) and branching (NF); however, this higher performance was more accentuated in organic farming. Again, like T2, *Serrano* had a vast root diameter (AD) and thicker roots (>2.5).

Table 6. Mean values (n = 3) of root measurements from accessions: Average root diameter (AD, mm), branching density as number of total forks by length (BD, cm^{-1}), number of forks (NF, number), root density (RD, g cm^{-3}), surface area (SA, cm^2), total length (TL, cm), total volume (TV, cm^3), total length of roots with diameter less than 0.5 mm (<0.5, cm), total length of roots with diameter between 0.5 mm and 2.5 mm (0.5–2.5, cm), total length of roots with diameter higher than 2.5 mm (>2.5, mm) in two farming systems (organic; org and conventional; con) at two sampling times (T1: summer, T2: autumn).

Sample	Root Traits	Farming System	*BOL-58*	*Serrano*	*Bola*	*Piquillo*	Average
T1	AD	Org	0.88 ^{a}	1.06 ^{b}	0.77 ^{a}	0.80 ^{a}	0.88 ^{NS}
		Con	0.75 ^{ns}	0.79 ^{ns}	0.88 ^{ns}	0.96 ^{ns}	0.85 ^{NS}
	BD	Org	01.08 ^{ns}	00.84 ^{ns}	00.93 ^{ns}	00.94 ^{ns}	00.95 *
		Con	00.90 ^{b}	00.76 ^{ab}	00.59 ^{a}	00.74 ^{ab}	00.75
	NF	Org	896.50 ^{ns}	539.00 ^{ns}	1001.00 ^{ns}	715.33 ^{ns}	787.96 ^{NS}
		Con	987.00 ^{ab}	588.00 ^{a}	1375.67 ^{b}	1230.67 ^{b}	1045.33 ^{NS}
	RD	Org	0.36 ^{ns}	0.35 ^{ns}	0.30 ^{ns}	0.31 ^{ns}	0.33 *
		Con	0.16 ^{ns}	0.17 ^{ns}	0.15 ^{ns}	0.16 ^{ns}	0.16
	SA	Org	229.96 ^{ns}	214.40 ^{ns}	249.69 ^{ns}	182.88 ^{ns}	219.23
		Con	263.43 ^{a}	190.60 ^{a}	658.16 ^{b}	499.66 ^{b}	402.96 *
	TL	Org	827.97 ^{ns}	634.98 ^{ns}	1042.61 ^{ns}	728.04 ^{ns}	808.40
		Con	1097.45 ^{a}	767.85 ^{a}	2356.56 ^{c}	1650.75 ^{b}	1468.15 *
	TV	Org	5.09 ^{ns}	5.85 ^{ns}	4.76 ^{ns}	3.70 ^{ns}	4.85
		Con	5.15 ^{a}	3.80 ^{a}	14.91 ^{b}	12.20 ^{b}	9.01 *
	<0.5	Org.	411.63 ^{ns}	258.57 ^{ns}	558.74 ^{ns}	353.67 ^{ns}	395.65
		Con.	550.34 ^{ab}	386.05 ^{a}	732.84 ^{b}	550.65 ^{ab}	554.97 *
	0.5–2.5	Org.	403.25 ^{ns}	350.15 ^{ns}	473.91 ^{ns}	365.93 ^{ns}	398.31
		Con.	539.32 ^{a}	374.29 ^{a}	1608.41 ^{c}	1088.20 ^{b}	902.56 *
	>2.5	Org.	12.43 ^{a}	25.11 ^{b}	07.62 ^{a}	06.78 ^{a}	12.98 *
		Con.	06.43 ^{a}	07.32 ^{a}	13.29 ^{b}	10.79 ^{ab}	09.45
T2	AD	Org	0.77 ^{a}	0.99 ^{b}	0.69 ^{a}	0.81 ^{a}	0.82
		Con	0.93 ^{ns}	1.06 ^{ns}	0.93 ^{ns}	0.83 ^{ns}	0.94 *
	BD	Org	01.84 ^{b}	01.15 ^{a}	01.63 ^{b}	01.60 ^{b}	01.56 *
		Con	01.60 ^{b}	00.87 ^{a}	01.01 ^{a}	01.14 ^{a}	01.15
	NF	Org	5567.33 ^{a}	2408.50 ^{a}	4940.33 ^{a}	9037.67 ^{b}	5488.46 *
		Con	3913.67 ^{ns}	1373.50 ^{ns}	1759.33 ^{ns}	2862.00 ^{ns}	2477.13
	RD	Org	0.50 ^{a}	0.88 ^{b}	0.28 ^{a}	0.40 ^{a}	0.52 ^{NS}
		Con	0.40 ^{a}	0.74 ^{b}	0.29 ^{a}	0.40 ^{a}	0.45 ^{NS}
	SA	Org	742.55 ^{a}	645.90 ^{a}	668.90 ^{a}	1450.86 ^{b}	877.05 *
		Con	695.48 ^{ns}	506.02 ^{ns}	465.67 ^{ns}	646.71 ^{ns}	578.47
	TL	Org	2976.85 ^{a}	2102.00 ^{a}	3034.65 ^{a}	5622.49 ^{b}	3434.00 *
		Con	2449.62 ^{ns}	1554.31 ^{ns}	1592.53 ^{ns}	2452.94 ^{ns}	2012.35
	TV	Org	14.88 ^{a}	16.51 ^{a}	12.17 ^{a}	30.60 ^{b}	18.54 ^{NS}
		Con	22.23 ^{ns}	14.40 ^{ns}	10.97 ^{ns}	14.62 ^{ns}	15.56 ^{NS}
	<0.5	Org.	1860.15 ^{a}	1212.32 ^{a}	1935.92 ^{a}	3267.68 ^{b}	2069.02 *
		Con.	1378.57 ^{ns}	841.44 ^{ns}	702.66 ^{ns}	1426.58 ^{ns}	1087.31
	0.5–2.5	Org.	1006.41 ^{a}	771.91 ^{a}	1020.79 ^{a}	2151.75 ^{b}	1237.71 *
		Con.	997.92 ^{ns}	580.75 ^{ns}	839.86 ^{ns}	937.97 ^{ns}	839.12
	>2.5	Org.	107.48 ^{ab}	117.57 ^{ab}	74.77 ^{a}	198.18 ^{b}	124.50 *
		Con.	69.95 ^{ab}	129.53 ^{b}	48.82 ^{a}	85.88 ^{ab}	83.55

* Mean values with different lower-case letters within rows indicate significant differences among rhizospheres including bulk soil based on the Student–Newman–Keuls multiple range test at p-value < 0.05, ns indicates no significant differences. Asterisk in last column (average) indicates significant differences between farming systems for pairs of data from the same substrate at p-value < 0.05, NS in the last column indicates no significant differences between farming systems for pairs of data from the same substrate.

2.3. *Correlations and Exploratory Factor Analysis*

2.3.1. Correlation among Root, Biomass, and Rhizosphere Traits

To better track root–soil interactions, a correlation analysis was carried out (Figure 2). Some interesting correlations between plant traits and rhizosphere parameters were observed (Figure 2). Interestingly, these correlations varied depending on the sampling time and farming system. In the first place, the respiration of the rhizospheres using Gal as the substrate was positively correlated at T1 and in both farming systems with many root parameters, mainly related to the root length and weight. This situation changed at T2, where the correlations were different for each farming system; in the organic field, respiration induced with galactose was negatively correlated with higher plant biomass, whereas in the conventional field, plant parameters were uncorrelated with plant traits.

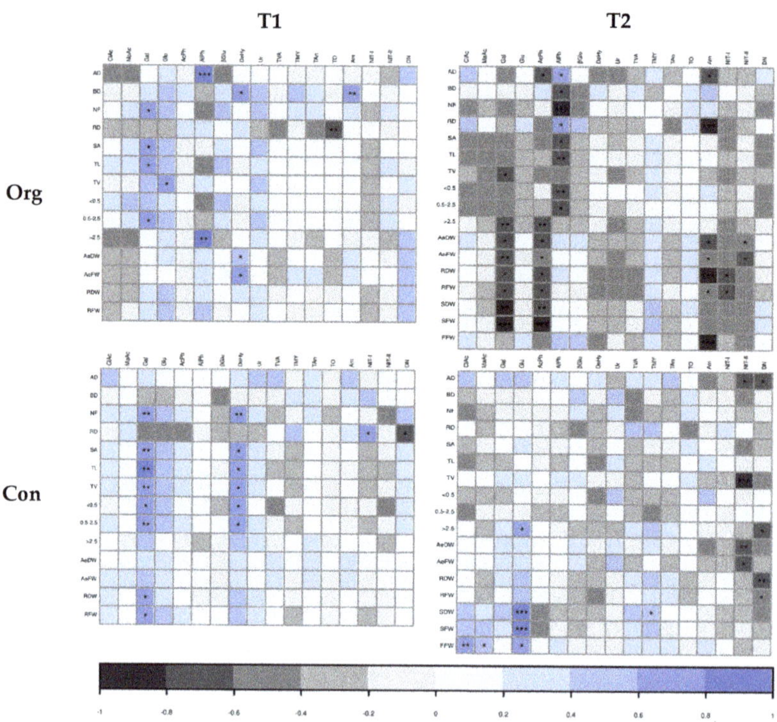

Figure 2. Heatmap showing correlations among plant traits and rhizosphere traits: citric acid (CiAc), malic acid (MaAc), galactose (Gal), glucose (Glu), acid phosphatase (AcPh), alkaline phosphatase (AlPh), β-glucosidase (βGlu), dehydrogenase (DeHy), and urease (Ur), total viable aerobic mesophilic bacterial count (TVA), total molds and yeast count (TMY), total anaerobic bacteria (TAn), total oligotrophic bacterial count (TO), ammonification potential (Am), nitrification I potential (NIT-I), nitrification II potential (NIT-II), denitrification potential (DN) in two farming systems (organic; org and conventional; con) at two sample times (T1: summer, T2: autumn). Pearson's multiple correlation coefficient (−1 to 1 range, dark-gray to blue scale), in which the significance level of the *t*-test at 5%, 1%, and 0.1% was evaluated. The significance levels were indicated by *, **, and ***, respectively.

At T1, AlPh was positively correlated only in the organic field, with a diameter of the root (AD and >2.5). DeHy activity was positively correlated with different plant traits, depending on the system. In the organic field, it was positively correlated to BD and biomass traits, whereas in the conventional field, it was correlated with NF and length parameters (TL, TV, <0.5 and 0.5–2.5). At T2, the only enzymatic activities that significantly correlated with root traits were AcPh and AlPh, and only in organic soil. AcPh activity was correlated negatively with lower biomass, whereas AlPh activity was correlated negatively with lower diameter classes of root length but positively with AD and RD.

In general, the microbial count was not correlated with any plant parameter at any time or farming system, but a correlation was observed with the higher oligotrophic microorganisms (TO) with less dense roots (RD) (Figure 2).

The nitrogen cycle parameters showed very few correlations with plant traits at T1 (Figure 2). Nevertheless, a negative correlation between Am potential and BD for organic farming, a positive correlation between RD and NIT-I potential, and a negative correlation between RD and DN potential were observed. At T2, more significant correlations were observed. In this case, AM potential was correlated with thinner (AD) and less dense roots (RD). This happened only in the case of the organic field, where Am, NIT-I, and NIT-II were also somehow correlated with lower values of biomass and production (AeDW, AeFW, RDW, RFW, and FFW). This negative correlation was only maintained in the conventional farming system for shoot biomass (AeDW and AeFW) and the NIT-II potential of the soil. In this farming system, interestingly, the root diameter (AD), the length of the thick roots (>2.5), and the root biomass (RDW and RFW) were negatively correlated with DN potential.

2.3.2. Factor Analysis of Rhizosphere and Bulk Soil

To resume all possible correlations among soil traits, an exploratory factor analysis was carried out, where factor load was considered significant at levels higher than 0.4. For the analysis, bulk samples were taken for both farming systems and sampling times; all root and biomass traits were excluded from the analysis. At T1, fifteen traits were factorizable (through the overall Measure of Sampling Adequacy, or MSA, at least 0.6 [29], Table 7) in three factors. At T2, sixteen traits were factorizable, one more (DeHy) than at T1.

Table 7. Factor loads for rhizosphere and bulk soil analyzed parameters at two sampling times. Citric acid (CiAc), malic acid (MaAc), galactose (Gal), glucose (Glu), acid phosphatase (AcPh), β-glucosidase (βGlu), dehydrogenase (DeHy), and urease (Ur), total viable aerobic mesophilic bacterial count (TVA), total molds and yeast count (TMY), total anaerobic bacteria (TAn), total oligotrophic bacterial count (TO), ammonification potential (Am), nitrification I potential (NIT-I), nitrification II potential (NIT-II), denitrification potential (DN) in two farming systems (organic: org or conventional: con) at two sampling times (T1, summer and T2, autumn).

	T1			T2		
Trait	**F1.1**	**F2.1**	**F3.1**	**F1.2**	**F2.2**	**F3.2**
CiAc	0.83	−0.36	0.19	−0.04	0.91	0.16
MaAc	0.57	−0.38	−0.03	0.01	0.92	−0.03
Gal	0.77	0.14	−0.06	0.45	0.26	0.55
Glu	0.71	0.12	−0.17	0.50	0.22	0.09
AcPh	−0.35	0.19	0.66	0.60	−0.06	0.07
βGlu	−0.50	0.29	0.40	0.73	0.33	0.12
DeHy	-	-	-	−0.40	0.06	0.19
Ur	−0.05	−0.04	0.72	0.97	0.10	0.06
TVA	−0.15	0.73	0.37	0.27	−0.03	0.91
TMY	0.12	0.43	0.32	0.24	0.18	0.14
TAn	−0.25	0.47	0.32	−0.17	0.19	0.45
TO	0.05	0.81	−0.16	0.81	−0.08	0.28
Am	0.06	0.01	0.51	0.68	−0.26	0.24
NIT-I	−0.57	0.29	0.37	0.44	0.28	0.03
NIT-II	0.13	−0.14	−0.42	0.87	−0.19	0.26
DN	0.05	−0.16	−0.01	−0.39	0.20	−0.48
% of Var.	30.21	14.38	12.26	35.89	15.46	10.67
Eigenvalues	4.53	2.16	1.84	5.74	2.47	1.71

At T1, the main loads for the first factor (which explains the 30.21% variation, F1.1) were all the induced respiration substrates (CiAc, MaAc, Gal, and Glu) with positive loads, and βGlu activity and NIT-I with negative loads (Table 7). The main loads for the second factor (14.38% of the variation, F2.1) were all the microbial counts with positive loads (Table 7). For the third factor (12.26% of variance explained, F3.1), the main loads were obtained with AcPh, βGlu, Ur, and Am potentials with positive loads and NIT-II with a negative load. When observing the projection (Figure 3) of the rhizosphere's scores of each accession and bulk sample in each factor (Table S1, Supplementary Data), first (F1.1) and second (F2.1) factors grouped each FS, being conventionally characterized by higher respiration rates and lower βGlu activity and NIT-I potential (Figure 3). For *Serrano* rhizosphere samples with scores that were not differentiated between farming systems, these F1.1 and F2.1 traits had a strong genotype influence, therefore being grouped very closely regardless of the farming system.

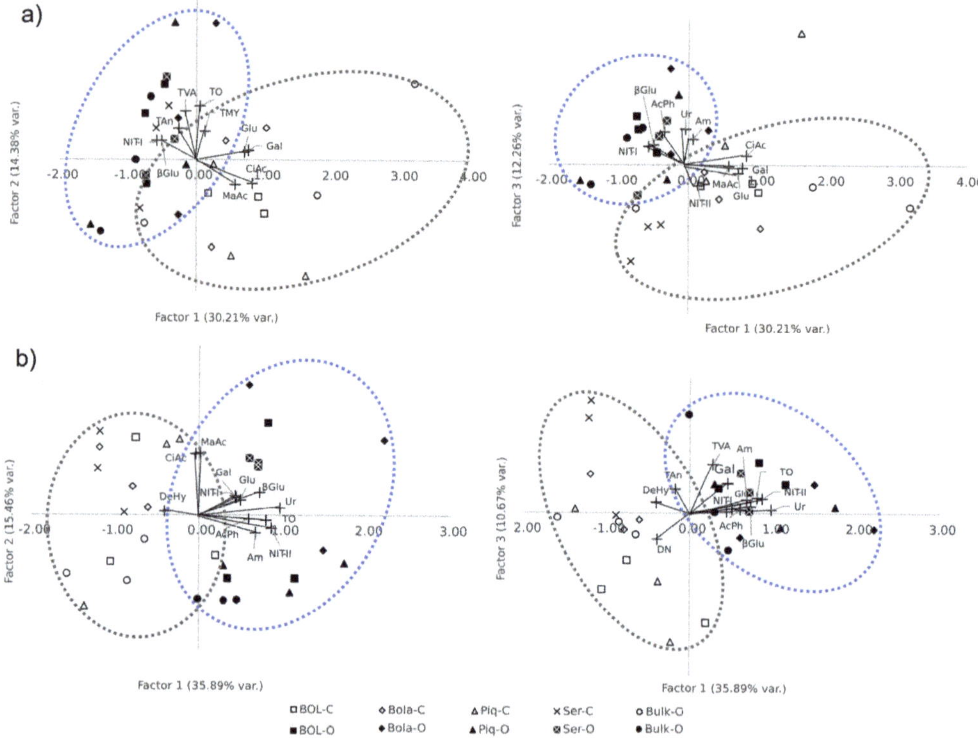

Figure 3. Varietal projection on the new factorial space. Scorings of accessions, bulks, and trait loads. (**a**) in vegetative stage (T1), left is first and second factors projection; right is first and third factors projection. (**b**) in late fruiting stage (T2), left is first and second factors projection; right is first and third factors projection. Blue dotted line group are mostly organic farming system samples, and gray dotted line group are mostly conventional farming system samples.

At T2, the first factor (35.89% of the variation, F1.2; Table 7) was characterized by positive loads Gal, Glu, AcPh, βGlu, Ur, TO, Am, NIT-I, and NIT-II, and DeHy with a negative load (Table 7). The second factor (F2.2) had only two but very high loads, CiAc and MaAc. Finally, the third factor (F3.2) had three traits with positive loads (Gal, TVA, and Tan) and one with negative and significant loads (DN). At this time, again, we were unable to differentiate the F1.2 scores among farming systems (Figure 3). Additionally, with F2.2, carboxylic acid's induced respiration weight interestingly differentiated all bulk and

Serrano samples. F3.2 was able to differentiate a few accessions from the bulk samples; in the organic farming system, *BOL-58* and *Serrano* had positive scores, and in the conventional FS, *BOL-58* and bulk had negative scores (Table S1, Supplementary Data) (Figure 3).

3. Discussion

3.1. The Farming System Conditions the Status of the Soil

The results showed the complexity and dynamics of the soils tested. In this experiment, we selected two similar fields that were specifically chosen to minimize the effects of the climatological conditions and physicochemical properties of the soil. Both soils differ mainly in the management of the crop, with the application of chemical fertilizers and pesticides in the case of conventional farming and only organic matter amendments and organic agriculture-authorized products in the case of organic cultivation. The results showed clear differences in the studied soil traits, which we were able to differentiate among both soils from the very beginning of the experiment. This clearly showed the imprint of the historic records of the soil, which were somehow maintained through the season [30]. Olayemi et al. [31] in a 6-year study on loamy-silty soils of a semiarid climate from Colorado (USA) concluded that soil biological communities are generally enriched and more diverse under continuous organic residue retention, resulting in higher soil biodiversity and a range of critical soil functions mediated by soil organisms. The results obtained here for the different soil traits studied agree with that idea.

First, the induced respiration rate measured through the colorimetric method may be an indicator of the level of microorganism biomass and diversity in certain soils [27,32–34]. In addition, the characterization of soil microbial catabolic diversity through substrate-induced respiration could be a way to monitor soil biological resistance and resilience [35]. However, microbial respiration and carbon utilization are not static but variable throughout the year, probably due to seasonal variations in the characteristics of the studied ecosystems [36]. Therefore, the differences among the induced respiration results in the two sampling times in this experiment are not surprising at all, as sampling was carried out both at the beginning (T1: summer) and at the end (T2: autumn) of the warmest period of the year in Valencia (average T°, min and max = 24.3°, 9.3–37 °C). Previous studies carried out in drylands showed that soil microbial respiration can adapt to the environmental temperature through the physiological adjustment of individual or entire microbial populations [37]. In addition, this response of microbial communities to different temperatures could be used to predict climate-induced changes in carbon fluxes [38]. It has been found that increased temperature reduces total microbial biomass, but at the same time, the response to temperature is dependent upon substrate quality [39]. Differences in the temperature sensitivities of taxa and the taxonomic composition of communities determine community-assembled bacterial growth [40].

In the present experiment, the induced respiration rate in the organic soil was lower at the beginning of the experiment, whereas at T2, the respiration rates from the soils of the two systems were alike. The fact that the respiration rates were more stable in the organic field than in the conventional may indicate that the microbial populations are buffered throughout the warm season, probably due to their specific microbial profile, which may differ from the microbial community in the conventional plot. Despite the general trend observed through time according to the farming system, it is important to note that not all substrates behaved the same. Creamer et al. [41] demonstrated, after testing eight substrates in 81 soils, that the substrate behavior was dependent upon combinations of land-use, pH, and soil organic matter. Specifically, they reported greater utilization of carboxylic acid-based substrates in arable sites, which concords with our results, with higher respiration rates with CiAc and MaAc, especially in the conventional field. In addition, the soils assayed had a pH of 8.2, which has been reported as negative for the use of Gal but positive for the use of organic acids [41].

Secondly, other important indicators of soil quality are soil enzymes [42]. The different soil enzymatic activities are the result of proliferating microorganisms and the accumulation of enzyme action. The main sources of accumulated enzymes are the cells of microorganisms, and a small part may come from organic plant and animal residues. Dehydrogenase and β-glucosidase are generally used as indicators for microbial activity. Dehydrogenase is involved in intracellular oxidation-reduction processes, and β-glucosidase, as an extracellular enzyme, is fundamental in the hydrolysis and degradation of soil carbohydrates, releasing glucose. This represents the important contribution of energy to soil microorganisms. Alkaline and acid phosphatase are two non-specific enzymes that catalyze the hydrolysis of glycerophosphates and differ by their optimal pH for action, 11 and 6, respectively, and are involved in the release of P from organic forms. Finally, urease activity in soil may be associated with living cells, dead cells, or cell debris, or may even be immobilized in humic clays and colloids. Urease hydrolyzes urea into ammonium, a usable form of N by plants, and carbon dioxide, participating actively in the nitrogen cycle and then in the fertility of the soil [43]. The enzymatic activities observed in the soils analyzed here remained stable throughout the studied period, and were similar to those reported by other authors such as Jat et al. [44], who studied soils from the rhizosphere and bulk soils of cereal crops with different management in India. On the other hand, other authors reported that in Western Spain, for acid phosphatases, levels were more likely to be 10–40 μmol g^{-1} h^{-1} [45].

In addition to soil properties, soil management has been described as a driver of the soil enzymatic activity [46]. The results in this paper showed significant differences among the two-farming systems for all tested enzymes, except for dehydrogenase. Interestingly, of all the enzymes evaluated in this experiment, DHA was the only one found to have exclusive intracellular activity. This may indicate similar microbial mass among conventional and organic fields but different profiles of microorganisms. Moreover, as the fundamental enzyme for the carbon cycle in soil, on average, β-glucosidase activity was proximally 30% higher in organic soil, evidencing the importance of the presence of organic carbon in the soil, which was higher due to the manure fertilization in the organic field [47]. In fact, its higher content in organic matter has been continuously correlated with higher enzymatic activity. AlPh, AcPh, and urease were also significantly higher in the organic than conventional farming systems; these enzymes are influenced not only by the organic matter of the soil but also by the fertilization status. Higher fertilization has been correlated with lower activity of those enzymes, although in the case of acid phosphatase, it has been described that increasing levels of N produce higher activity [48].

Thirdly, microbial counts offer an opportunity to quantify the microbial biomass and its profile. In concordance with previous enzymatic activity, microbial counts were more abundant in the organic than conventional farming systems, with dependence on the community profile and the sampling time. For instance, total viable aerobic (TVA) counts were higher in organic than conventional at both sampling times, whereas anaerobic microbial counts were higher in organic only at T1 and for oligotrophs only at T2. Higher levels of microorganism diversity are usually reported in organic farming due to the higher amount of organic matter [49]. Additionally, shifts over time in microbial communities are common [50]. On the contrary, counts of molds and yeast were similar regardless of the farming system or sampling time. Other authors have pointed out that these communities are relatively stable, with a typical profile depending on the soil type and climate and a few genera dominating over the others [51,52].

Finally, as in the case of microbial counts, the results presented here also indicate differences in the N cycling dynamics among farming systems. Microorganisms have been controlling the Earth's nitrogen cycle since life originated [53]. The nitrogen cycle refers to the dynamic process of circulating this element cyclically through the soil and the atmosphere, allowing the transformation of nitrogen into forms accessible to the metabolism of microorganisms, plants, and animals. Nitrogen fixation is the process of reducing molecular nitrogen to ammonia. The group of atmospheric nitrogen fixers consists of numerous organisms, including (i) aerobic nitro-gen-fixers from the genera *Azotobacter*, *Beijerinkia*, *Derexia* and *Azotomones*; (ii) strict anaerobic bacteria, such as those of the genus *Clostridium*;

and (iii) symbiotic fixing bacteria such as *Rhizobium* for the Fabaceae family or *Frankia* for non-legume angiosperms. Mineralization is the process of transforming organic nitrogen into ammonia. The functional group of mineralizers is broad and includes fungi and bacteria. In our experiment, total molds and yeast and total anaerobic bacteria correlate with the potential mineralization rate. Some fungal genera such *Mucor* and *Rhizopus* or *Aspergillus* and *Penicillium* have been described as N mineralizers, whereas mineralizing bacteria could be represented by genera such as *Pseudomonas, Clostridium, Serratia, Bacillus, Escherichia* and *Micrococcus* [54]. In the nitrification process, microorganisms convert ammonium to nitrate to obtain energy. In our experiment, total oligo-tropic counts and anaerobic correlates with the nitrification process. Typical nitrificating bacteria belong to the family *Nitrobacteriaceae: Nitrosomonas, Nitrosobolus, Nitrosospira, Nitrosococcus,* and *Nitrosovibrio*, oxidizing ammonium nitrogen to nitrite and *Nitrobacter*, *Nitrospira* and *Nitrococcus* that oxidize nitrite to nitrate. Both mineralization and nitrification were higher in the organic system than the conventional one, indicating a higher ability to recirculate N among the system. Higher mineralization and nitrification rates were observed in organic farming systems described by some authors [55–57] due to the use by microbial communities of SOM (soil organic matter). The nitrification process is known to be enhanced when soil is warm (20–30 °C), which explains the significant drop in the nitrification potential of the soils at T2 (milder temperatures), which was more intense in the conventional than organic farming system.

Denitrification occurs when N is lost through the conversion of nitrate to gaseous forms of N, such as nitric oxide, nitrous oxide, and dinitrogen gas. In this experiment, conventional soil suffered higher denitrification, but the type of gas produced was not identified. Nitrous oxide is a potent greenhouse gas with a global warming potential [58], whereas N_2 is inert. Lazcano et al. [59] pointed out the need to build up soil C stocks to contribute to N retention as microbial or stabilized organic N in the soil while increasing the abundance of denitrifying microorganisms and, thus, reducing the emissions of N_2O by favoring the completion of denitrification to produce dinitrogen gas.

3.2. Rhizosphere Performance Depends on the Genotypes as the Crop Evolves

Despite the great importance of the soil properties and the farming system on the studied soil parameters, in this experiment, it was also possible to observe the influence of the plants on the rhizosphere properties, especially at T2. This showed that during a plant's growth, it interacts with the surrounding environment in a very specific way. Rhizodeposits, root exudates, and root border cells shape microbial communities, pH, and other factors in the rhizosphere, thereby allowing plants to uptake a wider variety of nutrients for growth and inhibiting possible pathogens [15,60].

In this experiment, there were not many differences in the respiration rates between the rhizospheres of the accessions and the bulk soil, except for *Bola* and *Serrano*. Although other authors have also identified differences in rhizospheric catabolic activities at the species-dependent level [61], to our knowledge, this is the first time that such differences in respiration rates have been observed at the accession level.

Regarding enzymatic activity, generally, it is usually higher in the rhizosphere zone than in bulk soil due to the higher organic carbon deposition in this area, which creates favorable conditions for microbial activities [62]. Profuse vegetation, high root colonization, and no tillage have been correlated with greater soil enzyme activity [43,63,64]. Generally, rhizosphere soil is characterized by a higher amount of very labile carbon and lower contents of mineral nitrogen as well as other nutrients, with a 19–32 times higher number of microorganisms compared to bulk soil [65]. Contrary to what was expected, there were not a great deal of differences in this experiment among the samples on the rhizosphere or the bulk soil, except for the *Serrano* accession. In previous experiments, the action of the accessions on the enzymatic activity of the soils was more intense [66,67].

Exudates and other secondary metabolites have been described to alter the rhizosphere microbiota, as stated in a study by Hu et al. [68]. In our case, it was possible to observe this effect only for *Serrano* and *Piquillo*, which modified total viable counts at T2. The effect of certain plant growth-promoting rhizobacteria (PGPR) and the exudates of some accessions may be responsible for the increase in the total number of microorganisms observed, as other authors have described [12,13]. These results have shown that the rhizosphere communities modulated by the different exuded molecules make certain groups of microorganisms have more affinity for some genotypes than others [69]. Furthermore, in the case of *Piquillo* accession, N mineralization presented higher values in interactions in the conventional farming system. This ability to mobilize the N cycle seemed to be correlated with increasing aerial plant mass and fruit weight. There is clear evidence that plants are not passive conduits, taking up whatever N diffuses to their roots; instead, they can improve their N nutrition by (a) establishing symbiosis with soil microorganisms; (b) stimulating the activity of microorganisms in the root vicinity to increase N availability; and (c) increasing N conservation in soil by limiting microbial processes that lead to N losses, such as nitrification and denitrification, directly through the release of inhibitors from their roots [70,71].

3.3. Plant-Soil Interactions Are Complex and Multifactorial

Piquillo, *Serrano*, and *Bola* were selected to be part of this experiment due to their good performance for phosphorous acquisition in previous studies, either P uptake efficiency or P utilization efficiency. In those experiments, root length increased under P deficiency and fine roots were found to be correlated with P efficiency parameters [14]. Here, the capacity of these genotypes to alter the rhizosphere's microbial community and function has been described for the first time, although the impact of such alterations on the actual nutrition status of the plant remains to be studied.

Contrary to what was expected, root morphological traits showed few correlations with the rhizosphere, being significant only for galactose and glucose-induced respiration, AcPh and AlPh, and some steps of the nitrogen cycle. Furthermore, the correlations were not regularly seen through the different farming systems and sampling times. Therefore, we can suggest that root exudation was more important than root morphology in creating differences among the accession's rhizospheres. Unfortunately, the study of root exudates is difficult and still needs improvement [72].

Root exudation, root morphology, and mycorrhizal symbioses have been described as shaping belowground resource acquisition strategies in a species-dependent manner [73]. It seems that each plant species has its own strategy that favors one of the possible solutions over the others; for instance, the response of maize to P deficiency seems to be more dependent on root morphological changes than increasing root exudates [74]. The independence of the root morphology and the level of root exudates has also been described in studies such as that of Iannucci et al. [75], where they studied eight durum wheat genotypes for their root morphology, exudates, and soil community and only found a correlation between the last two. For future analyses, it would be convenient to study whether the root exudates are correlated with a certain root morphology, as some of our correlations may suggest, or whether they are totally independent.

4. Materials and Methods

4.1. Plant Material, Experimental Design, and Sampling

Four different pepper accessions, two Mediterranean (*Bola* and *Piquillo*) and two Latin-American (*BOL-58* and *Serrano*), were grown in an open field in the 2018 spring-summer season in two farming systems (FS): an organically managed field (FS-O) and a conventional field (FS-C), both located in Sagunto, Valencia, Spain, which belongs to a Mediterranean climatic area with hot and dry summers (Supplementary Table S2). Both fields were clay-loam with a pH of 8.22, EC 0.28 dS/m organic and 0.3 dS/m conventional, with a percentage of organic matter of 1.85% for organic and 1.65% for conventional. Both fields were furrow-irrigated with water from the same well. The fields were managed as in Ribes-Moya [5]. In the organic farm system, sheep manure (4 kg/m^2) was applied as fertilizer at the beginning of the season. In the conventional field, there was one application of vegetable humus (4 kg/m^2) and one application of a mix of nitrogen, phosphorus, and potassium (15-15-15) (50 g/m^2) before transplanting, plus three foliar applications of calcium nitrate (10 g/L), and one application of iron chelate (3 kg/1000 m^2) after transplanting. Pests and diseases were not treated in organic cultivation, whereas chlorpyrifos (48%, EC) and abamectin (1.8%, EC) were applied, combined with copper oxychloride (58.8% WP) as fungicide, in the conventional field. Adventitious plants were controlled mechanically at both sites. Three blocks of five plants per accession were randomly distributed in each field. Two phenological stages (PS) were also considered: at the end of the vegetative stage (T1) and the end of the fruiting stage (T2).

Rhizosphere soil samples (100 g) corresponding to each genotype were obtained as follows: at T1 and T2, the first 5 cm of topsoil of one plant per block was removed to avoid contamination from the surface. Then, the shovel was cleaned, the full plant was dug up to at least 20 cm depth, and the sample was taken carefully between the roots. Three bulk soil samples were taken with the same depth and procedure, but in the middle of the lanes, for each FS. All soil samples (rhizospheric and bulk) and plants (aerial and subterranean parts) were refrigerated and processed within 24 h after sampling.

4.2. Induced Respiration (IR)

The respiration analysis was carried out using a MicroResp® microplate-based system, according to the manufacturer's manual [27,76], with modifications. Following manual recommendations and availability, multiple carbon sources were used as substrates for IR: citric acid (CiAc), malic acid (MaAc), galactose (Gal), and glucose (Glu). Absorbance readings were made at 595 nm and converted into respiration rates (μg CO_2 released by g soil^{-1} h^{-1}). For CO_2 calculations, each microplate included a row of calibration wells, in which known amounts of inorganic Rx citric acid + sodium bicarbonate were added to produce known amounts of sodium citrate + carbon dioxide. The absorbance readings of the calibration wells in the plate were transformed into CO_2 liberation rates (respiration) thanks to a calibration curve of CO_2 liberation, with the same reaction created with Dansensor's Checkpoint O_2/CO_2®.

4.3. Enzymatic Activities (EA) Analysis

Acid and alkaline phosphomonoesterase (AcPh and AlPh, respectively) activity were measured based on the method of Tabatabai and Bermner [77] (phosphatases activity as μmol p-nitrophenol soil g^{-1} h^{-1}). It consisted of the spectrophotometric determination of the p-nitrophenol released when the soil is incubated at 37 °C for 1 h with a buffered solution (pH = 6.5 for acid, pH = 11 for alkaline) of p-nitrophenylphosphate. The amount of released p-nitrophenol was measured with a spectrophotometer at a 400 nm wavelength. To obtain the final concentrations, the raw absorbance data by sample were interpolated in the calibration of the standard curve.

B-glucosidase (BGlu) activity was measured as described in Tabatabai [78]. The colorimetric method is based on the determination of the p-nitrophenol obtained by the action of the enzyme β-glucosidase after incubating the soil with the substrate p-nitrophenol β-D-glucopyranoside at pH 6. The incubation was carried out at 37 °C for an hour, and the released p-nitrophenol was removed by filtration after the addition of $CaCl_2$ and THAM-NaOH pH 12. With the absorbance readings at 400 nm of the standard, the calibration curve was calculated and used to obtain the concentration in μmol p-nitrophenol soil g^{-1} h^{-1}. With the values obtained, net activity was calculated.

Dehydrogenase (DeHy) activity was measured as described in Trevors [79] and García et al. [80]. The enzymatic reaction is based on the spectrophotometric measurement at the wavelength of 540 nm of the iodonitrotetrazolium formazan (INTF), formed when the soil is incubated with 2-p-iodophenyl-3-p-nitrophenyl-5-phenyltetrazolium (INT) in the dark for 20 h at 20 °C. The absorbance for the sample was interpolated into the equation of the calibration curve, obtaining the amount of μmol INTF soil g^{-1} h^{-1}.

Urease (Ur) activity was measured according to Kandeler and Gerber [81] and modified by Kandeler et al. [82]. This colorimetric method is based on the determination of the ammonium released in the incubation of a soil solution at 37 °C for 2 h, where the ammonia produced by urease activity reacts with salicylate and dichloro isocyanide to give a bluish-green color. The absorbance at 610 nm was converted to the ammonia nitrogen concentration (μmol N-NH_4^+ soil g^{-1} h^{-1}).

4.4. Microbial Count (MC)

One gram of soil per sample was used to cultivate and count different soil microorganisms: total viable aerobic mesophilic bacteria count (TVA) was made with Plate Count Agar (Scharlau, Barcelona, Spain) at 28 °C for 48 h, total molds and yeast count (TMY) with Saboureaud Chloramphenicol Agar (Scharlau, Barcelona, Spain), at 28 °C for 3 d, total strictly anaerobic bacteria count (Tan) with Schaendler Agar (Conda Pronadisa, Madrid, Spain) at 37 °C in CO_2 atmosphere for 48 h, and total oligotrophic bacteria count (TO) with Oligotrophic Agar, composed of dipotassium phosphate (Panreac, Barcelona, Spain), magnesium sulfate heptahydrate (Panreac, Barcelona, Spain), peptone (Scharlau, Barcelona, Spain), glycerin (Scharlau, Barcelona, Spain), and bacteriologic agar (Labkem, Dublin, Ireland) [83] at 28 °C, 5 d. All the analyses were performed in duplicate. Results were expressed in log CFU/g of soil.

4.5. N_2 Cycle (NC)

Rhizosphere soil samples were tested for their capacity to perform different steps of the N_2 cycle. To evaluate the mineralization, nitrification I and II, and denitrification potential of the rhizospheres, 0.1 g of soil was added to different substrates (as follows) and incubated at 28 °C for 1 to 3 weeks. Then, measurements were performed each week of the NH_{4^+}, NO_{2^-}, NO_{3^-} and N_2 production, respectively.

Peptone water (Scharlau, Barcelona, Spain) was used as the substrate for the mineralization process (Am). NH_4^+ production was measured seven days after sowing using a colorimetric test strip method based on the Neßler reagent (Mquant TM Ammonium Test, Merk Darmstadt, Germany).

Ammonium sulfate solution was used to measure the conversion of ammonia into nitrate, from now on called nitrification I (NitI), after 14 days of incubation using test strips (Mquant TM, Nitrites test Merck Darmstadt, Germany). The concentration of nitrite ions was calculated by observing the color change produced by a reddish violet azodye, which is formed due to the diazotization of the nitrosating species and subsequent coupling with N-(1-naphthyl) ethylenediamine [84].

Nitrite solution was used to measure nitrification II with the same frequency as the previous one by means of test strips (Mquant TM Nitrates Test Merk Darmstadt, Germany). Nitrate ions are measured based on the same chemical reactions as in nitrification I. The nitrate ion is reduced to nitrite ion; these nitrite ions, in an acidic medium, form the nitrosating species and, when reacting with an aromatic amine, undergo diazotization, forming a diazonium salt. Later, it binds to N-(1-naphthyl) ethylenediamine, giving rise to a reddish violet azodye [84].

Nitrate solution was used to measure denitrification through the level of nitrogen gas produced at 7 days, 14 days, and 21 days, and was assessed with a Durham hood gas production expressed in a range from 0 to 4. The results were expressed in mg NH_{4^+}-N/g, mg NO_{2^-}-N/g, mg NO_{3^-}-N/g, and N_2 production in a 0–4 scale.

4.6. Evaluation of Plant Samples

Plant biomass (shoots, fruits, and roots parts) was weighted fresh and dry. All parameters were sampled at PS-1 and PS-2, except stump and fruit weight. These were only obtained at PS-2, as the plants were not big enough at PS-1, and the stump was not very differentiated and had no fruits: aerial part fresh weight (AeFW), aerial part dry weight

(AeDW), fruits fresh weight (FFW), root fresh weight (RFW), root dry weight (RDW), stump fresh weight (SFW), and stump dry weight (SDW).

Roots were washed, spread in a transparent sheet, and scanned to be measured. Images of scanned roots were measured with WinRHIZO-Pro 2003b, obtaining: average diameter (AD, cm), branching density (BD, number of forks by total length), number of forks (NF), root density (RD, dry weight (g) by volume (cm^3)), surface area (SA, cm^2), total length (TL, cm), total volume (TV, m^3), and length by diameter classes (0.5, 0.5–2.5, and >2.5, mm).

4.7. Statistical Analysis

To determine the differences between the main effects (accessions and bulk soil, FS, and interaction) and multivariate analysis of variance (MANOVA) for every trait, Student–Newman–Keuls (SNK) tests were carried out to compare accessions at each farming system and time using Statgraphics software V.18.1.13 (64-bits).

The rest of the analyses were carried out with R (R Core Team, Viena, Austria, ×64 V. 4.1.0 (18 May 2021) in Rstudio Team, V. 1.4.1106 (11 February 2021)): Pearson's multiple correlation coefficient analysis was carried out for each stage (full heatmap in Supplementary Data), and the packages used were: corrplot [85], svglite [86]. Exploratory factor analysis (EFA) was carried out at each phenological stage. For EFA, packages used were psych [87], tidyverse [88], mvnormtest [89], nFactors [90], EFA.MRFA [91], and dplyr [92]. Final numbers of factors were decided as a three-criteria decision-making path: number of eigenvalues greater than one, parallel analysis, and previous knowledge of the main effects (MANOVA's effects).

5. Conclusions

The present study contributes novel insights into how pepper genotypes, farming systems, and soils interact. Clear differences in the bulk and rhizosphere soils depending on the farming system indicate that it is a key factor in shaping the health status of soil, nutrient cycling capacity, and emissions. However, we are still far from understanding all possible changes and their consequences for food production and the global environment. Pepper genotypes changed the rhizosphere's functionality in a specific way, probably because of specific exudates. However, again, future studies are needed to discover those exudate profiles, their relationship with the proliferation of certain microorganisms, and their relationship with the root architecture. A combined study on plant genotypes and soil microbial profiles should be conducted in the future to create soil-tailored sustainable agriculture.

Supplementary Materials: The following supporting information can be downloaded at: https://www.mdpi.com/article/10.3390/plants12091873/s1, Table S1: Scorings of accessions, bulks, and trait loads; Table S2: Monthly temperatures and precipitation.

Author Contributions: Conceptualization, A.F. and A.R.-B.; methodology, A.F., A.J.-B., A.M.R.-M., C.P., M.D.R. and A.R.-B.; software, I.I.M.-M.; validation, A.J.-B., C.P. and M.D.R.; formal analysis, A.F., A.R.-B., A.M.R.-M. and I.I.M.-M.; investigation, I.I.M.-M. and C.P.M.; resources, A.R.-B., A.F. and M.D.R.; data curation, I.I.M.-M.; writing—original draft preparation, I.I.M.-M., A.F. and C.P.M.; writing—review and editing, I.I.M.-M., A.F. and A.J.-B.; visualization, I.I.M.-M.; supervision, A.F.; project administration, A.R.-B. and A.F.; funding acquisition, A.F. and A.R.-B. All authors have read and agreed to the published version of the manuscript.

Funding: This research was partially funded by MCIN with funding of the European Union NextGenerationEU (PRTR-C17.I1) and by Generalitat Valenciana, grant number: PRTR-C17.I1 AGROALNEXT/2022/027". IIMM work was supported by CONACYT-CONCYTEP pre-doctoral scholarship number 47274 by the Mexican government.

Data Availability Statement: R scripts and data used for statistical analysis and figures are available at: https://github.com/ivalethia/efa-root-soil, (30 April 2022).

Conflicts of Interest: The authors declare no conflict of interest. The funders had no role in the design of the study; in the collection, analyses, or interpretation of data; in the writing of the manuscript; or in the decision to publish the results.

References

1. Rodriguez-Burruezo, A.; González-Mas, M.D.C.; Nuez, F. Carotenoid composition and vitamin A value in ají (*Capsicum baccatum* L.) and rocoto (*C. pubescens* R. & P.), 2 pepper species from the Andean region. *J. Food Sci.* **2010**, *75*, S446–S453. [CrossRef] [PubMed]
2. FAOSTAT Statistic Division, Food and Agriculture Organization of the United Nations. Available online: http://www.fao.org/faostat/ (accessed on 26 October 2022).
3. MAPA Ministerio de Agricultura, Pesca y Alimentación. 2021. Available online: https://www.mapa.gob.es/es/ (accessed on 26 October 2022).
4. Willer, H.; Trávníček, J.; Meier, C.; Schlatter, B. (Eds.) *The World of Organic Agriculture 2021-Statistics and Emerging Trends*; Research Institute of Organic Agriculture (FiBL) and IFOAM–Organics International: Bonn, Germany, 2021.
5. Ribes-Moya, A.M.; Raigon, M.D.; Moreno-Peris, E.; Fita, A.; Rodriguez-Burruezo, A. Response to organic cultivation of heirloom Capsicum peppers: Variation in the level of bioactive compounds and effect of ripening. *PLoS ONE* **2018**, *13*, e0207888. [CrossRef] [PubMed]
6. Ribes-Moya, A.M.; Adalid, A.M.; Raigon, M.D.; Hellín, P.; Fita, A.; Rodriguez-Burruezo, A. Variation in flavonoids in a collection of peppers (*Capsicum* sp.) under organic and conventional cultivation: Effect of the genotype, ripening stage, and growing system. *J. Sci. Food Agric.* **2020**, *100*, 2208–2223. [CrossRef] [PubMed]
7. Tripodi, P.; Cardi, T.; Bianchi, G.; Migliori, C.A.; Schiavi, M.; Rotino, G.L.; Lo Scalzo, R. Genetic and environmental factors underlying variation in yield performance and bioactive compound content of hot pepper varieties (*Capsicum annuum*) cultivated in two contrasting Italian locations. *Eur. Food Res. Technol.* **2018**, *244*, 1555–1567. [CrossRef]
8. Zheng, Q.; Hu, Y.; Zhang, S.; Noll, L.; Böckle, T.; Dietrich, M.; Herbold, C.W.; Eichorst, S.A.; Woebken, D.; Richter, A.; et al. Soil multifunctionality is affected by the soil environment and by microbial community composition and diversity. *Soil Biol. Biochem.* **2019**, *136*, 107521. [CrossRef]
9. Yang, J.; Kloepper, J.W.; Ryu, C.M. Rhizosphere bacteria help plants tolerate abiotic stress. *Trends Plant Sci.* **2009**, *14*, 1–4. [CrossRef]
10. Kandasamy, S.; Loganathan, K.; Muthuraj, R.; Duraisamy, S.; Seetharaman, S.; Thiruvengadam, R.; Ponnusamy, B.; Ramasamy, S. Understanding the molecular basis of plant growth promotional effect of *Pseudomonas fluorescens* on rice through protein profiling. *Proteome Sci.* **2009**, *7*, 47. [CrossRef]
11. Kloepper, J.W.; Ryu, C.M.; Zhang, S. Induced systemic resistance and promotion of plant growth by *Bacillus* spp. *Phytopathology* **2004**, *94*, 1259–1266. [CrossRef]
12. Singh, B.K.; Millard, P.; Whiteley, A.S.; Murrell, J.C. Unravelling rhizosphere–microbial interactions: Opportunities and limitations. *Trends Microbiol.* **2004**, *12*, 386–393. [CrossRef]
13. Wang, X.; Whalley, W.R.; Miller, A.J.; White, P.J.; Zhang, F.; Shen, J. Sustainable Cropping Requires Adaptation to a Heterogeneous Rhizosphere. *Trends Plant Sci.* **2020**, *25*, 1194–1202. [CrossRef]
14. Pereira-Dias, L.; Gil-Villar, D.; Castell-Zeising, V.; Quiñones, A.; Calatayud, Á.; Rodríguez-Burruezo, A.; Fita, A. Main root adaptations in pepper germplasm (*Capsicum* spp.) to phosphorus low-input conditions. *Agronomy* **2020**, *10*, 637. [CrossRef]
15. Sánchez-Cañizares, C.; Jorrín, B.; Poole, P.S.; Tkacz, A. Understanding the holobiont: The interdependence of plants and their microbiome. *Curr. Opin. Microbiol.* **2017**, *38*, 188–196. [CrossRef]
16. Aira, M.; Gomez-Brandon, M.; Lazcano, C.; Bååth, E.; Dominguez, J. Plant genotype strongly modifies the structure and growth of maize rhizosphere microbial communities. *Soil Biol. Biochem.* **2010**, *42*, 2276–2281. [CrossRef]
17. Toljander, J.F.; Lindahl, B.D.; Paul, L.R.; Elfstrand, M.; Finlay, R.D. Influence of arbuscular mycorrhizal mycelial exudates on soil bacterial growth and community structure. *FEMS Microbiol. Ecol.* **2007**, *61*, 295–304. [CrossRef] [PubMed]
18. Edwards, C.A. *Earthworm Ecology*, 2nd ed.; CRC Press: London, UK, 2004.
19. Tate, R.L. *Soil Microbiology*; John Wiley & Sons Ltd.: New York, NY, USA, 2000.
20. Nelkner, J.; Henke, C.; Lin, T.W.; Pätzold, W.; Hassa, J.; Jaenicke, S.; Grosch, R.; Pühler, A.; Sczyrba, A.; Schlüter, A. Effect of Long-Term Farming Practices on Agricultural Soil Microbiome Members Represented by Metagenomically Assembled Genomes (MAGs) and Their Predicted Plant-Beneficial Genes. *Genes* **2019**, *10*, 424. [CrossRef]
21. Griffiths, B.S.; Philippot, L. Insights into the resistance and resilience of the soil microbial community. *FEMS Micro.-Biol.* **2013**, *37*, 112–129. [CrossRef] [PubMed]
22. Degens, B.P.; Harris, J.A. Development of a physiological approach to measuring the catabolic diversity of soil microbial communities. *Soil Biol. Biochem.* **1997**, *29*, 1309–1320. [CrossRef]
23. Nkongolo, K.K.; Narendrula-Kotha, R. Advances in monitoring soil microbial community dynamic and function. *J. Appl. Genet.* **2020**, *61*, 249–263. [CrossRef]

24. Degens, B.P.; Schipper, L.A.; Sparling, G.P.; Vojvodic-Vukovic, M. Decreases in organic C reserves in soils can reduce the catabolic diversity of soil microbial communities. *Soil Biol. Biochem.* **2000**, *32*, 189–196. [CrossRef]
25. Aon, M.A.; Cabello, M.N.; Sarena, D.E.; Colaneri, A.C.; Franco, M.G.; Burgos, J.L.; Cortassa, S.I. Spatio-temporal patterns of soil microbial and enzymatic activities in an agricultural soil. *Appl. Soil Ecol.* **2001**, *18*, 239–254. [CrossRef]
26. Lin, Q.; Brookes, P.C. An evaluation of the substrate-induced respiration method. *Soil Biol. Biochem.* **1999**, *31*, 1969–1983. [CrossRef]
27. Campbell, C.D.; Chapman, S.J.; Cameron, C.M.; Davidson, M.S.; Potts, J.M. A rapid microtiter plate method to measure carbon dioxide evolved from carbon substrate amendments so as to determine the physiological profiles of soil microbial communities by using whole soil. *Appl. Env. Microbiol.* **2003**, *69*, 3593–3599. [CrossRef] [PubMed]
28. Fierer, N.; Wood, S.A.; de Mesquita, C.P.B. How microbes can, and cannot, be used to assess soil health. *Soil Biol. Biochem.* **2021**, *153*, 108111. [CrossRef]
29. Kaiser, H.F. An index of factorial simplicity. *Psychometrika* **1974**, *39*, 31–36. [CrossRef]
30. De Brogniez, D.; Ballabio, C.; Stevens, A.; Jones, R.J.A.; Montanarella, L.; van Wesemael, B. A map of the topsoil organic carbon content of Europe generated by a generalized additive model. *Eur. J. Soil Sci.* **2015**, *66*, 121–134. [CrossRef]
31. Olayemi, O.P.; Schneekloth, J.P.; Wallenstein, M.D.; Trivedi, P.; Calderón, F.J.; Corwin, J.; Fonte, S.J. Soil macrofauna and microbial communities respond in similar ways to management drivers in an irrigated maize system of Colorado (USA). *Appl. Soil Ecol.* **2022**, *178*, 104562. [CrossRef]
32. Rowell, M.J. Colorimetric method for CO_2 measurement in soils. *Soil Biol. Biochem.* **1995**, *27*, 373–375. [CrossRef]
33. Cookson, W.R.; Murphy, D.V.; Roper, M.M. Characterizing the relationships between soil organic matter components and microbial function and composition along a tillage disturbance gradient. *Soil Biol. Biochem.* **2008**, *40*, 763–777. [CrossRef]
34. Sandén, T.; Zavattaro, L.; Spiegel, H.; Grignani, C.; Sandén, H.; Baumgarten, A.; Tiirola, M.; Mikkonen, A. Out of sight: Profiling soil characteristics, nutrients and bacterial communities affected by organic amendments down to one meter in a long-term maize experiment. *Appl. Soil Ecol.* **2019**, *134*, 54–63. [CrossRef]
35. Swallow, M.J.; Quideau, S.A. A method for determining community level physiological profiles of organic soil horizons. *Soil Sci. Soc. Am. J.* **2015**, *79*, 536–542. [CrossRef]
36. Babur, E.; Dindaroğlu, T.; Riaz, M.; Uslu, O.S. Seasonal variations in litter layers' characteristics control microbial respiration and microbial carbon utilization under mature pine, cedar, and beech forest stands in the Eastern Mediterranean Karstic Ecosystems. *Microb. Ecol.* **2022**, *84*, 153–167. [CrossRef] [PubMed]
37. Dacal, M.; Bradford, M.A.; Plaza, C.; Maestre, F.T.; García-Palacios, P. Soil microbial respiration adapts to ambient temperature in global drylands. *Nat. Ecol. Evol.* **2019**, *3*, 232–238. [CrossRef] [PubMed]
38. Bradford, M.A.; McCulley, R.L.; Crowther, T.; Oldfield, E.E.; Wood, S.A.; Fierer, N. Cross-biome patterns in soil microbial respiration predictable from evolutionary theory on thermal adaptation. *Nat. Ecol. Evol.* **2019**, *3*, 223–231. [CrossRef]
39. Ali, R.S.; Poll, C.; Kandeler, E. Dynamics of soil respiration and microbial communities: Interactive controls of temperature and substrate quality. *Soil Biol. Biochem.* **2018**, *127*, 60–70. [CrossRef]
40. Wang, C.; Morrissey, E.M.; Mau, R.L.; Hayer, M.; Piñeiro, J.; Mack, M.C.; Marks, J.C.; Bell, S.L.; Miller, S.N.; Schwartz, E.; et al. The temperature sensitivity of soil: Microbial biodiversity, growth, and carbon mineralization. *ISME J.* **2021**, *15*, 2738–2747. [CrossRef] [PubMed]
41. Creamer, R.E.; Stone, D.; Berry, P.; Kuiper, I. Measuring respiration profiles of soil microbial communities across Europe using MicroResp™ method. *Appl. Soil Ecol.* **2016**, *97*, 36–43. [CrossRef]
42. Attademo, A.M.; Sanchez-Hernandez, J.C.; Lajmanovich, R.C.; Repetti, M.R.; Peltzer, P.M. Enzyme Activities as Indicators of Soil Quality: Response to Intensive Soybean and Rice Crops. *Water Air Soil Pollut.* **2021**, *232*, 295. [CrossRef]
43. Dotaniya, M.L.; Aparna, K.; Dotaniya, C.K.; Singh, M.; Regar, K.L. Role of soil enzymes in sustainable crop production. In *Enzymes in Food Biotechnology*; Academic Press: Cambridge, MA, USA, 2019.
44. Jat, H.S.; Datta, A.; Choudhary, M.; Sharma, P.C.; Dixit, B.; Jat, M.L. Soil enzymes activity: Effect of climate smart agriculture on rhizosphere and bulk soil under cereal based systems of north-west India. *Eur. J. Soil Biol.* **2021**, *103*, 103292. [CrossRef]
45. Sun, Y.; Goll, D.S.; Ciais, P.; Peng, S.; Margalef, O.; Asensio, D.; Sardans, J.; Peñuelas, J. Spatial Pattern and Environmental Drivers of Acid Phosphatase Activity in Europe. *Front. Big Data* **2020**, *2*, 51. [CrossRef]
46. Bandick, A.K.; Dick, R.P. Field management effects on soil enzyme activities. *Soil Biol. Biochem.* **1999**, *31*, 1471–1479. [CrossRef]
47. Mangalassery, S.; Mooney, S.J.; Sparkes, D.L.; Fraser, W.T.; Sjögersten, S. Impacts of zero tillage on soil enzyme activities, microbial characteristics and organic matter functional chemistry in temperate soils. *Eur. J. Soil Biol.* **2015**, *68*, 9–17. [CrossRef]
48. Margalef, O.; Sardans, J.; Maspons, J.; Molowny-Horas, R.; Fernández-Martínez, M.; Janssens, I.A.; Richter, A.; Ciais, P.; Obersteiner, M.; Peñuelas, J. The effect of global change on soil phosphatase activity. *Glob. Chang. Biol.* **2021**, *27*, 5989–6003. [CrossRef] [PubMed]
49. Habteselassie, M.; Woodruff, L.; Norton, J.; Ouyang, Y.; Sintim, H. Changes in microbial communities in soil treated with organic or conventional N sources. *J. Environ. Qual.* **2022**, *51*, 1144–1154. [CrossRef] [PubMed]
50. Zhao, J.; Ni, T.; Li, Y.; Xiong, W.; Ran, W.; Shen, B.; Shen, Q.; Zhang, R. Responses of Bacterial Communities in Arable Soils in a Rice-Wheat Cropping System to Different Fertilizer Regimes and Sampling Times. *PLoS ONE* **2014**, *9*, e85301. [CrossRef] [PubMed]

51. Liu, Y.R.; Delgado-Baquerizo, M.; Wang, J.T.; Hu, H.W.; Yang, Z.; He, J.Z. New insights into the role of microbial community composition in driving soil respiration rates. *Soil Biol. Biochem.* **2018**, *118*, 35–41. [CrossRef]
52. Tkacz, A.; Bestion, E.; Bo, Z.; Hortala, M.; Poole, P.S. Influence of plant fraction, soil, and plant species on microbiota: A multikingdom comparison. *Mbio* **2020**, *11*, e02785-19. [CrossRef]
53. Canfield, D.E.; Glazer, A.N.; Falkowski, P.G. The Evolution and Future of Earth's Nitrogen Cycle. *Science* **2010**, *330*, 192–196. [CrossRef]
54. Xiao, D.; He, X.; Wang, G.; Xu, X.; Hu, Y.; Chen, X.; Zhang, W.; Su, Y.; Wang, K.; Soromotin, A.V.; et al. Network analysis reveals bacterial and fungal keystone taxa involved in straw and soil organic matter mineralization. *Appl. Soil Ecol.* **2022**, *173*, 104395. [CrossRef]
55. Henneron, L.; Kardol, P.; Wardle, D.A.; Cros, C.; Fontaine, S. Rhizosphere control of soil nitrogen cycling: A key component of plant economic strategies. *New Phytol.* **2020**, *228*, 1269–1282. [CrossRef]
56. Yue, H.; Banerjee, S.; Liu, C.; Ren, Q.; Zhang, W.; Zhang, B.; Tian, X.; Wei, G.; Shu, D. Fertilizing-induced changes in the nitrifying microbiota associated with soil nitrification and crop yield. *Sci. Total Environ.* **2022**, *841*, 156752. [CrossRef]
57. Lin, H.-C.; Huber, J.A.; Gerl, G.; Hülsbergen, K.-J. Nitrogen balances and nitrogen-use efficiency of different organic and conventional farming systems. *Nutr. Cycl. Agroecosystems* **2016**, *105*, 1–23. [CrossRef]
58. Masson-Delmotte, V.; Zhai, P.; Pirani, A.; Connors, S.L.; Péan, C.; Berger, S.; Caud, N.; Chen, Y.; Goldfarb, L.; Gomis, M.I.; et al. (Eds.) *IPCC, 2021: Climate Change 2021: The Physical Science Basis. Contribution of Working Group I to the Sixth Assessment Report of the Intergovernmental Panel on Climate Change*; Cambridge University Press: Cambridge, UK; New York, NY, USA, 2021.
59. Lazcano, C.; Zhu-Barker, X.; Decock, C. Effects of Organic Fertilizers on the Soil Microorganisms Responsible for N2O Emissions: A Review. *Microorganisms* **2021**, *9*, 983. [CrossRef] [PubMed]
60. Vives-Peris, V.; de Ollas, C.; Gómez-Cadenas, A.; Pérez-Clemente, R.M. Root exudates: From plant to rhizosphere and beyond. *Plant Cell Rep.* **2019**, *39*, 3–17. [CrossRef]
61. Brolsma, K.M.; Vonk, J.A.; Mommer, L.; Van Ruijven, J.; Hoffland, E.; De Goede, R.G. Microbial catabolic diversity in and beyond the rhizosphere of plant species and plant genotypes. *Pedobiologia* **2017**, *61*, 43–49. [CrossRef]
62. Hirte, J.; Leifeld, J.; Abiven, S.; Oberholzer, H.R.; Mayer, J. Below ground carbon inputs to soil via root biomass and rhizodeposition of field-grown maize and wheat at harvest are independent of net primary productivity. *Agric. Ecosyst. Environ.* **2018**, *265*, 556–566. [CrossRef]
63. Bergstrom, D.W.; Monreal, C.M.; Tomlin, A.D.; Miller, J.J. Interpretation of soil enzyme activities in a comparison of tillage practices along a topographic and textural gradient. *Can. J. Soil Sci.* **2000**, *80*, 71–79. [CrossRef]
64. Choudhary, M.; Jat, H.S.; Datta, A.; Yadav, A.K.; Sapkota, T.B.; Mondal, S.; Meena, R.P.; Sharma, P.C.; Jat, M.L. Sustainable ntensification influences soil quality, biota, and productivity in cereal-based agroecosystems. *Appl. Soil Ecol.* **2018**, *126*, 189–198. [CrossRef]
65. Kuzyakov, Y. Factors affecting rhizosphere priming effects. *J. Plant Nutr. Soil Sci.* **2002**, *165*, 382–396. [CrossRef]
66. Fita, A.; Garcia-Martinez, M.D.; Raigon, M.D.; Lerma, M.D.; Moreno, E.; Rodriguez-Burruezo, A. Peppers: Soil dynamics, root architecture and fruit quality. In *International Congress. STRATEGIES for Organic and Low Input Agricultures and Their Food Systems*; Solibam Congress: Nantes, France, 2014.
67. Ribes-Moya, A.M.; Morales-Manzo, I.I.; Aguilar, C.L.; Raigón, M.D.; Rodríguez-Burruezo, A. Estudio preliminar de la actividad enzimática fosfatasa alcalina y catalasa en cultivos ecológico y convencional de ecotipos de pimiento (*Capsicum* sp.). In *XXVII Jornadas Técnicas de SEAE. VI Congreso Valenciano de Agricultura Ecológica*; Sociedad Española de Agricultura Ecológica: Gandía, España, 2019; pp. 244–254.
68. Hu, L.; Robert, C.A.M.; Cadot, S.; Zhang, X.; Ye, M.; Li, B.; Manzo, D.; Chervet, N.; Steinger, T.; van der Heijden, M.G.A.; et al. Root exudate metabolites drive plant-soil feedbacks on growth and defense by shaping the rhizosphere microbiota. *Nat. Commun.* **2018**, *9*, 2738. [CrossRef]
69. Berendsen, R.L.; Pieterse, C.M.J.; Bakker, P.A.H.M. The rhizosphere microbiome and plant health. *Trends Plant Sci.* **2012**, *17*, 478–486. [CrossRef]
70. Moreau, D.; Bardgett, R.D.; Finlay, R.D.; Jones, D.L.; Philippot, L. A plant perspective on nitrogen cycling in the rhizosphere. *Funct. Ecol.* **2019**, *33*, 540–552. [CrossRef]
71. Schmidt, J.E.; Kent, A.D.; Brisson, V.L.; Gaudín, A.C.M. Agricultural management and plant selection interactively affect rhizosphere microbial community structure and nitrogen cycling. *Microbiome* **2019**, *7*, 146. [CrossRef] [PubMed]
72. Oburger, E.; Jones, D.L. Sampling root exudates—Mission impossible? *Rhizosphere* **2018**, *6*, 116–133. [CrossRef]
73. Wen, Z.; White, P.J.; Shen, J.; Lambers, H. Linking root exudation to belowground economic traits for resource acquisition. *New Phytol.* **2021**, *233*, 1620–1635. [CrossRef] [PubMed]
74. Wen, Z.; Li, H.; Shen, J.; Rengel, Z. Maize responds to low shoot P concentration by altering root morphology rather than increasing root exudation. *Plant Soil* **2017**, *416*, 377–389. [CrossRef]
75. Iannucci, A.; Canfora, L.; Nigro, F.; De Vita, P.; Beleggia, R. Relationships between root morphology, root exudate compounds and rhizosphere microbial community in durum wheat. *Appl. Soil Ecol.* **2021**, *158*, 103781. [CrossRef]
76. Microresp. James Hutton Ltd. Scotland, UK, V3.2. 2022. Available online: www.microresp.com (accessed on 6 October 2022).
77. Tabatabai, M.A.; Bremner, J.M. Use of p-nitrophenyl phosphate for assay of soil phosphate activity. *Soil Biol. Biochem.* **1969**, *1*, 301–307. [CrossRef]

78. Tabatabai, M.A. Soil enzymes. In *Methods of Soil Analysis, Part 2, Chemical and Microbiological Properties*, 2nd ed.; Page, A.L., Miller, R.H., Keeney, D.R., Eds.; American Society of Agronomy—Soil Science Society of America: Madison, WI, USA, 1982; pp. 903–947.
79. Trevors, J.T.; Mayfield, C.I.; Innis, W.E. Measurement of electron transport systema (ETS) activity in soil. *Microb. Ecol.* **1982**, *8*, 163–168. [CrossRef]
80. García, C.; Hernandez, T.; Costa, F.; Ceccanti, B.; Masciandaro, G. The dehydrogenase activity of soil as an ecological marker in process of perturbed system regeneration. In *Proceedings of the XI International Symposium of Environmental Biochemistry*; Gallardo-Lancho, J., Ed.; CSIC: Salamanca, Spain, September 1993; pp. 89–100.
81. Kandeler, E.; Gerber, H. Short-term assay of soil urease activity using colorimetric determination of ammonium. *Biol. Fert. Soils.* **1988**, *6*, 68–72. [CrossRef]
82. Kandeler, E.; Stemmer, M.; Klimanek, E.M. Response of soil microbial biomass, urease and xylanase within particle size fraction to long-term soil management. *Soil. Biol. Biochem.* **1999**, *31*, 261–273. [CrossRef]
83. Alef, K.; Nannipieri, P. *Methods in Applied Soil Microbiology and Biochemistry*; Academic Press: Cambridge, MA, USA, 1995.
84. Capitán, L.F.; Avidad, R.; Fernández, M.D.; Ariza, A. Sensor de Un Solo Uso Para la Detección y Determinación de Nitrito En Aguas. Patent 2185462, 19 June 2004.
85. Wei, T.; Simko, V. R Package 'Corrplot': Visualization of a Correlation Matrix (Version 0.92). 2021. Available online: https://github.com/taiyun/corrplot (accessed on 13 March 2022).
86. Wickham, H.; Henry, L.; Pedersen, T.; Luciani, T.; Decorde, M.; Lise, V. Svglite: An 'SVG' Graphics Device. R Package Version 2.1.0. 2022. Available online: https://CRAN.R-project.org/package=svglite (accessed on 14 March 2022).
87. Revelle, W. Psych: Procedures for Personality and Psychological Research. R Package Version 2.2.9. 2022. Available online: https://CRAN.R-project.org/package=psych (accessed on 13 March 2022).
88. Wickham, H.; Averick, M.; Bryan, J.; Chang, W.; McGowan, L.D.; François, R.; Grolemund, G.; Hayes, A.; Henry, L.; Hester, J.; et al. Welcome to the tidyverse. *J. Open Source Softw.* **2019**, *4*, 1686. [CrossRef]
89. Jarek, S. Mvnormtest: Normality Test for Multivariate Variables. R Package Version 0.1-9. 2012. Available online: https://CRAN.R-project.org/package=mvnormtest (accessed on 13 March 2022).
90. Raiche, G.; Magis, D. nFactors: Parallel Analysis and Other Non Graphical Solutions to the Cattell Scree Test. R Package Version 2.4.1.1. 2022. Available online: https://CRAN.R-project.org/package=nFactors (accessed on 15 March 2023).
91. Navarro-Gonzalez, D.; Lorenzo-Seva, U. EFA.MRFA: Dimensionality Assessment Using Minimum Rank Factor Analysis. R Package Version 1.1.2. 2021. Available online: https://CRAN.R-project.org/package=EFA.MRFA (accessed on 15 March 2023).
92. Wickham, H.; François, R.; Henry, L.; Müller, K. Dplyr: A Grammar of Data Manipulation. R Package Version 1.0.10. 2022. Available online: https://CRAN.R-project.org/package=dplyr (accessed on 1 February 2023).

plants

MDPI

Article

Grapefruit Root and Rhizosphere Responses to Varying Planting Densities, Fertilizer Concentrations and Application Methods

John M. Santiago [1], Davie M. Kadyampakeni [2], John-Paul Fox [1], Alan L. Wright [3], Sandra M. Guzmán [4], Rhuanito Soranz Ferrarezi [5] and Lorenzo Rossi [1,*]

1 Indian River Research and Education Center, Horticultural Sciences Department, Institute of Food and Agricultural Sciences, University of Florida, Fort Pierce, FL 34945, USA
2 Citrus Research and Education Center, Soil, Water and Ecosystem Sciences Department, Institute of Food and Agricultural Sciences, University of Florida, Lake Alfred, FL 33850, USA
3 Indian River Research and Education Center, Soil, Water and Ecosystem Sciences Department, Institute of Food and Agricultural Sciences, University of Florida, Fort Pierce, FL 34945, USA
4 Indian River Research and Education Center, Agricultural and Biological Engineering Department, Institute of Food and Agricultural Sciences, University of Florida, Fort Pierce, FL 34945, USA
5 Department of Horticulture, University of Georgia, Athens, GA 30602, USA
* Correspondence: l.rossi@ufl.edu; Tel.: +1-(772)-577-7341

Abstract: Huanglongbing (HLB) disease has caused a severe decline in citrus production globally over the past decade. There is a need for improved nutrient regimens to better manage the productivity of HLB-affected trees, as current guidelines are based on healthy trees. The aim of this study was to evaluate the effects of different fertilizer application methods and rates with different planting densities on HLB-affected citrus root and soil health. Plant material consisted of 'Ray Ruby' (*Citrus* × *paradisi*) grapefruit trees grafted on 'Kuharske' citrange (*Citrus* × *sinensis* × *Citrus trifoliata*). The study consisted of 4 foliar fertilizer treatments, which included 0×, 1.5×, 3× and 6× the University of Florida Institute of Food and Agriculture (UF/IFAS) recommended guidelines for B, Mn and Zn. Additionally, 2 ground-applied fertilizer treatments were used, specifically controlled-release fertilizer (CRF1): 12−3−14 + B, Fe, Mn and Zn micronutrients at 1× UF/IFAS recommendation, and (CRF2): 12−3−14 + 2× Mg + 3× B, Fe, Mn and Zn micronutrients, with micronutrients applied as sulfur-coated products. The planting densities implemented were low (300 trees ha^{-1}), medium (440 trees ha^{-1}) and high (975 trees ha^{-1}). The CRF fertilizer resulted in greater soil nutrient concentrations through all of the time sampling points, with significant differences in soil Zn and Mn. Grapefruit treated with ground-applied CRF2 and 3× foliar fertilizers resulted in the greatest bacterial alpha and beta diversity in the rhizosphere. Significantly greater abundances of Rhizobiales and Vicinamibacterales were found in the grapefruit rhizosphere of trees treated with 0× UF/IFAS foliar fertilizer compared to higher doses of foliar fertilizers.

Keywords: citrus greening; *Citrus paradisi*; flatwoods; plant nutrition; rhizosphere

Citation: Santiago, J.M.; Kadyampakeni, D.M.; Fox, J.-P.; Wright, A.L.; Guzmán, S.M.; Ferrarezi, R.S.; Rossi, L. Grapefruit Root and Rhizosphere Responses to Varying Planting Densities, Fertilizer Concentrations and Application Methods. *Plants* **2023**, *12*, 1659. https://doi.org/10.3390/plants12081659

Academic Editor: Hisato Kunitake

Received: 28 February 2023
Revised: 11 April 2023
Accepted: 12 April 2023
Published: 15 April 2023

1. Introduction

Much citrus production worldwide has significantly declined due to a disease known as huanglongbing (HLB, or citrus greening) [1]. In Florida, citrus production has been reduced by more than 70% since HLB was first detected in 2005 [2–4]. The disease is associated with *Candidatus* Liberibacter asiaticus (CLas), bacteria transmitted to tree hosts by a vector called the Asian Citrus Psyllid (*Diaphorina citri*, ACP) [5–7]. As the bacteria colonize the sieve tubes within the phloem, callose deposition occurs, resulting in the plugging of the phloem, thus inhibiting nutrient uptake [1]. Symptoms following infection can include leaf chlorosis, leaf drop, reduced canopy density, smaller fruit size, root dieback, lack of juice quality and reduced yield [8]. Grapefruit, in particular, are more prone to root dieback when infected with HLB compared to specialty citrus, such as lemon and lime [9]. Most of the grapefruit production in Florida takes place in the Indian River district, a 200-mile-long area that borders the Atlantic Ocean. Since the introduction of HLB, grapefruit production has undergone an 85% decline [10].

There are several approaches toward dealing with HLB and alleviating symptoms, including the development of disease-tolerant rootstocks [7,11], the implementation of vector control [12] and the use of soil amendments [13] to improve soil health. The effect of different methods and rates of fertilizer application on citrus health and yield have also been studied [4,14–16]. There are several ways of applying nutrients to citrus: foliar applications, ground-applied granular fertilizers, fertigation and banding [14]. The mobility of nutrients in soil and plant tissues is a notable factor in deciding which application method is most appropriate [15]. For instance, when applying micronutrients (generally performed in smaller quantities), the efficiency of root uptake may be compromised when soil conditions are not favorable, and thus the foliar application of micronutrients may serve as a better alternative due to greater efficiency [17]. When applying micronutrients to the ground, variation in soil characteristics, such as soil pH, drainage and moisture-holding capacity, can significantly impact mobility, solubility and the root uptake of nutrients [15].

The development of improved nutrient guidelines has proven to be beneficial since most of the established guidelines are based on healthy noninfected citrus trees rather than HLB-affected trees [18]. Several studies have examined the effects of various fertilizer types and rates on citrus health and yield [4,9,19–22]. The fruit yield increased in 'Valencia' sweet orange trees (*Citrus* × *sinensis*) when evaluating varying rates of manganese (Mn) via foliar application at 3× the recommended rate [23]. Additionally, when examining the effect of foliar application rates on mandarins, increased rates of boron (B) and zinc (Zn) above the recommended guidelines resulted in greater fruit yield and quality [24]. The effect of controlled-release fertilizer (CRF) formulations has also been tested on HLB-affected citrus, with CRF treatments resulting in a greater concentration of nitrogen (N), calcium (Ca), sulfur (S) and B in leaves compared to soluble dry granular fertilizers [10]. Furthermore, [25] found that CRF formulations resulted in significantly high yields in HLB-affected sweet orange trees relative to conventional fertilizer sources.

The application of various types and rates of fertilizers may also have subsequent effects on the microbial communities in the soil, notably those that reside within the rhizosphere (a portion of the soil that encompasses the roots of plants). When studying the effects of various N fertilizer treatments on wheat health and rhizosphere composition, Ref. [26] found that the abundance of dominant soil bacteria was significantly altered through changes in pH from long-term fertilization. Similarly, Ref. [27] found that tea orchards treated with a long-term application of organic fertilizer resulted in significant shifts in soil pH correlated with substantial differences in rhizosphere bacterial diversity. Potential shifts in rhizosphere community composition from fertilizer may be crucial for plant health, as they consist of plant-growth-promoting rhizobacteria (PGPR), which can improve plant growth through symbiotic relationships [28]. Some bacteria, such as *Bacillus subtilis*, can assist plant hosts directly in the defense against pathogens [29]. Furthermore, PGPR can make plant essential nutrients available for root uptake, as they are capable of solubilizing nutrients into forms that can be utilized by plants [30]. Further insight into the interactions shared between fertilizer regimens, rhizosphere microbial composition and root health is still needed, specifically toward citrus. This study aimed to determine the impact of foliar and ground-applied fertilizer treatments on grapefruit root health and rhizosphere composition at varying planting densities. It is predicted that grapefruit treated with greater nutrient concentrations will result in a more diverse rhizosphere bacterial community composition.

2. Results

2.1. Soil Nutrient Concentrations

Soil nutrient concentrations were influenced by planting density treatments, but no clear patterns were established (Table 1). In September 2020, trees planted in a high density (975 trees ha^{-1}) had significantly greater soil Mg (20%, Figure 1A), whereas no significant differences were observed for Ca (Figure 2A). In September 2020, trees planted in a high density had significantly greater soil Zn (31%, Figure 3A) than trees planted in a medium density (440 trees ha^{-1}). However, trees planted in a medium density had significantly greater soil P (17%, Figure 1A) compared to trees planted in a high density. In January 2021, trees planted in a low density (300 trees ha^{-1}) had significantly greater soil Zn (23%, Figure 3A) compared to trees planted in a high density. In May 2021, trees planted in a high density had significantly greater soil Mn (29%, Figure 3A) and Zn (28%, Figure 3A) compared to trees planted in a low density. In September 2021, trees planted in a high density had significantly greater soil P (25%, Figure 1A) and B (14%, Figure 4A) compared to trees planted in a medium density. In January 2022, trees planted in a high density had significantly greater soil Zn (36%, Figure 3A) compared to trees planted in a medium density.

Table 1. Corresponding *p*-values and r values (Pearson's coefficient) of correlations examined between soil nutrient concentrations and planting density treatments from all time points of sampling, which include September 2020, January 2021, May 2021, September 2021 and January 2022.

Nutrient	r	*p*-Value
P	−0.05	0.4
K	0.03	0.25
Mg	−0.01	0.98
Ca	0.12	0.006
B	0.08	0.08
Zn	−0.06	0.18
Mn	0.06	0.9
Fe	0.12	0.08
Cu	0.01	0.79

Ground-applied CRF treatments influenced soil nutrient concentrations; however, similarly to what was reported for planting density, no patterns were established (Table 2). Soil samples fertilized with CRF1 treatment had significantly greater K (41%, Figure 1B) and B (34%, Figure 4B) than those fertilized with the CRF2 treatment in September 2020. However, soil receiving CRF2 treatment had significantly greater Mn (56%, Figure 3B) than the CRF1 treatment. In January 2021, soil receiving CRF2 treatment had significantly greater P (16%, Figure 1B) and Zn (21%, Figure 3B) than those fertilized with CRF1. In May 2021, soil fertilized with the CRF2 treatment had significantly greater Mg (11%, Figure 1B), Mn (75%, Figure 3B), Zn (26%, Figure 3B) and B (23%, Figure 4B) than those fertilized with CRF1. In September 2021, soil fertilized with CRF2 had significantly greater Zn (41%, Figure 3B), Mn (100%, Figure 3B) and Cu (11%, Figure 3B) than those fertilized with CRF2. In January 2022, soil fertilized with CRF2 had significantly greater B (23%, Figure 4B) than soil fertilized with CRF1. No significant differences were detected for Ca (Figure 2B).

Table 2. Corresponding *p*-values and r values (Pearson's coefficient) of correlations examined between soil nutrient concentrations and ground-applied CRF treatments from all time points of sampling, which include September 2020, January 2021, May 2021, September 2021 and January 2022.

Nutrient	r	*p*-Value
K	−0.05	0.33
P	0.09	0.08
Mg	0.1	0.03
Ca	−0.21	5.12×10^{-6}
B	0.13	0.004
Zn	0.17	0.0002
Mn	0.4	2.20×10^{-16}
Fe	0.23	3.17×10^{-7}
Cu	0.17	0.0002

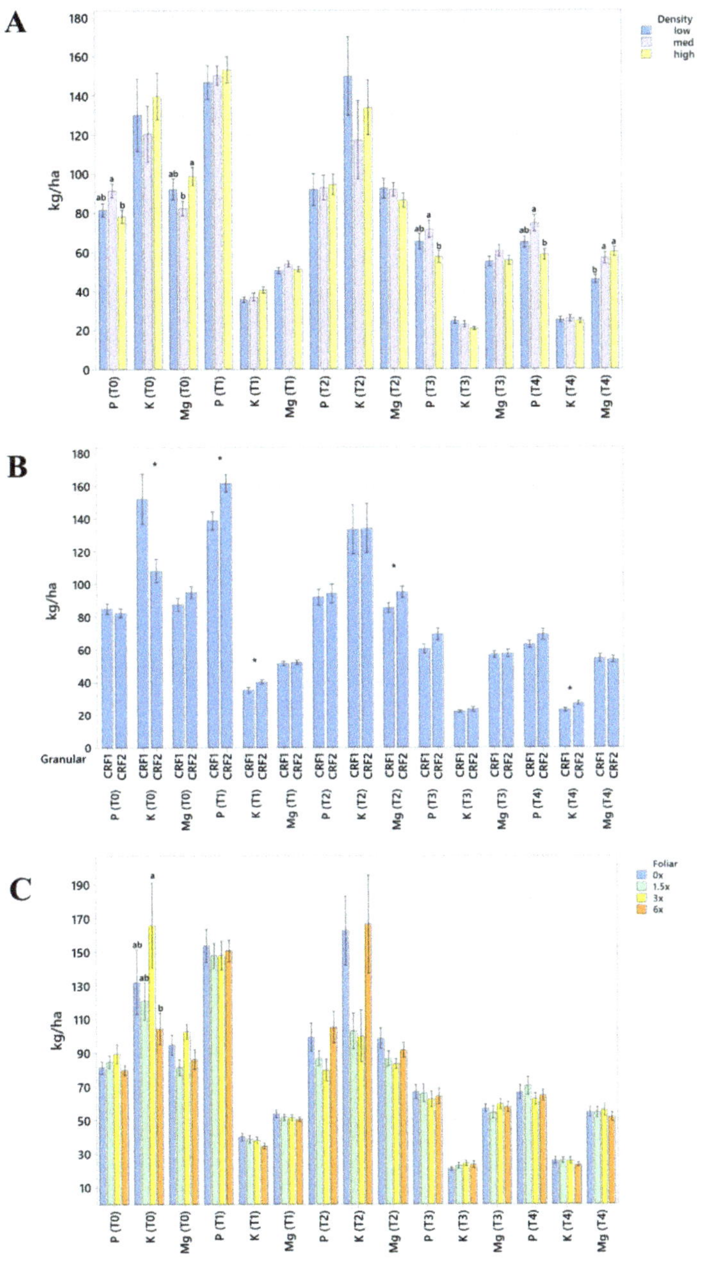

Figure 1. Soil macronutrient concentrations of 'Ray Ruby' grapefruit grafted on 'Kuharske' citrange planted in flatwood soils located in Fort Pierce, FL, USA, in response to (**A**) planting density, (**B**) ground-applied fertilizer and (**C**) foliar fertilizer treatments during September 2020 (T0), January 2021 (T1), May 2021 (T2), September 2021 (T3) and January 2022 (T4). Bars are ± standard deviation of the mean. Treatments with * and different letters were considered to be significantly different ($p < 0.05$).

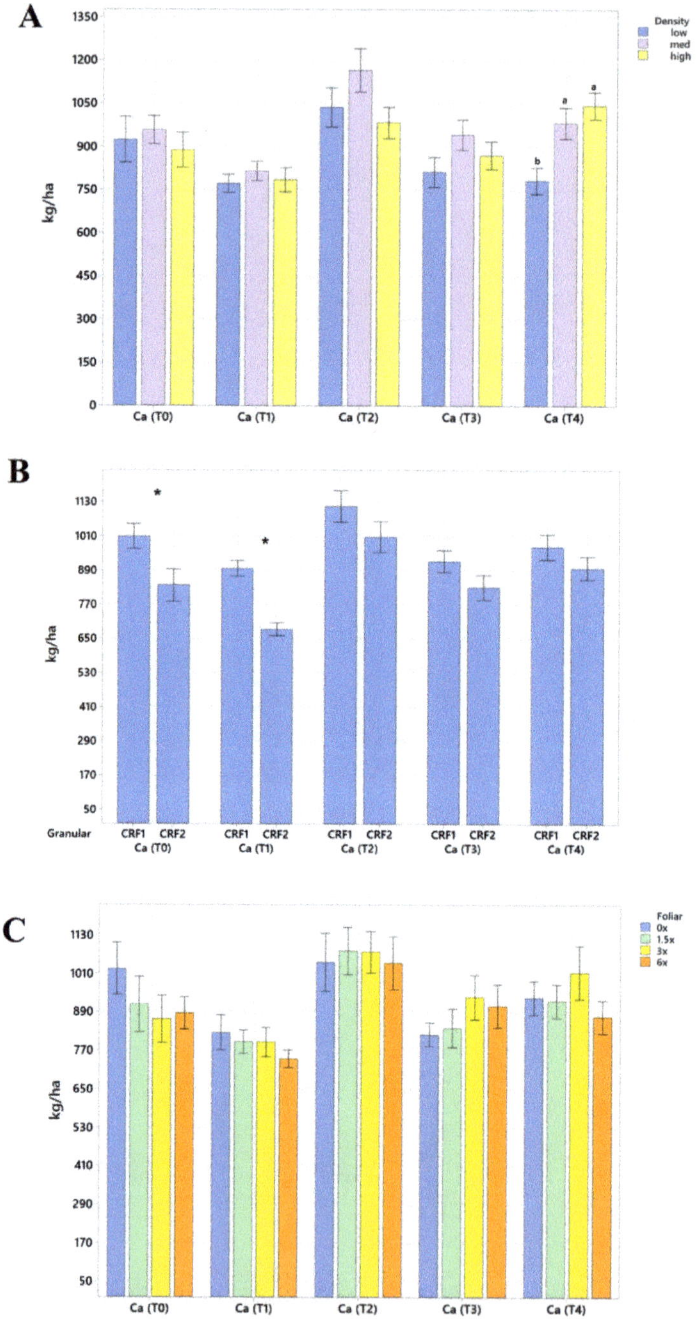

Figure 2. Soil Ca concentrations of 'Ray Ruby' grapefruit grafted on 'Kuharske' citrange planted in flatwood soils located in Fort Pierce, FL, USA, in response to (**A**) planting density, (**B**) ground-applied fertilizer and (**C**) foliar fertilizer treatments during September 2020 (T0), January 2021 (T1), May 2021 (T2), September 2021 (T3) and January 2022 (T4). Bars are ± standard deviation of the mean. Treatments with * and different letters were considered to be significantly different ($p < 0.05$).

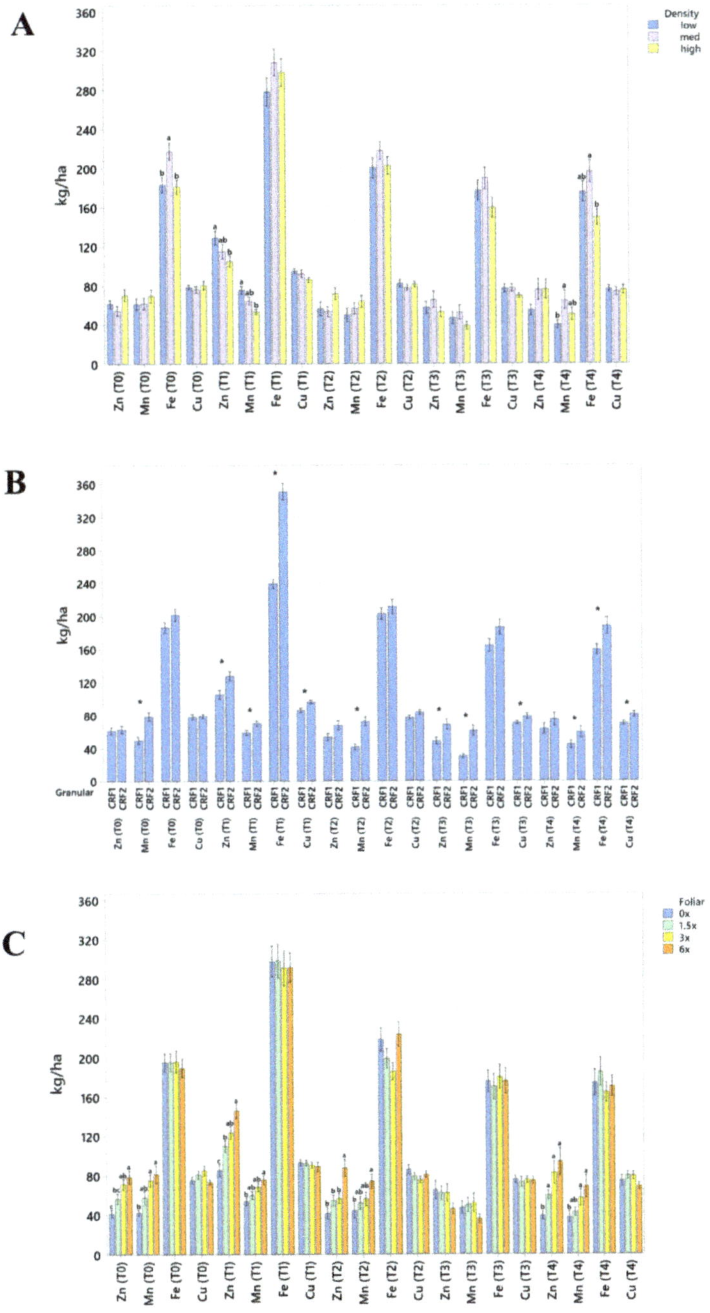

Figure 3. Soil micronutrient concentrations of 'Ray Ruby' grapefruit grafted on 'Kuharske' citrange planted in flatwood soils located in Fort Pierce, FL, USA, in response to (**A**) planting density, (**B**) ground-applied fertilizer and (**C**) foliar fertilizer treatments during September 2020 (T0), January 2021 (T1), May 2021 (T2), September 2021 (T3) and January 2022 (T4). Bars are ± standard deviation of the mean. Treatments with * and different letters were considered to be significantly different ($p < 0.05$).

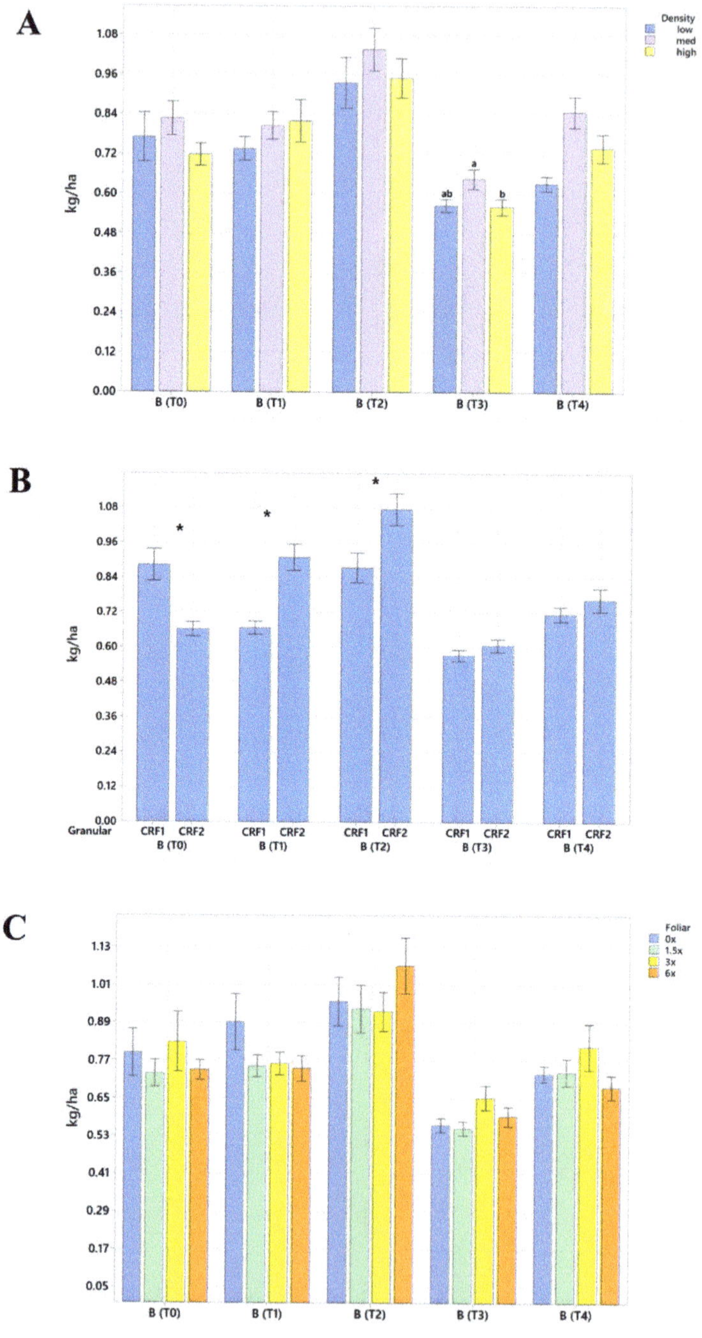

Figure 4. Soil boron concentrations of 'Ray Ruby' grapefruit grafted on 'Kuharske' citrange planted in flatwood soils located in Fort Pierce, FL, USA, in response to (**A**) planting density, (**B**) ground-applied fertilizer and (**C**) foliar fertilizer treatments during September 2020 (T0), January 2021 (T1), May 2021 (T2), September 2021 (T3) and January 2022 (T4). Bars are ± standard deviation of the mean. Treatments with * and different letters were considered to be significantly different ($p < 0.05$).

As previously reported for planting densities and ground-applied fertilizers, foliar fertilizer treatments affected soil nutrient concentrations. However, no clear trends were observed (Table 3) and no significant differences were detected in soil macronutrients, except for K in September 2020 (Figure 1C). Soil samples from 6× foliar treatments had significantly greater Zn (88% and 39%, Figure 3C) and Mn (99% and 47%, Figure 3C) concentrations than treatments fertilized with 0× and 1.5× foliar spray in September 2020. In January 2021, soil receiving 6× foliar sprays had significantly greater Zn (71% and 32.71%, Figure 3C) and Mn (41% and 26%, Figure 3C) concentrations than 0× and 1.5× foliar treatments. In May 2021, soil receiving 6× foliar sprays had significantly greater Zn (110% and 60%, Figure 3C) concentrations than those fertilized with 0× and 1.5× sprays. In January 2022, soil receiving 6× foliar sprays had significantly greater Zn (110%, 30% and 54%, Figure 3C) concentrations than those fertilized with 0×, 1.5× and 3× foliar sprays. Neither Ca or B showed significant differences (Figures 2C and 4C).

Table 3. Corresponding p-values and r values (Pearson's coefficient) of correlations examined between soil nutrient concentrations and foliar fertilizer treatments from all time points of sampling, which include September 2020, January 2021, May 2021, September 2021 and January 2022.

Nutrient	r	p-Value
K	−0.03	0.94
P	0.01	0.98
Mg	−0.03	0.39
Ca	−0.02	0.65
B	0.02	0.56
Zn	0.29	9.76×10^{-11}
Mn	0.24	9.19×10^{-8}
Fe	−0.09	0.84
Cu	−0.06	0.13

2.2. Root Nutrient Concentrations and Size

Root nutrient concentrations and size were significantly affected by planting density. During September 2020, trees planted in a high density resulted in significantly greater total root length (60%, Tables 4 and 5) than trees planted in a low density. In September 2021, the root concentrations of Mg (52%, Figure 5A) were significantly greater in trees planted in a high density than those planted in a low density. In May 2021, the root concentrations of N (18%, Figure 6A), Mn (73%, Figure 7A) and Zn (59%, Figure 7A) were significantly greater in trees planted in a high density compared to the trees planted in a medium density. Additionally, B was significantly higher in high-density plantings compared to those in a low planting density (45%, Figure 8A). In January 2022, trees planted in a high density had significantly greater Zn (59%, Figure 7A) and Mn (73%, Figure 7A) in their roots than those grown in a medium density. No significant differences were observed in root density in response to planting density (Figure 9A).

Root nutrient concentrations and measurements were significantly affected by ground-applied CRF treatments. In September 2020, grapefruit treated with the CRF2 treatment resulted in roots with a significantly greater root density (40%, Figure 9B) and grapefruit fertilized with the CRF1 treatment was significantly greater than those fertilized with the CRF2. In May 2021, the root concentrations of K (17.32%, Figure 6B) were significantly greater in trees fertilized with the CRF1 treatment compared to the CRF2. However, the root concentrations of P (11%, Figure 5B) and Mg (17%, Figure 5B) were significantly greater in trees fertilized with the CRF2 treatment compared to the CRF1. Additionally, the total root volume (122%, Tables 6 and 7) and total root area (63%, Tables 8 and 9) of grapefruit fertilized with the CRF2 treatment were significantly greater than those fertilized with the CRF1. In September 2021, the root concentrations of Mn (52%, Figure 7B) and Zn (35%, Figure 7B) were significantly greater in trees fertilized with CRF2 compared to CRF1. In January 2022, the root concentrations of P (20%, Figure 5B) and Mg (17%, Figure 5B) were significantly greater in trees fertilized with CRF2 compared to CRF1. However, the root concentrations of K (17%, Figure 6B) were significantly greater in trees fertilized with CRF1 compared to CRF2. Ground-applied fertilizer treatments were not observed to have a significant effect on the root B concentrations at all sampling times (Figure 8B). No significant differences were observed in root nutrient concentrations in response to foliar fertilizer treatments (Figures 5C, 6C, 7C and 8C); additionally, no significant difference in root density in response to foliar fertilizer treatments was observed (Figure 9C).

Table 4. Total root length of 'Ray Ruby' grapefruit grafted on 'Kuharske' citrange planted in flatwood soils located in Fort Pierce, FL, USA, and treated with ground-applied fertilizer, foliar fertilizer and different planting densities from September 2020 to January 2022.

Root Length (mm)		Sept. 2020	Jan. 2021	May 2021	Sept. 2021	Jan. 2022
Ground-applied [1]	CRF1	489 ± 120	622.6 ± 161	653.7 ± 143	545.5 ± 112	706.0 ± 95
	CRF2	572 ± 90	432.9 ± 93	450.2 ± 103	509.0 ± 79	655.2 ± 111
Foliar [2]	0×	504 ± 180	343.2 ± 143	539.3 ± 194	469.7 ± 118	617.5 ± 121
	1.5×	476 ± 101	431.0 ± 78	456.0 ± 100	405.8 ± 95	600.5 ± 124
	3×	401 ± 121	406.5 ± 131	329.8 ± 104	523.1 ± 155	685.1 ± 171
	6×	772 ± 194	1120.8 ± 318	994.8 ± 298	726.6 ± 171	817.8 ± 173
Planting density [3]	High	738 ± 128	651.6 ± 142	655.2 ± 183	729.2 ± 149	790.8 ± 156
	Low	398 ± 144	529.0 ± 227	559.2 ± 194	382.3 ± 79	508.6 ± 68
	Medium	402 ± 72	442.5 ± 176	593.5 ± 202	495.6 ± 113	747.8 ± 137

[1] Controlled release fertilizer blends 1 (CRF1): 12N-1.31 P-11.62K and micronutrients at 1× the UF/IFAS recommendation with micronutrients as sulfates (12-3-14 1× Micro) and CRF2: enhanced 12N-1.31 P-11.62K with 2× Mg and 2.5× the UF/IFAS recommendation with micronutrients as sulfur-coated products (#12-3-14 2.5× Micro). [2] Foliar fertilizer treatments, which included 0×, 1.5×, 3× and 6× the UF/IFAS recommendation. [3] Low (300 trees ha^{-1}), medium (440 trees ha^{-1}) and high (975 trees ha^{-1}).

Table 5. *p*-values obtained via a three-way ANOVA followed by Tukey's post hoc test, corresponding to the total root length of 'Ray Ruby' grapefruit grafted on 'Kuharske' citrange planted in flatwood soils located in Fort Pierce, FL, USA, and treated with ground-applied fertilizer, foliar fertilizer and different planting densities from September 2020 to January 2022. Bold values were considered to be significantly different ($p < 0.05$).

Root Length *p* Values		Sept. 2020	Jan. 2021	May 2021	Sept. 2021	Jan. 2022
Ground-applied [1]	CRF1—CRF2	0.28	0.74	0.35	0.78	0.75
Foliar [2]	1.5×–0×	0.84	0.44	0.90	0.98	0.99
	3×–0×	0.99	0.58	0.99	0.99	0.98
	6×–0×	0.28	0.98	0.99	0.52	0.76
	3×–1.5×	0.72	0.99	0.92	0.91	0.97
	6×–1.5×	0.72	0.71	0.85	0.32	0.73
	6×–3×	0.18	0.82	0.99	0.71	0.92
Planting density [3]	Low—High	0.04	0.77	0.93	0.06	0.27
	Med—High	0.32	0.90	0.86	0.22	0.94
	Med—Low	0.62	0.96	0.66	0.79	0.42

[1] Controlled-release fertilizer blends 1 (CRF1): 12N-1.31 P-11.62K and micronutrients at 1× the UF/IFAS recommendation with micronutrients as sulfates (12-3-14 1× Micro) and CRF2: enhanced 12N-1.31 P-11.62K with 2× Mg and 2.5× the UF/IFAS recommendation with micronutrients as sulfur-coated products (#12-3-14 2.5× Micro). [2] Foliar fertilizer treatments, which included 0×, 1.5×, 3× and 6× the UF/IFAS recommendation. [3] Low (300 trees ha^{-1}), medium (440 trees ha−1) and high (975 trees ha^{-1}).

Table 6. Total root volume of 'Ray Ruby' grapefruit grafted on 'Kuharske' citrange planted in flatwood soils located in Fort Pierce, FL, USA, and treated with ground-applied fertilizer, foliar fertilizer and different planting densities from September 2020 to January 2022.

Root Volume (mm^3)		Sept. 2020	Jan. 2021	May 2021	Sept. 2021	Jan. 2022
Ground-applied [1]	CRF1	491.8 ± 135	540.8 ± 117	467.1 ± 101	1139.7 ± 232	1949.4 ± 271
	CRF2	732.7 ± 329	439.1 ± 101	316.0 ± 73	1143.3 ± 170	1777.7 ± 285

Table 6. *Cont.*

Root Volume (mm^3)		Sept. 2020	Jan. 2021	May 2021	Sept. 2021	Jan. 2022
Foliar [2]	0×	430.9 ± 199	367.4 ± 131	386.1 ± 138	1044.9 ± 256	1691.2 ± 367
	1.5×	374.4 ± 110	344.8 ± 70	349.2 ± 75	908.5 ± 227	1647.4 ± 316
	3×	856.8 ± 512	607.2 ± 176	285.6 ± 99	1163.3 ± 336	1861.2 ± 443
	6×	692.7 ± 290	755.9 ± 98	672.0 ± 229	1479.3 ± 329	2225.4 ± 452
Planting density [3]	High	916.5 ± 435	637.0 ± 171	397.5 ± 102	1527.2 ± 312	2131.2 ± 438
	Low	485.4 ± 210	456.4 ± 134	423.4 ± 150	835.7 ± 155	1409.2 ± 161
	Medium	391.6 ± 119	389.4 ± 90	449.9 ± 136	1109.8 ± 249	2072.9 ± 371

[1] Controlled release fertilizer blends 1 (CRF1): 12N-1.31 P-11.62K and micronutrients at 1× the UF/IFAS recommendation with micronutrients as sulfates (12-3-14 1 × Micro) and CRF2: enhanced 12N-1.31 P-11.62K with 2 × Mg and 2.5× the UF/IFAS recommendation with micronutrients as sulfur-coated products (#12-3-14 2.5 × Micro). [2] Foliar fertilizer treatments, which included 0×, 1.5×, 3× and 6× the UF/IFAS recommendation. [3] Low (300 trees ha^{-1}), medium (440 trees ha−1) and high (975 trees ha^{-1}).

Table 7. *p*-values obtained via a three-way ANOVA followed by Tukey's post hoc test, corresponding to the total root volume of 'Ray Ruby' grapefruit grafted on 'Kuharske' citrange planted in flatwood soils located in Fort Pierce, FL, USA, and treated with ground-applied fertilizer, foliar fertilizer and different planting densities from September 2020 to January 2022.

Root Volume *p* Values		Sept. 2020	Jan. 2021	May 2021	Sept. 2021	Jan. 2022
Ground-applied [1]	CRF1—CRF2	0.82	0.67	0.03	0.98	0.66
Foliar [2]	1.5×–0×	0.92	0.69	0.65	0.99	0.99
	3×–0×	0.92	0.29	0.88	0.99	0.99
	6×–0×	0.77	0.99	0.97	0.69	0.78
	3×–1.5×	0.99	0.83	0.98	0.92	0.97
	6×–1.5×	0.99	0.62	0.86	0.49	0.72
	6×–3×	0.98	0.26	0.98	0.86	0.92
Planting density [3]	Low—High	0.22	0.79	0.63	0.09	0.29
	Med—High	0.35	0.97	0.98	0.35	0.97
	Med—Low	0.97	0.69	0.51	0.73	0.38

[1] Controlled release fertilizer blends 1 (CRF1): 12N-1.31 P-11.62K and micronutrients at 1× the UF/IFAS recommendation with micronutrients as sulfates (12-3-14 1 × Micro) and CRF2: enhanced 12N-1.31 P-11.62K with 2 × Mg and 2.5× the UF/IFAS recommendation with micronutrients as sulfur-coated products (#12-3-14 2.5 × Micro). [2] Foliar fertilizer treatments, which included 0×, 1.5×, 3× and 6× the UF/IFAS recommendation. [3] Low (300 trees ha^{-1}), medium (440 trees ha−1) and high (975 trees ha^{-1}).

Table 8. Total root area of 'Ray Ruby' grapefruit grafted on 'Kuharske' citrange planted in flatwood soils located in Fort Pierce, FL, USA, and treated with ground-applied fertilizer, foliar fertilizer and different planting densities from September 2020 to January 2022.

Root Area (mm^2)		Sept. 2020	Jan. 2021	May 2021	Sept. 2021	Jan. 2022
Ground-applied [1]	CRF1	1598.3 ± 428	1463.1 ± 328	1869.0 ± 420	2763.6 ± 566	4111.4 ± 576
	CRF2	1837.3 ± 498	1452.4 ± 312	1308.2 ± 301	2664.3 ± 402	3773.9 ± 621
Foliar [2]	0×	1392.9 ± 499	1199.5 ± 457	1580.4 ± 556	2440.6 ± 605	3540.3 ± 771
	1.5×	1240.9 ± 330	1173.8 ± 247	1392.2 ± 300	2126.1 ± 512	3485.3 ± 693
	3×	1770.8 ± 733	1586.1 ± 427	1057.7 ± 354	2733.5 ± 799	3962.2 ± 963
	6×	2388.7 ± 825	2202.7 ± 319	2771.5 ± 973	3633.9 ± 831	4722.7 ± 974
Planting density [3]	High	2425.8 ± 668	2144.6 ± 532	1644.4 ± 464	3687.8 ± 754	4556.2 ± 949
	Low	1433.9 ± 575	1208.7 ± 252	1686.7 ± 593	1977.3 ± 383	2962.8 ± 363
	Medium	1205.9 ± 260	1039.3 ± 221	1800.6 ± 578	2598.5 ± 587	4364.4 ± 787

[1] Controlled-release fertilizer blends 1 (CRF1): 12N-1.31 P-11.62K and micronutrients at 1× the UF/IFAS recommendation with micronutrients as sulfates (12-3-14 1× Micro) and CRF2: enhanced 12N-1.31 P-11.62K with 2× Mg and 2.5× the UF/IFAS recommendation with micronutrients as sulfur-coated products (#12-3-14 2.5× Micro). [2] Foliar fertilizer treatments, which included 0×, 1.5×, 3× and 6× the UF/IFAS recommendation. [3] Low (300 trees ha^{-1}), medium (440 trees ha−1) and high (975 trees ha^{-1}).

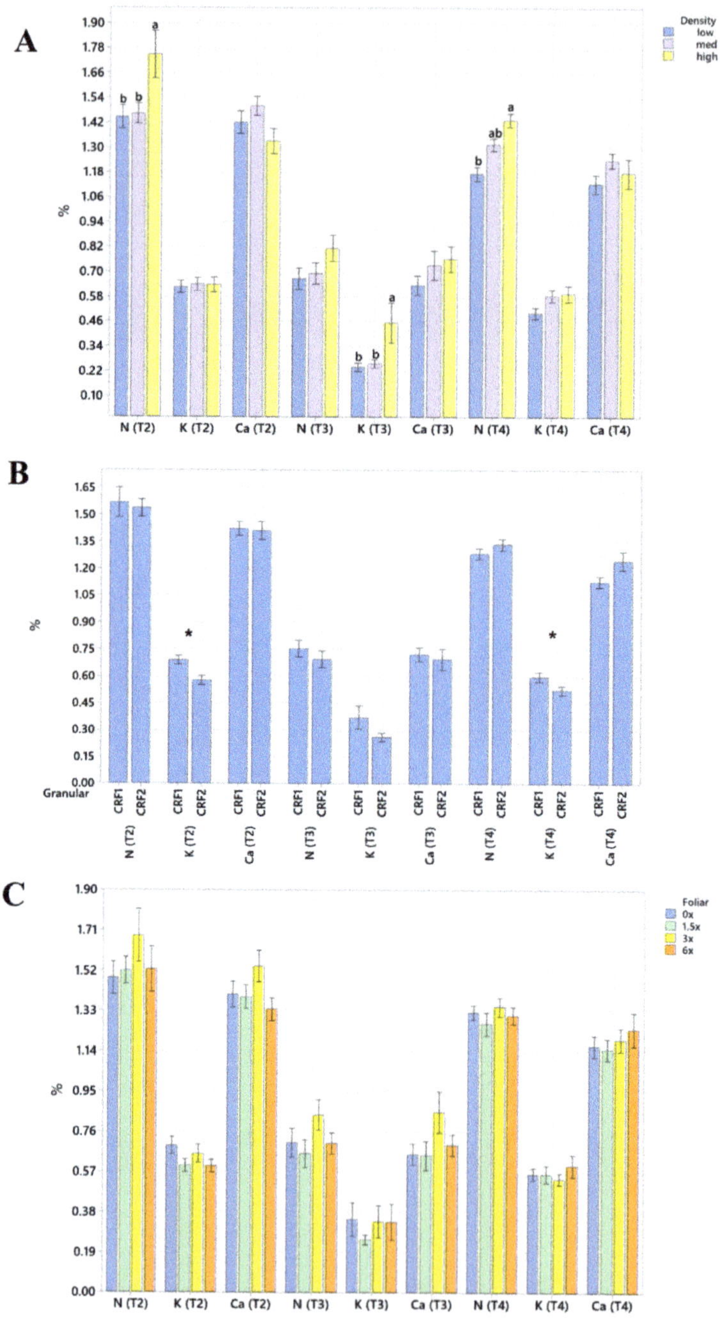

Figure 5. Root P, Mg and S concentrations of 'Ray Ruby' grapefruit grafted on 'Kuharske' citrange planted in flatwood soils located in Fort Pierce, FL, USA. Graphs indicate response to (**A**) planting density, (**B**) ground-applied fertilizer and (**C**) foliar fertilizer treatments during May 2021 (T2), September 2021 (T3) and January 2022 (T4). Bars are ± standard deviation of the mean. Treatments with * and different letters were considered to be significantly different ($p < 0.05$).

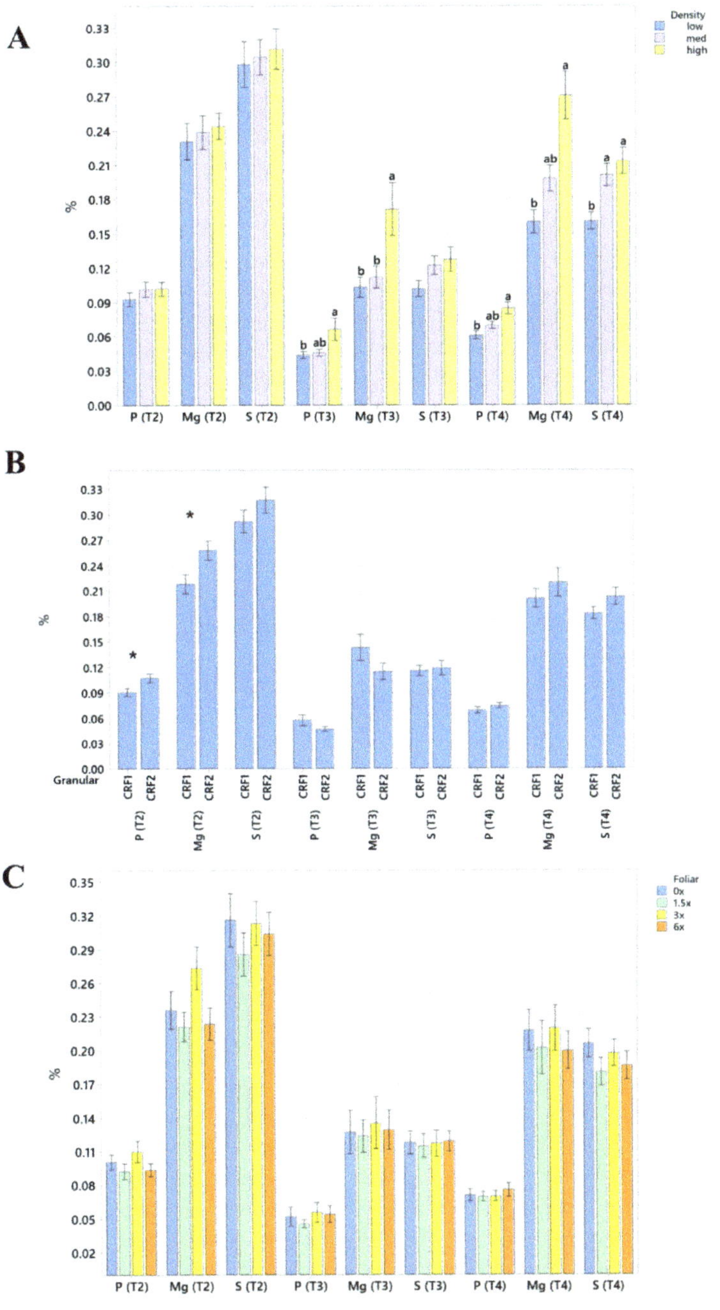

Figure 6. Root N, K and Ca concentrations of 'Ray Ruby' grapefruit grafted on 'Kuharske' citrange planted in flatwood soils located in Fort Pierce, FL, USA. Graphs indicate response to (**A**) planting density, (**B**) ground-applied fertilizer and (**C**) foliar fertilizer treatments during May 2021 (T2), September 2021 (T3) and January 2022 (T4). Bars are ± standard deviation of the mean. Treatments with * and different letters were considered to be significantly different ($p < 0.05$).

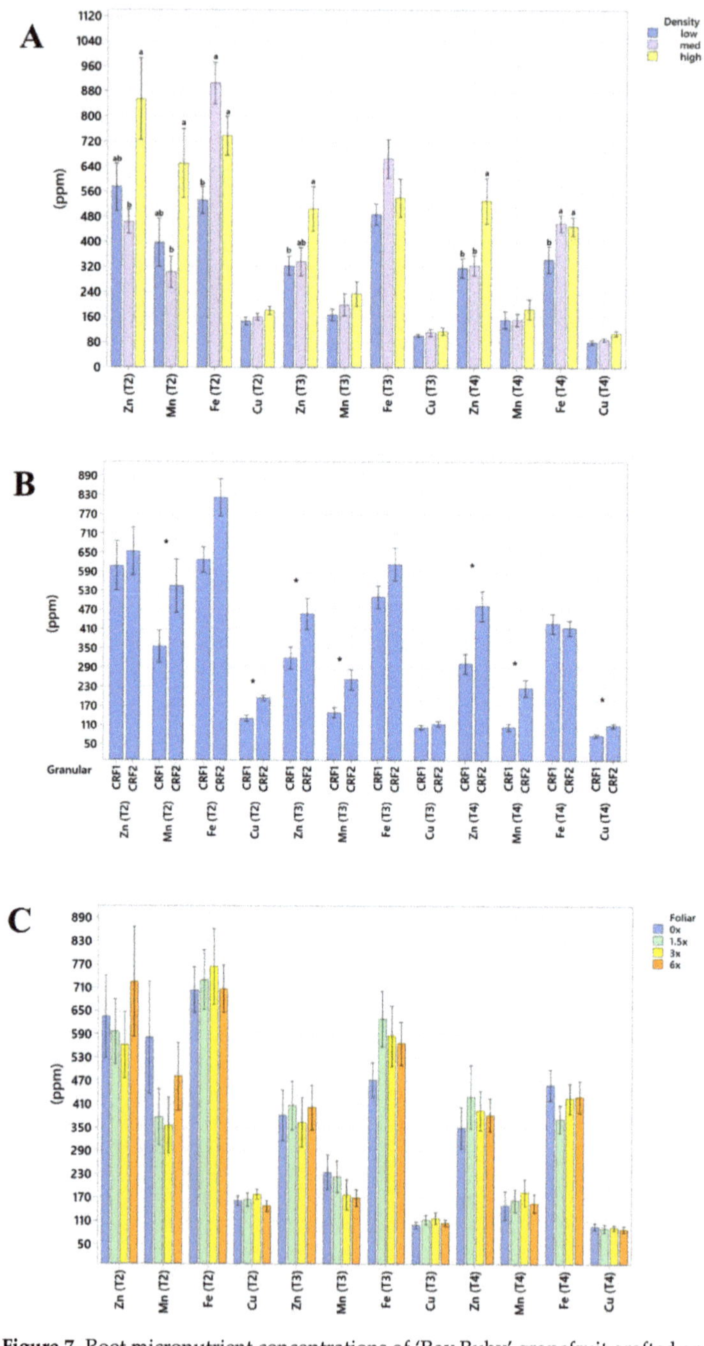

Figure 7. Root micronutrient concentrations of 'Ray Ruby' grapefruit grafted on 'Kuharske' citrange planted in flatwood soils located in Fort Pierce, FL, USA. Graphs indicate response to (**A**) planting density, (**B**) ground-applied fertilizer and (**C**) foliar fertilizer treatments during May 2021 (T2), September 2021 (T3) and January 2022 (T4). Bars are ± standard deviation of the mean, and treatments with * and different letters were considered to be significantly different ($p < 0.05$).

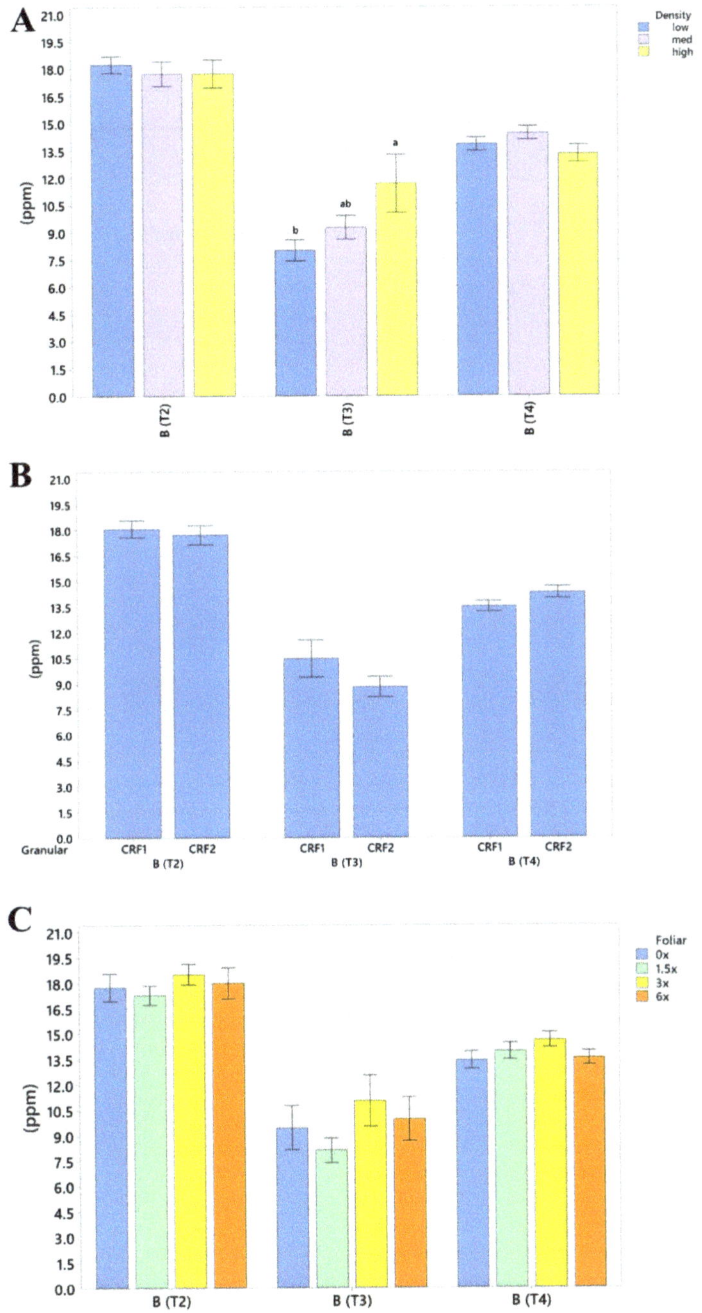

Figure 8. Root B concentrations of 'Ray Ruby' grapefruit grafted on 'Kuharske' citrange planted in flatwood soils located in Fort Pierce, FL, USA. Graphs indicate response to (**A**) planting density, (**B**) ground-applied fertilizer and (**C**) foliar fertilizer treatments during May 2021 (T2), September 2021 (T3) and January 2022 (T4). Bars are ± standard deviation of the mean, and treatments with different letters were considered to be significantly different ($p < 0.05$).

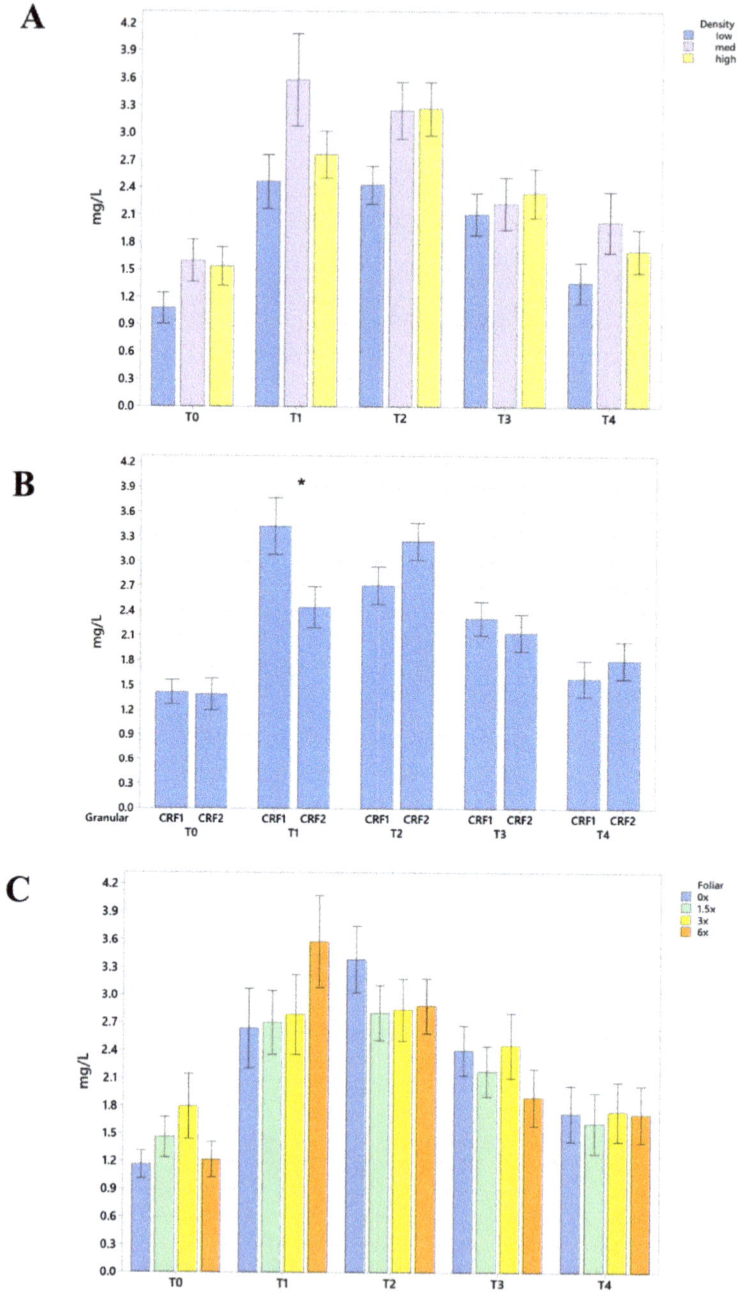

Figure 9. Root density of 'Ray Ruby' grapefruit grafted on 'Kuharske' citrange planted in flatwood soils located in Fort Pierce, FL, USA, in response to (**A**) planting density, (**B**) ground-applied fertilizer and (**C**) foliar fertilizer treatments during September 2020 (T0), January 2021 (T1), May 2021 (T2), September 2021 (T3) and January 2022 (T4). Bars are ± standard deviation of the mean, and treatments with * were considered to be significantly different ($p < 0.05$).

Table 9. *p*-values obtained via a three-way ANOVA followed by Tukey's post hoc test, corresponding to the total root area of 'Ray Ruby' grapefruit grafted on 'Kuharske' citrange planted in flatwood soils located in Fort Pierce, FL, USA, and treated with ground-applied fertilizer, foliar fertilizer and different planting densities from September 2020 to January 2022. Bold values were considered to be significantly different ($p < 0.05$).

Root Area *p* Values		Sept. 2020	Jan. 2021	May 2021	Sept. 2021	Jan. 2022
Ground-applied [1]	CRF1—CRF2	0.71	0.38	0.03	0.88	0.69
Foliar [2]	1.5×–0×	0.94	0.37	0.73	0.98	0.99
	3×–0×	0.98	0.16	0.93	0.98	0.98
	6×–0×	0.52	0.99	0.99	0.59	0.77
	3×–1.5×	0.99	0.89	0.98	0.91	0.97
	6×–1.5×	0.86	0.30	0.83	0.40	0.72
	6×–3×	0.67	0.14	0.97	0.79	0.92
Planting density [3]	Low—High	0.13	0.93	0.81	0.08	0.28
	Med—High	0.36	0.74	0.88	0.28	0.95
	Med—Low	0.85	0.53	0.53	0.76	0.39

[1] Controlled-release fertilizer blends 1 (CRF1): 12N-1.31 P-11.62K and micronutrients at 1× the UF/IFAS recommendation with micronutrients as sulfates (12-3-14 1× Micro) and CRF2: enhanced 12N-1.31 P-11.62K with 2× Mg and 2.5× the UF/IFAS recommendation with micronutrients as sulfur-coated products (#12-3-14 2.5× Micro). [2] Foliar fertilizer treatments, which included 0×, 1.5×, 3× and 6× the UF/IFAS recommendation. [3] Low (300 trees ha^{-1}), medium (440 trees ha−1) and high (975 trees ha^{-1}).

2.3. *Rhizosphere Microbiome Diversity*

Data were log-transformed to reduce error rates caused by rarefaction and later utilized for alpha and beta diversity analyses of the rhizosphere bacterial community through the R package "Phyloseq" v1.24.0 (McMurdie and Holmes 2013). Rhizosphere bacterial alpha diversity did vary according to treatments according to the Shannon index. Grapefruit trees treated with CRF 2 had a greater bacterial alpha diversity than those treated with CRF 1 (Figure 10A). Grapefruit trees treated with 3× foliar fertilizer had a greater bacterial alpha diversity compared to those treated with 0×, 1.5× and 6× foliar fertilizer (Figure 11A).

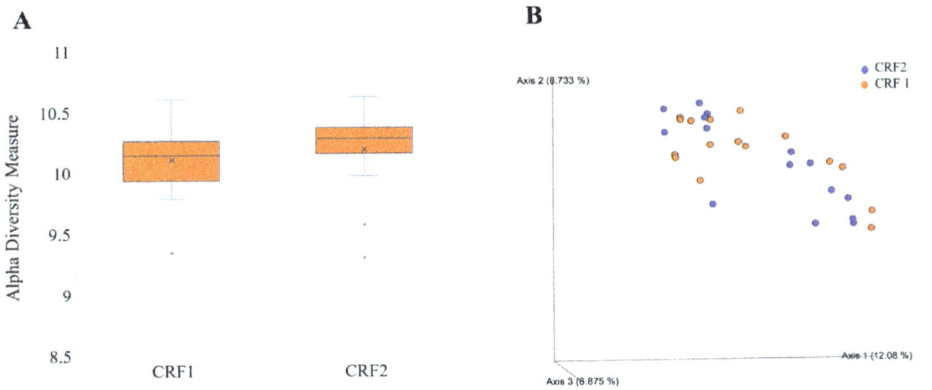

Figure 10. Alpha (**A**) and beta diversity (**B**) in rhizosphere bacteria of 'Ray Ruby' grapefruit grafted on 'Kuharske' citrange planted in flatwood soils located in Fort Pierce, FL, USA, and grown with different ground-applied fertilizer treatments. Alpha diversity was measured using the Shannon index of rhizosphere bacteria among treatments. Plotted in Figure (**A**) are boxes (interquartile), the median (line within each box), the mean (× within each box), and whiskers (lowest and greatest values). Principal coordinates analysis (PCoA) based on the Bray–Curtis dissimilarity matrix of rhizosphere bacterial samples can be found in Figure (**B**), where colors indicate treatment and include CRF1 (orange) and CRF2 (blue).

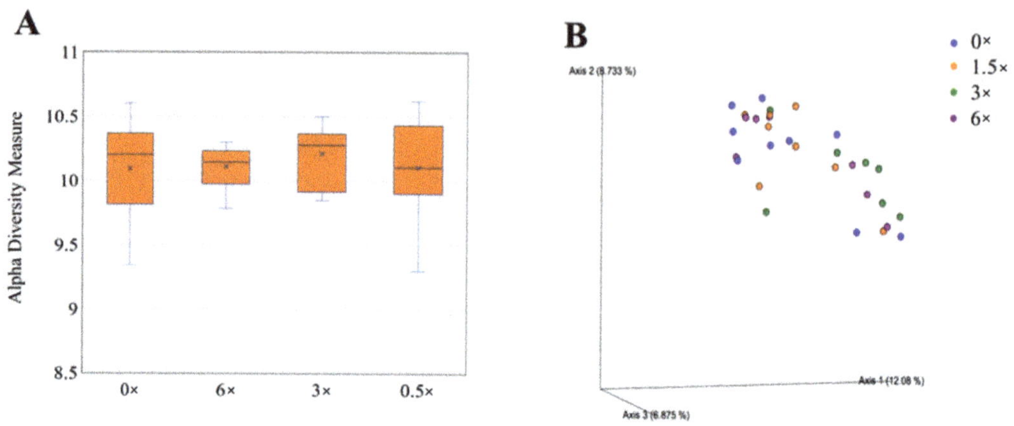

Figure 11. Alpha (**A**) and beta diversity (**B**) in rhizosphere bacteria of 'Ray Ruby' grapefruit grafted on 'Kuharske' citrange planted in flatwood soils located in Fort Pierce, FL, USA, and treated with various doses of foliar fertilizer applications. Alpha diversity was measured using the Shannon index of rhizosphere bacteria among treatments. Plotted in Figure (**A**) are boxes (interquartile), the median (line within each box), the mean (× within each box), and whiskers (lowest and greatest values). Principal coordinates analysis (PCoA) based on the Bray–Curtis dissimilarity matrix of rhizosphere bacterial samples can be found in Figure (**B**), where colors indicate treatment and include 0× (blue), 1.5× (orange), 3× (green) and 6× (purple) the UF/IFAS recommended.

Beta diversity analyses included principal coordinate analysis (PCoA) on Bray–Curtis distances. An ANOSIM test was performed to determine significant differences in beta diversity between treatments. Rhizosphere bacterial beta diversity did vary according to treatments according to the Shannon index. Grapefruit trees treated with CRF 2 had greater bacterial beta diversity than those treated with CRF 1 (Figure 10B). Grapefruit trees treated with 3× foliar fertilizer had greater bacterial beta diversity compared to those treated with 0×, 1.5× and 6× foliar (Figure 11B).

There was variation in the relative abundance of bacterial taxonomic orders among treatments. Grapefruit trees treated with 0× foliar fertilizer had a rhizosphere bacterial community with a significantly greater abundance of Reyranellales ($p < 0.05$), Rhizobiales ($p < 0.05$) and Rickettsiales ($p < 0.05$) compared to those from the 1.5× foliar treatment. Additionally, grapefruit trees at 0× foliar fertilizer had a rhizosphere bacterial community with a significantly greater abundance of Nitrosotaleales ($p < 0.05$), Vicinamibacterales ($p < 0.05$), Tistrellales ($p < 0.05$) and Solirubrobacterales ($p < 0.05$) compared to those from the 3× foliar fertilizer treatment. Furthermore, grapefruit trees with 0× foliar fertilizer had a rhizosphere bacterial community with a significantly greater abundance of Nitrososphaerales ($p < 0.05$), Clostridiales ($p < 0.05$), Caulobacterales ($p < 0.05$) and Rickettsiales ($p < 0.05$) compared to 6× foliar treatment.

3. Discussion

Planting density affected root nutrient concentrations and root size, with notably significantly greater root concentrations of Mn and Zn concentrations in a high density than in a medium density in May 2021 and January 2022. When planting grapefruit in higher densities, the greater presence of roots within a given soil area may have allowed for greater root interception of essential nutrients. Similarly, Gezahegn et al. found that closely spaced fava bean plants had increased root elongation, increasing moisture and nutrient uptake [31]. Additionally, the greater presence and activity of grapefruit roots from high-density plantings may also increase the amount of root exudates in these soils, potentially contributing to observed increases in root concentrations of Mn and Zn. Root exudates function to recruit microbes from the bulk soil to the rhizosphere for a multitude of tasks, whereby some of which range from plant defense to nutrient acquisition [32,33]. The type and amount of root exudates released vary according to the plant's growth stage and the root characteristics, such as root physiology and morphology [34,35].

Changes in root nutrient concentrations and measured parameters were more influenced by ground-applied CRF treatments than foliar fertilizer treatments, due to application sites being closer to the root zone than foliar application sites in the canopy. Long-term ground-applied CRF use can have more pronounced effects on soil characteristics, such as pH (commonly increased soil acidification), when compared to foliar fertilizers, affecting the availability of plant essential nutrients [36,37]. Furthermore, long-term excessive fertilizer applications have also been shown to alter total organic carbon (TOC), basic cation content and soil physical properties, further emphasizing the effect of ground-applied CRF on overall root health [38,39]. The phenology of citrus trees with several vegetive flushes during the year, in combination with the sub-tropical weather of Florida (with an abundance of rain during the summer and fall seasons) and the deterioration in tree health caused by HLB disease, may have contributed toward the inconsistency in nutrient concentrations, and the lack of observed patterns during the time periods of the study.

When comparing the impact of ground-applied CRF on root health, the CRF2 treatment had an overall greater beneficial effect compared to the CRF1 treatment, notably in May 2021, with significant increases in the total root area and total root volume of grapefruit. Similarly, [40] found that 2× the dose of micronutrients (Mn, Zn and B) in HLB-affected sweet orange trees led to a higher median root lifespan. Greater concentrations of micronutrients (3× B, Fe, Mn and Zn) in the CRF2 treatment may have also contributed toward better root health. Nutrient supply is imperative to disease control because nutrients influence plant resistance, pathogen vigor, growth and associated factors [15].

Both foliar and ground-applied fertilizer treatments had a greater impact on soil micronutrient concentrations compared to planting density at all timepoints, as confirmed through Pearson correlation coefficients. This was expected, as micronutrients were the main component that changed through the foliar and ground-applied fertilizer treatments.

Excess amounts of micronutrients provided by the CRF2 treatment may have improved HLB-affected root health by reducing the activity of pathogenic organisms, such as CLas. Greater concentrations of micronutrients, such as Mn and Zn, have been shown to exhibit antimicrobial functions [41]. For instance, Zn has been shown to exhibit host-pathogen interactions, as increased concentrations have been shown to suppress the growth of potential phytopathogens [42,43], thus promoting soil and plant health. Higher doses of foliar Mn in HLB-affected trees reduced symptom severity [16].

Significantly greater abundances of Rhizobiales and Vicinamibacterales may have been recruited from excess root exudates released from stressed HLB-affected grapefruit treated with 0× UF/IFAS foliar fertilizer compared to the other trees treated with higher doses of foliar fertilizers. Rhizobiales are a bacterial order of interest as they can interact with host plants to produce auxins, vitamins and N fixation, and protect the plants against stress [44]. Additionally, Vicinamibacterales have also been classified as PGPR, as they have been associated with increased plant available nutrient concentrations (specifically N and P) in rice rhizosphere soil [45]. Typically, greater amounts of exudates are released from roots during periods of stress for the purpose of recruiting microorganisms, specifically plant-growth-promoting rhizobacteria (PGPR) to assist in the acquisition of plant essential resources that would otherwise be unavailable [46–48]. Similarly, host plants grown under nutrient-deficient conditions have been shown to increase the synthesis of a root exudate known as Strigolactones to promote mycorrhizal fungal recruitment for nutrient acquisition [49,50]. Moreover, flavonoids released from the roots of nutrient-deficient legumes have been shown to stimulate bacterial root infection, leading to the establishment of nodules that promote N fixation [51].

4. Materials and Methods

4.1. Experimental Design

The study site is located at the University of Florida Institute of Food and Agricultural Sciences (UF/IFAS) Indian River Research and Education Center in Fort Pierce, FL (latitude 27.435342°, longitude -80.445197°, altitude 10 m). Plant material consisted of 'Ray Ruby' grapefruit trees (*Citrus × paradisi*) grafted on 'Kuharske' citrange (*Citrus × sinensis × Citrus trifoliata*). Trees were planted in September 2013 in flatwood soils (Pineda sands classified as loamy, siliceous, active, hyperthermic Arenic Glossaqualfs) which are poorly drained and consist of 96% sand, 2.5% silt and 1.5% clay and have an argillic soil layer at 90 cm below the soil surface. The average soil pH was 5.8 and the cation exchange capacity (CEC) was 3.5 c_{mol} kg^{-1}. Trees were grown on raised beds roughly 1 m tall to facilitate drainage, with swales 15 m between beds. Irrigation was delivered using 39.7 L h^{-1} microjet sprinklers (Maxijet, Dundee, FL, USA). The original experiment, which only focused on aboveground parameters, has been described in [9] and was arranged in a split-split-plot design with three factors, each consisting of plant densities, ground-applied controlled-release fertilizers (applied in February, May and September) and foliar-applied fertilizer combinations (applied in March, June and October). The three planting densities implemented were low (300 trees ha^{-1}), medium (440 trees ha^{-1}) and high (975 trees ha^{-1}). The study consisted of two ground-applied fertilizer treatments, specifically, controlled-release fertilizer blends 1 (CRF1): 12.00N-1.31P-11.62K and micronutrients at 1× the UF/IFAS recommendation with micronutrients as sulfates (12-3-14 1× Micro) and CRF2: enhanced 12.00N-1.31P-11.62K with 2× Mg and 2.5× the UF/IFAS recommendation with micronutrients as sulfur-coated products (#12-3-14 2.5× Micro, Tables 10 and 11). Additionally, there were four foliar fertilizer treatments, which included 0×, 1.5×, 3× and 6× the UF/IFAS recommendation [14]. Root and soil parameter sampling was performed every 4 months from September 2020 to January 2022. Rhizosphere samples were collected in January 2021.

Table 10. Details about treatment factors and arrangements for experimental design.

Experimental Design	Factor and Level
Main plot	CRF application in the soil • 12-3-14 + micronutrients at 1× UF/IFAS recommendation • 12-3-14 + 2× Mg + micronutrients (3× UF/IFAS B, Fe, Mn and Zn)
Subplot	Three plant densities • Single-row low density (SR/LD): 300 trees per ha • Single-row high density (SR/HD): 440 trees per ha • Double-row high density (DR/HD): 975 trees per ha
Sub-subplot	Four foliar treatments • No supplemental nutrients applied (0×) • 1.5 times the recommended doses of B, Mn and Zn [1] (1.5×) • 3.0 times the recommended doses of B, Mn and Zn [1] (3.0×) • 6.0 times the recommended doses of B, Mn and Zn [1] (6.0×)

[1] B = 0.28 kg/ha, Zn = 5.6 kg/ha and Mn = 4.2 kg/ha: (UF/IFAS recommendation).

Table 11. Controlled-release fertilizer formula.

	12-3-14 + Micronutrient at 1× UF/IFAS Recommendation		12-3-14 +2 × Mg + Micronutrients (3× UF/IFAS B, Fe, Mn and Zn)	
	Nutrient (%)	Amount (kg ha^{-1})	Nutrient (%)	Amount (kg ha^{-1})
N	12	180	12	180
P_2O_5	3	45	3	45
K_2O	14	209	14	209
Ca	1	15	1	15
Mg	1.2	18	2.4	36

Table 11. Cont.

	12-3-14 + Micronutrient at 1× UF/IFAS Recommendation		12-3-14 +2 × Mg + Micronutrients (3× UF/IFAS B, Fe, Mn and Zn)	
	Nutrient (%)	Amount (kg ha^{-1})	Nutrient (%)	Amount (kg ha^{-1})
S	13	194	15	228
B	0.1	0.7	0.1	1.7
Cu	0	0.6	0.1	1.5
Fe	0.4	5.9	1	15
Mn	0.6	8.4	1.4	21
Mo	0	0	0	0.2
Zn	0.4	5.9	1	15

4.2. Soil Nutrient Analysis

A soil auger (One-Piece Auger model #400.48, AMS, Inc., American Falls, ID) 7 cm in diameter and 10 cm in depth was used to collect soil samples for nutrient concentrations. A single core was taken within the irrigated zone from one tree per experimental plot (total of 96) across eight experimental blocks. Soil samples were analyzed for extractable N, P, K, Mg, Ca, S, B, Zn, Mn, Fe and Cu.

Soil samples were dried overnight at 80 °C, and nutrient concentrations were determined using Mehlich III extraction [52]. A total of 25 mL of Mehlich III extractant solution (0.2 M CH_3COOH + 0.015 M NH_4F + 0.013 M HNO_3 + 0.001 M EDTA + 0.25 M H_4NO_3) was pipetted into extraction tubes containing 2.5 ± 0.05 g of soil. Soil nutrient concentrations were measured using inductively coupled plasma optical emission spectroscopy (ICP-OES, Spectro Ciros CCD, Fitzburg, MA, USA) [14].

4.3. Root Parameter Analysis

Root parameters were measured using a minirhizotron system (CID Bio-Science CI-602, CID Bioscience, Inc. Camas, WA, USA). Each tube consisted of three scannable windows at different depths (0–19 cm, 19–39 cm and 39–59 cm), and all three windows in every tube were scanned for each sampling point. After images were acquired, average root length and density were calculated using commercial software (RootSnap™ Version 1.3.2.25, CID Bio-Science, Camas, WA, USA).

4.4. Deoxyribonucleic Acid (DNA) Isolation and Quantification

Rhizosphere samples were taken shortly after bulk soil sampling, which consisted of rhizosphere soil (located around the roots) being lightly shaken from the roots and placed in 50 mL sterile tubes. Approximately, 50 g of the soil was collected and stored at −20 °C before DNA extraction. Approximately, 15 mL of 1× sterile phosphate-buffered saline (800 mL distilled water, 8 g NaCl, 0.2 g KCl, 1.44 g Na_2PO_4 and 0.24 g KH_2PO_4) was added to the sample and shaken by hand for 15 s. Roots were removed with forceps and discarded, the remaining soil was centrifuged at 3000 g for 15 min and the supernatant was discarded. Soil DNA was extracted from 0.25 g of the soil pellet using the DNeasy PowerSoil Kit (Qiagen Inc., Germantown, MD, USA) according to the manufacturer's instructions. A fluorometer (Qubit, Thermofisher Scientific, Wilmington, NC, USA) was used to quantify the extracted DNA and determine whether the DNA was concentrated enough for sequencing (>1 ng/µL). Rhizosphere DNA was amplified with 515Fa/926R universal bacterial [53] primers and sequenced at the Genomics and Microbiome Core Facility at Rush University, Chicago, IL, USA.

4.5. Rhizosphere Microbiome Diversity and Statistical Analysis

After sequencing, bioinformatic data were processed using DADA2 [54] within the Qiime 2 [55] package. Raw sequences were demultiplexed. DADA2 was used to filter chimeras, primers and adapters and assemble pair-ended sequences. Taxonomy was assigned to amplicon sequence variants (ASVs) with the reference dataset SILVA 128 database for 16S rRNA using a naïve Bayes classifier in Qiime2 [56]. Alpha and beta diversity analyses of the bacterial community were performed on log-normalized data to avoid an increase in error rates due to rarefaction [57] with the R package "Phyloseq" v1.24.0 [58]. Alpha diversity analyses included the number of observed ASVs (Shannon index), whereas beta diversity analyses included principal coordinate analysis (PCoA) on weighted UniFrac distances. A two-way analysis of similarities (ANOSIM) test was performed to determine significant differences in beta diversity between treatments.

4.6. Plant and Soil Data Statistical Analysis

An analysis of variance (three-way ANOVA) was performed using statistical software (R Version 3.6.0, RStudio, Boston, MA, USA). The main effect means were separated using Tukey's honestly significant difference post hoc test. Differences were considered to be significant when p-values were less than or equal to 0.05. Additionally, Pearson correlation coefficients were computed.

5. Conclusions

In this study, we examined the impact of planting densities, ground-applied controlled-release fertilizers and foliar fertilizer dosages on both soil and plant parameters, including rhizosphere community composition. The use of the CRF2 ground-applied fertilizer resulted in higher soil nutrient concentrations through all time points of the study, notably with significant differences in Zn and Mn. Additionally, increased micronutrient concentrations provided by the CRF2 ground-applied fertilizer may have provided additional nutrients required to assist the root health of HLB-affected grapefruit trees. Furthermore, HLB-affected grapefruit at 0× the UF/IFAS recommended foliar fertilizer resulted in a significantly greater abundance of Vicinamibacterales and Rhizobiales, whereby both of which are PGPR that may have been recruited to assist grapefruit health under nutrient-deprived conditions.

Author Contributions: All authors contributed to the study conception and design. Material preparation, data collection and analysis were performed by J.M.S. and J.-P.F. The first draft of the manuscript was written by J.M.S. and all authors commented on the previous versions of the manuscript. All authors have read and agreed to the published version of the manuscript.

Funding: This research was funded by the UF/IFAS Citrus Initiative, U.S. Department of Agriculture (USDA)-APHIS MAC, agreement no. AP19PPQSandT00C116. This work was also supported by the U.S. Department of Agriculture National Institute of Food and Agriculture, Hatch project #FLA-IRC-005743.

Institutional Review Board Statement: Not applicable.

Informed Consent Statement: Not applicable.

Data Availability Statement: The datasets generated during and/or analyzed during the current study have been deposited to NCBI. The BioProject accession number is PRJNA937287.

Acknowledgments: We thank Tom James and Randy Burton for the treatment application and research field maintenance.

Conflicts of Interest: The authors have no relevant financial or non-financial interests to disclose.

References

1. Bodaghi, S.; Meyering, B.; Pugina, G.; Bowman, K.D.; Albrecht, U. Different Sweet Orange–Rootstock Combinations Infected by *Candidatus* Liberibacter asiaticus under Greenhouse Conditions: Effects on the Scion. *HortScience* **2022**, *57*, 56–64. [CrossRef]
2. Albrecht, U.; Mccollum, G.; Bowman, K. Influence of rootstock variety on Huanglongbing disease development in field-grown sweet orange (*Citrus sinensis* [L.] Osbeck) trees. *Sci. Hortic.* **2012**, *138*, 71–80. [CrossRef]
3. Kwakye, S.; Kadyampakeni, D.; Morgan, K.; Vashisth, T.; Wright, A. Effects of Iron Rates on Growth and Development of Young Huanglongbing-affected Citrus Trees in Florida. *HortScience* **2022**, *57*, 1092–1098. [CrossRef]

4. Hallman, L.M.; Kadyampakeni, D.M.; Ferrarezi, R.S.; Wright, A.L.; Ritenour, M.A.; Johnson, E.G.; Rossi, L. Impact of Ground Applied Micronutrients on Root Growth and Fruit Yield of Severely Huanglongbing-Affected Grapefruit Trees. *Horticulturae* **2022**, *8*, 763. [CrossRef]
5. Johnson, E.G.; Wu, J.; Bright, D.B.; Graham, J.H. Association of 'CandidatusLiberibacter asiaticus' root infection, but not phloem plugging with root loss on huanglongbing-affected trees prior to appearance of foliar symptoms. *Plant Pathol.* **2014**, *63*, 290–298. [CrossRef]
6. Zheng, Z.; Chen, J.; Deng, X. Historical Perspectives, Management, and Current Research of Citrus HLB in Guangdong Province of China, Where the Disease has been Endemic for Over a Hundred Years. *Phytopathology®* **2018**, *108*, 1224–1236. [CrossRef] [PubMed]
7. Zapien-Macias, J.M.; Ferrarezi, R.S.; Spyke, P.D.; Castle, W.S.; Gmitter, F.G.; Grosser, J.W.; Rossi, L. Early Performance of Recently Released Rootstocks with Grapefruit, Navel Orange, and Mandarin Scions under Endemic Huanglongbing Conditions in Florida. *Horticulturae* **2022**, *8*, 1027. [CrossRef]
8. Bowman, K.D.; Mccollum, G.; Albrecht, U. Performance of 'Valencia' orange (*Citrus sinensis* [L.] Osbeck) on 17 rootstocks in a trial severely affected by huanglongbing. *Sci. Hortic.* **2016**, *201*, 355–361. [CrossRef]
9. Phuyal, D.; Nogueira, T.A.R.; Jani, A.D.; Kadyampakeni, D.M.; Morgan, K.T.; Ferrarezi, R.S. 'Ray Ruby' Grapefruit Affected by Huanglongbing I. Planting Density and Soil Nutrient Management. *HortScience* **2020**, *55*, 1411–1419. [CrossRef]
10. Hallman, L.M.; Kadyampakeni, D.M.; Fox, J.-P.; Wright, A.L.; Rossi, L. Root-Shoot Nutrient Dynamics of Huanglongbing-Affected Grapefruit Trees. *Plants* **2022**, *11*, 3226. [CrossRef]
11. Kunwar, S.; Meyering, B.; Grosser, J.; Gmitter, F.G.; Castle, W.S.; Albrecht, U. Field performance of 'Valencia' orange trees on diploid and tetraploid rootstocks in different huanglongbing-endemic growing environments. *Sci. Hortic.* **2022**, *309*, 111635. [CrossRef]
12. Cocuzza, G.E.M.; Alberto, U.; Hernández-Suárez, E.; Siverio, F.; Di Silvestro, S.; Tena, A.; Carmelo, R. A review on *Trioza erytreae* (African citrus psyllid), now in mainland Europe, and its potential risk as vector of huanglongbing (HLB) in citrus. *J. Pest Sci.* **2017**, *90*, 1–17. [CrossRef]
13. Li, B.; Wang, S.; Zhang, Y.; Qiu, D. Acid Soil Improvement Enhances Disease Tolerance in Citrus Infected by Candidatus Liberibacter asiaticus. *Int. J. Mol. Sci.* **2020**, *21*, 3614. [CrossRef] [PubMed]
14. Kadyampakeni, D.M.; Morgan, K.T.; Nkedi-Kizza, P.; Kasozi, G.N. Nutrient Management Options for Florida Citrus: A Review of NPK Application and Analytical Methods. *J. Plant Nutr.* **2015**, *38*, 568–583. [CrossRef]
15. Kadyampakeni, D.M.; Chinyukwi, T. Are macronutrients and micronutrients therapeutic for restoring performance of trees affected by citrus greening? A discussion of current practices and future research opportunities. *J. Plant Nutr.* **2022**, *44*, 2949–2969. [CrossRef]
16. Zambon, F.T.; Kadyampakeni, D.M.; Grosser, J.W. Ground Application of Overdoses of Manganese Have a Therapeutic Effect on Sweet Orange Trees Infected with Candidatus Liberibacter asiaticus. *HortScience* **2019**, *54*, 1077–1086. [CrossRef]
17. Atta, A.A.; Morgan, K.T.; Kadyampakeni, D.M.; Mahmoud, K.A. The Effect of Foliar and Ground-Applied Essential Nutrients on Huanglongbing-Affected Mature Citrus Trees. *Plants* **2021**, *10*, 925. [CrossRef]
18. Ghimire, L.; Grosser, J.; Vashisth, T. Differences in Nutrient Uptake Can Influence the Performance of Citrus Rootstocks under Huanglongbing Conditions. *HortScience* **2023**, *58*, 40–46. [CrossRef]
19. Esteves, E.; Kadyampakeni, D.M.; Zambon, F.; Ferrarezi, R.S.; Maltais-Landry, G. Magnesium fertilization has a greater impact on soil and leaf nutrient concentrations than nitrogen or calcium fertilization in Florida orange production. *Nutr. Cycl. Agroecosystems* **2022**, *122*, 73–87. [CrossRef]
20. Morgan, K.T.; Wheaton, T.A.; Castle, W.S.; Parsons, L.R. Response of Young and Maturing Citrus Trees Grown on a Sandy Soil to Irrigation Scheduling, Nitrogen Fertilizer Rate, and Nitrogen Application Method. *HortScience* **2009**, *44*, 145–150. [CrossRef]
21. Ferrarezi, R.S.; Jani, A.D.; James, H.T.; Gil, C.; Ritenour, M.A.; Wright, A.L. Sweet Orange Orchard Architecture Design, Fertilizer, and Irrigation Management Strategies under Huanglongbing-endemic Conditions in the Indian River Citrus District. *HortScience* **2020**, *55*, 2028–2036. [CrossRef]
22. Phuyal, D.; Nogueira, T.A.R.; Jani, A.D.; Kadyampakeni, D.M.; Morgan, K.T.; Ferrarezi, R.S. 'Ray Ruby' Grapefruit Affected by Huanglongbing II. Planting Density, Soil, and Foliar Nutrient Management. *HortScience* **2020**, *55*, 1420–1432. [CrossRef]
23. Morgan, K.T.; Rouse, R.E.; Ebel, R.C. Foliar Applications of Essential Nutrients on Growth and Yield of 'Valencia' Sweet Orange Infected with Huanglongbing. *HortScience* **2016**, *51*, 1482–1493. [CrossRef]
24. Al-Obeed, R.S.; Ahmed, M.A.-A.; Kassem, H.A.; Al-Saif, A.M. Improvement of "Kinnow" mandarin fruit productivity and quality by urea, boron and zinc foliar spray. *J. Plant Nutr.* **2018**, *41*, 609–618. [CrossRef]
25. Vashisth, T.; Grosse, J. Comparison of Controlled Release Fertilizer (CRF) for Newly Planted Sweet Orange Trees under Huanglongbing Prevalent Conditions. *J. Hortic.* **2018**, *5*, 244. [CrossRef]
26. Ren, N.; Wang, Y.; Ye, Y.; Zhao, Y.; Huang, Y.; Fu, W.; Chu, X. Effects of Continuous Nitrogen Fertilizer Application on the Diversity and Composition of Rhizosphere Soil Bacteria. *Front. Microbiol.* **2020**, *11*, 1948. [CrossRef] [PubMed]
27. Lin, W.; Lin, M.; Zhou, H.; Wu, H.; Li, Z.; Lin, W. The effects of chemical and organic fertilizer usage on rhizosphere soil in tea orchards. *PLoS ONE* **2019**, *14*, e0217018. [CrossRef]

28. Malik, L.; Sanaullah, M.; Mahmood, F.; Hussain, S.; Siddique, M.H.; Anwar, F.; Shahzad, T. Unlocking the potential of co-applied biochar and plant growth-promoting rhizobacteria (PGPR) for sustainable agriculture under stress conditions. *Chem. Biol. Technol. Agric.* **2022**, *9*, 58. [CrossRef]
29. Samaras, A.; Kamou, N.; Tzelepis, G.; Karamanoli, K.; Menkissoglu-Spiroudi, U.; Karaoglanidis, G.S. Root Transcriptional and Metabolic Dynamics Induced by the Plant Growth Promoting Rhizobacterium (PGPR) Bacillus subtilis Mbi600 on Cucumber Plants. *Plants* **2022**, *11*, 1218. [CrossRef]
30. Zeng, Q.; Ding, X.; Wang, J.; Han, X.; Iqbal, H.M.N.; Bilal, M. Insight into soil nitrogen and phosphorus availability and agricultural sustainability by plant growth-promoting rhizobacteria. *Environ. Sci. Pollut. Res.* **2022**, *29*, 45089–45106. [CrossRef]
31. Gezahegn, A.M.; Tesfaye, K.; Sharma, J.J.; Belel, M.D. Determination of optimum plant density for faba bean (*Vicia faba* L.) on vertisols at Haramaya, Eastern Ethiopia. *Cogent Food Agric.* **2016**, *2*, 1224485. [CrossRef]
32. Rolfe, S.A.; Griffiths, J.; Ton, J. Crying out for help with root exudates: Adaptive mechanisms by which stressed plants assemble health-promoting soil microbiomes. *Curr. Opin. Microbiol.* **2019**, *49*, 73–82. [CrossRef] [PubMed]
33. Williams, A.; Langridge, H.; Straathof, A.L.; Muhamadali, H.; Hollywood, K.A.; Goodacre, R.; Vries, F.T. Root functional traits explain root exudation rate and composition across a range of grassland species. *J. Ecol.* **2022**, *110*, 21–33. [CrossRef]
34. Chen, S.; Waghmode, T.R.; Sun, R.; Kuramae, E.E.; Hu, C.; Liu, B. Root-associated microbiomes of wheat under the combined effect of plant development and nitrogen fertilization. *Microbiome* **2019**, *7*, 136. [CrossRef]
35. Preece, C.; Peñuelas, J. A Return to the Wild: Root Exudates and Food Security. *Trends Plant Sci.* **2020**, *25*, 14–21. [CrossRef]
36. Hartemink, A.E.; Barrow, N.J. Soil pH–nutrient relationships: The diagram. *Plant Soil* **2023**. [CrossRef]
37. Kumar Bhatt, M.; Labanya, R.; Joshi, H.C. Influence of Long-term Chemical fertilizers and Organic Manures on Soil Fertility—A Review. *Univers. J. Agric. Res.* **2019**, *7*, 177–188. [CrossRef]
38. Tang, Y.; Wang, X.; Yang, Y.; Gao, B.; Wan, Y.; Li, Y.C.; Cheng, D. Activated-Lignite-Based Super Large Granular Slow-Release Fertilizers Improve Apple Tree Growth: Synthesis, Characterizations, and Laboratory and Field Evaluations. *J. Agric. Food Chem.* **2017**, *65*, 5879–5889. [CrossRef]
39. Dhaliwal, S.S.; Naresh, R.K.; Mandal, A.; Walia, M.K.; Gupta, R.K.; Singh, R.; Dhaliwal, M.K. Effect of manures and fertilizers on soil physical properties, build-up of macro and micronutrients and uptake in soil under different cropping systems: A review. *J. Plant Nutr.* **2019**, *42*, 2873–2900. [CrossRef]
40. Atta, A.A.; Morgan, K.T.; Hamido, S.A.; Kadyampakeni, D.M. Effect of Essential Nutrients on Roots Growth and Lifespan of Huanglongbing Affected Citrus Trees. *Plants* **2020**, *9*, 483. [CrossRef]
41. Da Silva, J.R.; De Alvarenga, F.V.; Boaretto, R.M.; Lopes, J.R.S.; Quaggio, J.A.; Coletta Filho, H.D.; Mattos, D. Following the effects of micronutrient supply in HLB-infected trees: Plant responses and 'Candidatus Liberibacter asiaticus' acquisition by the Asian citrus psyllid. *Trop. Plant Pathol.* **2020**, *45*, 597–610. [CrossRef]
42. Wu, D.; Ma, Y.; Yang, T.; Gao, G.; Wang, D.; Guo, X.; Chu, H. Phosphorus and Zinc Are Strongly Associated with Belowground Fungal Communities in Wheat Field under Long-Term Fertilization. *Microbiol. Spectr.* **2022**, *10*, e00110-22. [CrossRef] [PubMed]
43. Graham, J.H.; Johnson, E.G.; Myers, M.E.; Young, M.; Rajasekaran, P.; Das, S.; Santra, S. Potential of Nano-Formulated Zinc Oxide for Control of Citrus Canker on Grapefruit Trees. *Plant Dis.* **2016**, *100*, 2442–2447. [CrossRef] [PubMed]
44. Erlacher, A.; Cernava, T.; Cardinale, M.; Soh, J.; Sensen, C.; Grube, M.; Berg, G. Rhizobiales as functional and endosymbiontic members in the lichen symbiosis of *Lobaria pulmonaria* L. *Front. Microbiol.* **2015**, *6*, 53. [CrossRef]
45. Liu, H.; Brettell, L.; Qiu, Z.; Singh, B. Microbiome-Mediated Stress Resistance in Plants. *Trends Plant Sci.* **2020**, *25*, 733–743.
46. Vives-Peris, V.; Molina, L.; Segura, A.; Gómez-Cadenas, A.; Pérez-Clemente, R.M. Root exudates from citrus plants subjected to abiotic stress conditions have a positive effect on rhizobacteria. *J. Plant Physiol.* **2018**, *228*, 208–217. [CrossRef]
47. Lombardi, N.; Vitale, S.; Turrà, D.; Reverberi, M.; Fanelli, C.; Vinale, F.; Marra, R.; Ruocco, M.; Pascale, A.; D'Errico, G.; et al. Root Exudates of Stressed Plants Stimulate and Attract *Trichoderma* Soil Fungi. *Mol. Plant-Microbe Interact.* **2018**, *31*, 982–994. [CrossRef]
48. Gargallo-Garriga, A.; Preece, C.; Sardans, J.; Oravec, M.; Urban, O.; Peñuelas, J. Root exudate metabolomes change under drought and show limited capacity for recovery. *Sci. Rep.* **2018**, *8*, 12696. [CrossRef]
49. López-Ráez, J.A.; Pozo, M.J.; García-Garrido, J.M. Strigolactones: A cry for help in the rhizosphere. *Botany* **2011**, *89*, 513–522. [CrossRef]
50. Kim, B.; Westerhuis, J.; Smilde, A.; Flokova, K.; Suleiman, A.; Kuramae, E.; Bouwmeester, H.; Zancarini, A. Effect of strigolactones on recruitment of the rice root-associated microbiome. *FEMS Microbiol. Ecol.* **2022**, *98*, fiac010.
51. Dong, W.; Song, Y. The Significance of Flavonoids in the Process of Biological Nitrogen Fixation. *Int. J. Mol. Sci.* **2020**, *21*, 5926. [CrossRef] [PubMed]
52. Mylavarapu, R.; Obreza, T.; Morgan, K.; Hochmuth, G.; Nair, V.; Wright, A. Extraction of Soil Nutrients Using Mehlich-3 Reagent for Acid-Mineral Soils of Florida. *EDIS* **2014**, *2014*. [CrossRef]
53. Caporaso, J.G.; Lauber, C.L.; Walters, W.A.; Berg-Lyons, D.; Lozupone, C.A.; Turnbaugh, P.J.; Fierer, N.; Knight, R. Global patterns of 16S rRNA diversity at a depth of millions of sequences per sample. *Proc. Natl. Acad. Sci. USA* **2011**, *108*, 4516–4522. [CrossRef] [PubMed]
54. Callahan, B.J.; McMurdie, P.J.; Rosen, M.J.; Han, A.W.; Johnson, A.J.A.; Holmes, S.P. DADA2: High-resolution sample inference from Illumina amplicon data. *Nat. Methods* **2016**, *13*, 581–583. [CrossRef]

55. Caporaso, J.G.; Kuczynski, J.; Stombaugh, J.; Bittinger, K.; Bushman, F.D.; Costello, E.K.; Fierer, N.; Peña, A.G.; Goodrich, J.K.; Gordon, J.I.; et al. QIIME allows analysis of high-throughput community sequencing data. *Nat. Methods* **2010**, *7*, 335–336. [CrossRef]
56. Bokulich, N.A.; Joseph, C.M.L.; Allen, G.; Benson, A.K.; Mills, D.A. Next-Generation Sequencing Reveals Significant Bacterial Diversity of Botrytized Wine. *PLoS ONE* **2012**, *7*, e36357. [CrossRef]
57. McMurdie, P.J.; Holmes, S. Waste Not, Want Not: Why Rarefying Microbiome Data Is Inadmissible. *PLoS Comput. Biol.* **2014**, *10*, e1003531. [CrossRef]
58. McMurdie, P.J.; Holmes, S. phyloseq: An R Package for Reproducible Interactive Analysis and Graphics of Microbiome Census Data. *PLoS ONE* **2013**, *8*, e61217. [CrossRef]

plants

MDPI

Article

Physiological and Ultrastructural Responses to Excessive-Copper-Induced Toxicity in Two Differentially Copper Tolerant Citrus Species

Xin-Yu Li [1], Mei-Lan Lin [1], Fei Lu [1,2], Xin Zhou [1], Xing Xiong [1], Li-Song Chen [1] and Zeng-Rong Huang [1,2,*]

1 College of Resources and Environment, Fujian Agriculture and Forestry University, Fuzhou 350002, China
2 Key Lab of Soil Ecosystem Health and Regulation, Fujian Province University (Fujian Agriculture and Forestry University), Fuzhou 350002, China
* Correspondence: hzrapaul@126.com or huangzengrong@fafu.edu.cn

Abstract: Over-applied copper (Cu)-based agrochemicals are toxic to citrus trees. However, less information is available discussing the ultrastructural alterations in Cu-stressed citrus species. In the present study, seedlings of *Citrus sinensis* and *Citrus grandis* that differed in Cu-tolerance were sandy-cultured with nutrient solution containing 0.5 μM Cu (as control) or 300 μM Cu (as Cu toxicity) for 18 weeks. At the end of the treatments, the physiological parameters and ultrastructural features of the citrus leaves and roots were analyzed. The results indicate that Cu toxicity significantly decreased the ratio of shoot biomass to dry weight, the Cu translocation factor and the total chlorophyll of two citrus species. The anatomical and ultrastructural alterations verified that excessive Cu resulted in starch granules accumulated in the leaves and roots of the two citrus species. Under Cu toxicity, increased root flocculent precipitate and thickened root cell wall might reduce the Cu translocation from citrus roots to the shoots. Compared with *C. sinensis*, *C. grandis* maintained a relatively integral root cellular structure under Cu toxicity, which provided a structural basis for a higher Cu tolerance than *C. sinensis*. The present results increase our understanding of the physiological and ultrastructural responses to Cu toxicity in citrus species.

Keywords: anatomy; *Citrus grandis*; *Citrus sinensis*; copper toxicity; ultrastructure

Citation: Li, X.-Y.; Lin, M.-L.; Lu, F.; Zhou, X.; Xiong, X.; Chen, L.-S.; Huang, Z.-R. Physiological and Ultrastructural Responses to Excessive-Copper-Induced Toxicity in Two Differentially Copper Tolerant Citrus Species. *Plants* **2023**, *12*, 351. https://doi.org/10.3390/plants12020351

Academic Editor: Lorenzo Rossi

Received: 21 December 2022
Revised: 5 January 2023
Accepted: 10 January 2023
Published: 11 January 2023

1. Introduction

The soil copper (Cu) contamination induced by the over-application of Cu-based agrochemicals is a primary environmental and toxicological concern for sustainable agriculture [1,2]. In citrus orchard management, Cu-plus fungicides (such as Bordeaux mixture) and micronutrient fertilizers are frequently foliar-sprayed for pathogen controlling and nutrient adjustment, respectively [3,4]. The Cu-containing solution dripped from the trees or washed by precipitation increased the Cu concentration of the topsoil. Accordingly, old citrus orchards often had a significantly higher soil Cu concentration than younger ones [5,6]. Similar results were also reported in vineyards soils [7] and apple orchards [8].

It has been generally accepted that Cu concentrations (extracted with DPTA) above 5.0 μg/g [9] and 20.0 μg/g [10,11] represent excessive Cu levels in citrus orchard soil and citrus leaves, respectively. According to our previous investigation of the mineral nutrients in pummelo orchards in the Fujian province of China (309 samples), 28.3% of tested soils and 70.0% of pummelo leaves had excessive Cu [12], which induced potential Cu toxicity in citrus. Excessive Cu absorption by citrus roots resulted in oxidative stress, which inhibited root elongation and decreased the number of root branches [13]. Citrus leaves suffering from Cu toxicity are yellowish with lower chlorophyll content and photosynthetic rate [14]. Moreover, Cu phytotoxicity decreased the fruiting number of citrus species [15] and caused visible damage on the fruit surface, which reduced its profitability markedly.

Revealing the Cu-tolerant mechanisms of citrus species is crucial for optimized Cu-nutrient management and for improving environmental quality in the sustainable development of citrus production. For those purposes, we have investigated the Cu tolerance of citrus seedlings at the biochemical and physiological [16], transcriptional [17], proteomic [18] and metabolomic levels [19]. Our previous findings indicate that activated antioxidative defenses [19] promote Cu retention by the root cell wall [20] and secretion of root exudates [21] represented the most critical responses to Cu toxicity in citrus species. Despite versatile Cu-tolerant mechanisms of a biochemical or molecular basis being unveiled, evidence regarding ultrastructural alterations of citrus species under Cu toxicity is still missing.

The alterations of plant tissues and organs provide visible and direct evidence for a better understanding of plant physiological responses to Cu toxicity. For instance, Sánchez-Pardo et al. [22] revealed that a higher Cu tolerance of soybean than white lupin was associated with the size of palisade parenchyma and epidermal cells, which accounted for the Cu micro localization. Likewise, de Freitas et al. [23] have reported that Cu toxicity caused membrane and endoderm ruptures, inhibiting the root nutrient uptake and translocation from roots to shoots in *Inga subnuda* based on anatomical and ultrastructural analyses. Furthermore, using TEM, Minkina et al. [24] revealed that Cu toxicity disorganized the thylakoid membranes of leaves and damaged the integrity of the root cell wall and cytoplasmic membrane in barley (*Hordeum sativum*). The existing studies identified that the ultrastructural alteration of plant tissues exposing Cu toxicity varied with plant species and Cu tolerance. Therefore, the anatomical and ultrastructural investigations would offer vital insights into the Cu tolerance of citrus species.

The present study aimed to explore the anatomical and ultrastructural alterations in the leaves and roots of *C. sinensis* and *C. grandis* that were subjected to 300 μM Cu toxicity for 18 weeks. At the end of treatments, the physiological parameters, including biomass distribution, the Cu translocation and leaf total chlorophyll content, were measured. The anatomical features of citrus leaves and roots were examined under SEM. Additionally, the ultrastructural characteristics of citrus leaves and roots were observed by TEM. The results of the present study would extend our understanding of the physiological responses of citrus species conferring Cu toxicity on an ultrastructural basis.

2. Results

2.1. *Cu Toxicity Downregulated Seedling Height, Induced Leaf Chlorosis and Hampered Root Development of C. sinensis and C. grandis*

The citrus seedlings exposed to 300 μM Cu toxicity were significantly lower than the control seedlings in two citrus species (Figure 1A,D). The toxicity of 300 μM Cu decreased the leaf area and caused obvious leaf chlorosis compared with the control of two citrus species (Figure 1B,E). Moreover, the yellowish leaves were observed mainly on the upper leaves compared with the lower leaves under 300 μM Cu in both citrus species. The fresh roots of control seedlings were bright and vigorous. However, citrus roots that suffered from Cu toxicity were dark, with fewer root branches. Remarkably, the root development was inhibited by 300 μM Cu toxicity of two citrus species (Figure 1C,F).

Figure 1. The effects of Cu toxicity on seedling growth (**A**,**D**), leaf morphology (**B**,**E**) and root development (**C**,**F**) of two citrus species (*C. sinensis*: (**A**–**C**); *C. grandis*: (**D**–**F**)).

2.2. Cu Toxicity Altered the Biomass Distribution, Decreased the Cu Translocation Factor and Total Chlorophyll Content of *C. sinensis* and *C. grandis*

As shown in Table 1, 300 μM Cu toxicity decreased the ratio (shoot biomass/dry weight) by 7.42% and 4.12% in *C. sinensis* and *C. grandis* compared with the control, respectively. Contrastingly, the ratio (root biomass/dry weight) increased by 40.86% and 27.45% under 300 μM Cu toxicity in *C. sinensis* and *C. grandis* compared with the control. The comparison between two citrus species indicated that *C. sinensis* had a significantly lower ratio of shoot biomass to dry weight but a significantly higher ratio of root biomass to dry weight than *C. grandis,* both under control and 300 μM Cu toxicity. Moreover, the control seedlings of *C. grandis* had a noticeably higher Cu translocation from the lateral roots to the leaves than *C. sinensis*. However, the Cu translocation factors of two citrus species decreased significantly under Cu toxicity compared with the control. Under Cu toxicity, no significant difference was found between the two citrus species. Likewise, the total chlorophyll content of the two citrus species was decreased remarkably by Cu toxicity compared with the control without a significant difference between the two citrus species under Cu toxicity.

Table 1. The effects of excessive Cu on biomass distribution, Cu translocation factor (TF, defined as the ratio of Cu concentration in the leaves to the lateral roots) and total chlorophyll content of two citrus species. The values represent means ± SE of four replicates. Significant differences ($p \leq 0.05$) between treatments are indicated by different letters.

Species	Treatments	Shoot Biomass/Dry Weight (%)	Root Biomass/Dry Weight (%)	Cu TF (%)	Total Chlorophyll (μg/cm^2)
C. sinensis	Cu 0.5 μM	84.63 ± 0.26 b	15.37 ± 0.26 b	27.08 ± 0.08 b	73.22 ± 1.27 a
	Cu 300 μM	78.35 ± 0.90 c	21.65 ± 0.90 a	0.53 ± 0.04 c	66.16 ± 1.09 bc
C. grandis	Cu 0.5 μM	86.97 ± 0.85 a	13.03 ± 0.85 c	40.46 ± 3.65 a	68.73 ± 2.82 ab
	Cu 300 μM	83.39 ± 0.83 b	16.61 ± 0.83 b	0.52 ± 0.06 c	61.59 ± 2.89 c

2.3. Cu Toxicity Disrupted the Anatomical Structure of Leaves and Roots of *C. sinensis* and *C. grandis*

The SEM revealed that the epidermal cells of the leaves of *C. sinensis* from the control group were well-structured by a single layer in a rectangular-like shape. The palisade parenchyma could be found under the epidermis, consisting of layers of parenchymal cells in a regular column (Figure 2A). The spongy parenchyma adjacent to the lower leaf epidermis was irregularly distributed, containing a small number of starch granules (Figure 2B). Excessive Cu enlarged the epidermal cells and shrank the palisade parenchyma. Strikingly, the number of starch granules increased in the palisade and spongy parenchyma (Figure 2C,D). Compared with *C. sinensis*, *C. grandis* had a thicker epidermis and shorter palisade parenchyma under control. The number of starch granules in the palisade and spongy parenchyma was less than that in *C. sinensis* (Figure 2E,F). Cu toxicity disrupted the arrangement of spongy parenchyma and significantly increased the number of starch granules in the palisade and spongy parenchyma of *C. grandis* leaves (Figure 2G,H).

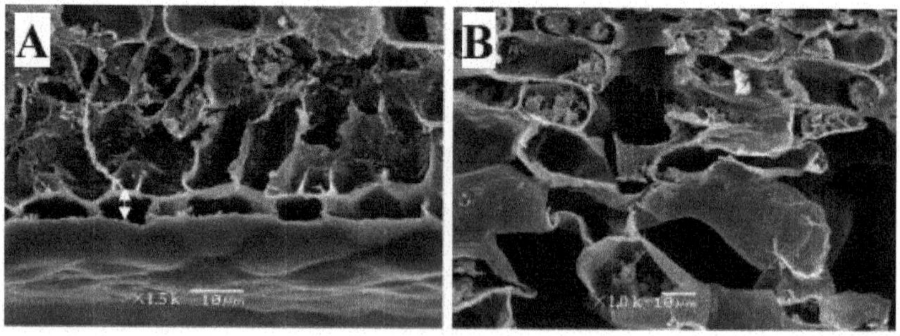

Figure 2. *Cont.*

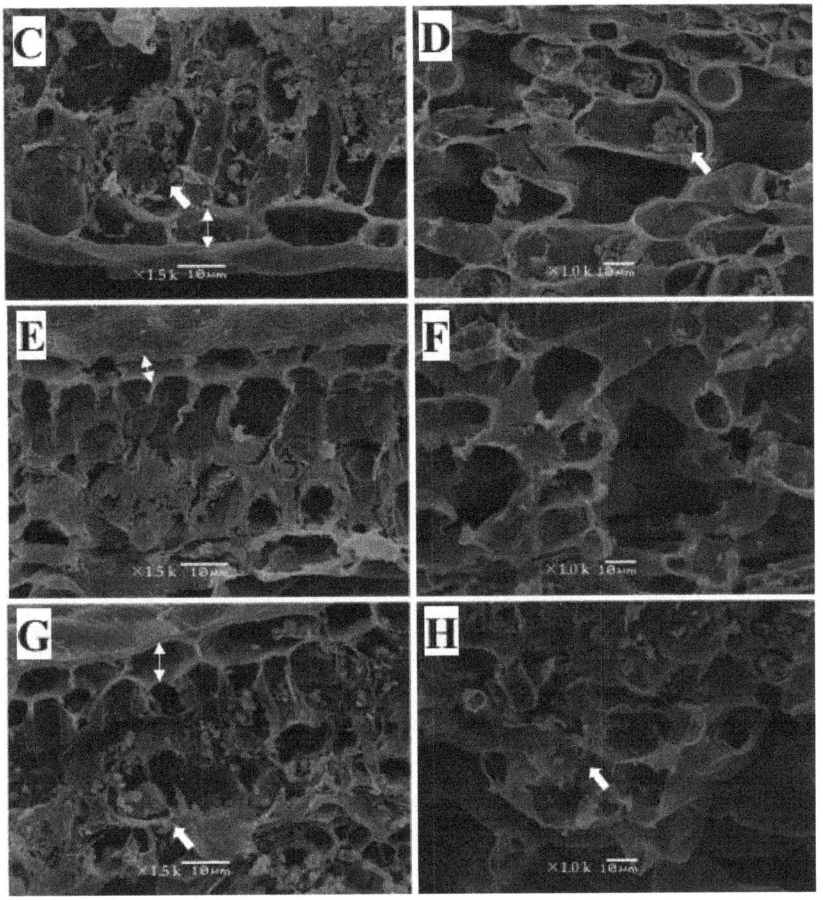

Figure 2. The SEM images of 0.5 μM (**A**,**B**,**E**,**F**) and 300 μM Cu-treated (**C**,**D**,**G**,**H**) leaves of two citrus species. *C. sinensis*: (**A–D**); *C. grandis*: (**E–H**). Images (**A**,**C**,**E**,**G**) show epidermis and palisade tissues of citrus leaves. Images (**B**,**D**,**F**,**H**) show the spongy tissue. The granules in cells labeled by arrows are starch granules. The two-way arrow represents the average thickness of the leaf epidermis. The scales in the figures represent 10 μm.

The cell layer in the root exodermis of control *C. sinensis* was well-structured in the transversal and longitudinal sections under SEM (Figures 3A and 4A). The distorted vessels and impaired structure at the cross-section of *C. sinensis* roots were found under Cu toxicity (Figure 3C). The xylem of control *C. sinensis* was almost empty (Figures 3B and 4B). However, the cross-section of the root xylem was covered by flocculent precipitate under Cu toxicity in *C. sinensis* (Figure 4D). Increased accumulation of starch granules was observed in roots of excessive Cu-treated *C. sinensis* (Figure 3D). Compared with *C. sinenis* under control, *C. grandis* had a thicker xylem and the inner ring was clear (Figure 4E,F). The accumulated starch granules, disrupted vessels and flocculent precipitate at the longitudinal section were also found to be in excess of Cu-treated *C. grandis* (Figures 3H and 4H). Compared with *C. sinensis* under Cu toxicity, *C. grandis* had a more apparent flocculent precipitate on the cross sections under SEM.

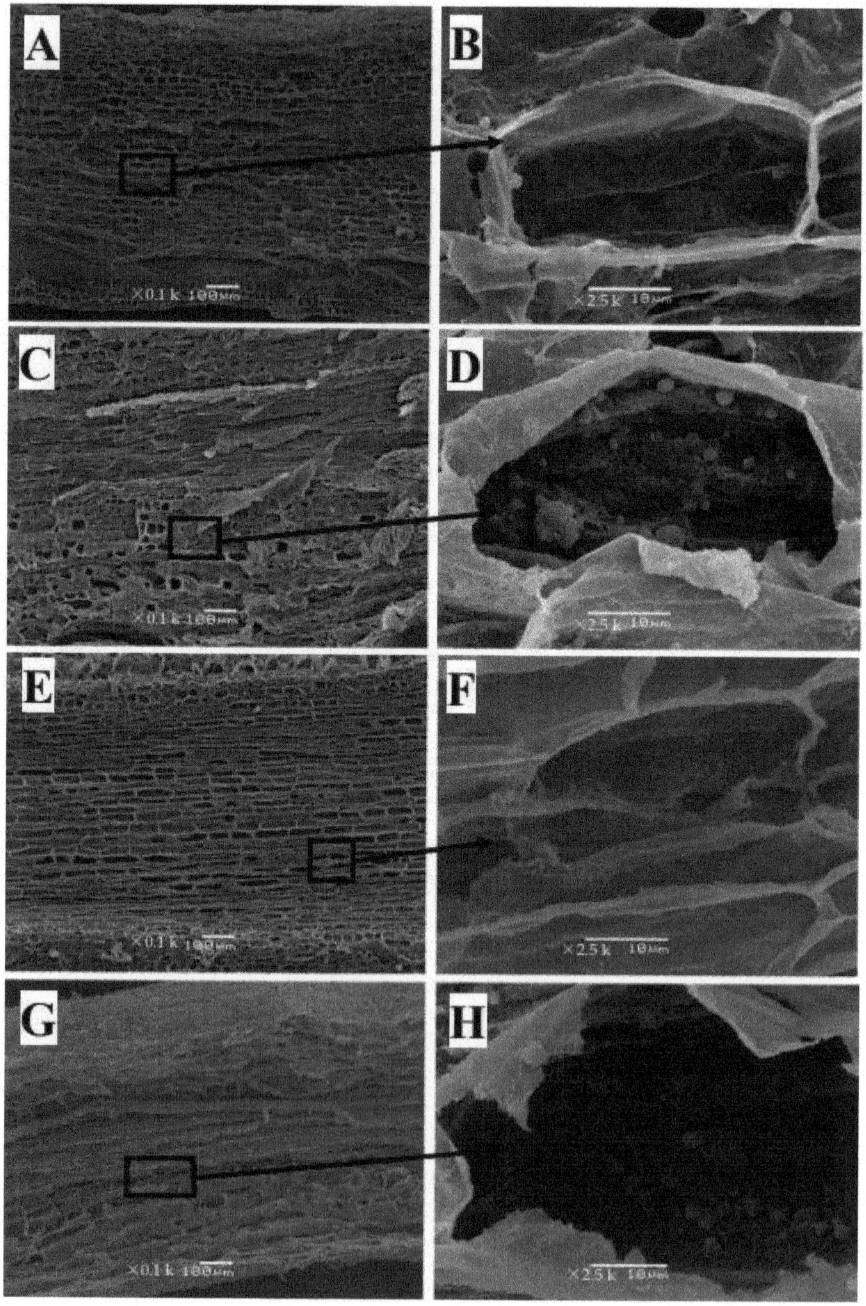

Figure 3. The SEM images of the longitudinal section of 0.5 μM (**A**,**B**,**E**,**F**) and 300 μM Cu-treated (**C**,**D**,**G**,**H**) roots of two citrus species. *C. sinensis*: (**A**–**D**); *C. grandis*: (**E**–**H**). The right figures are partial enlargements of the left figures. The granules in cells labeled by arrows are starch granules. The scales in figures (**A**,**C**,**E**,**G**) represent 100 μm. The scales in figures (**B**,**D**,**F**,**H**) represent 10 μm.

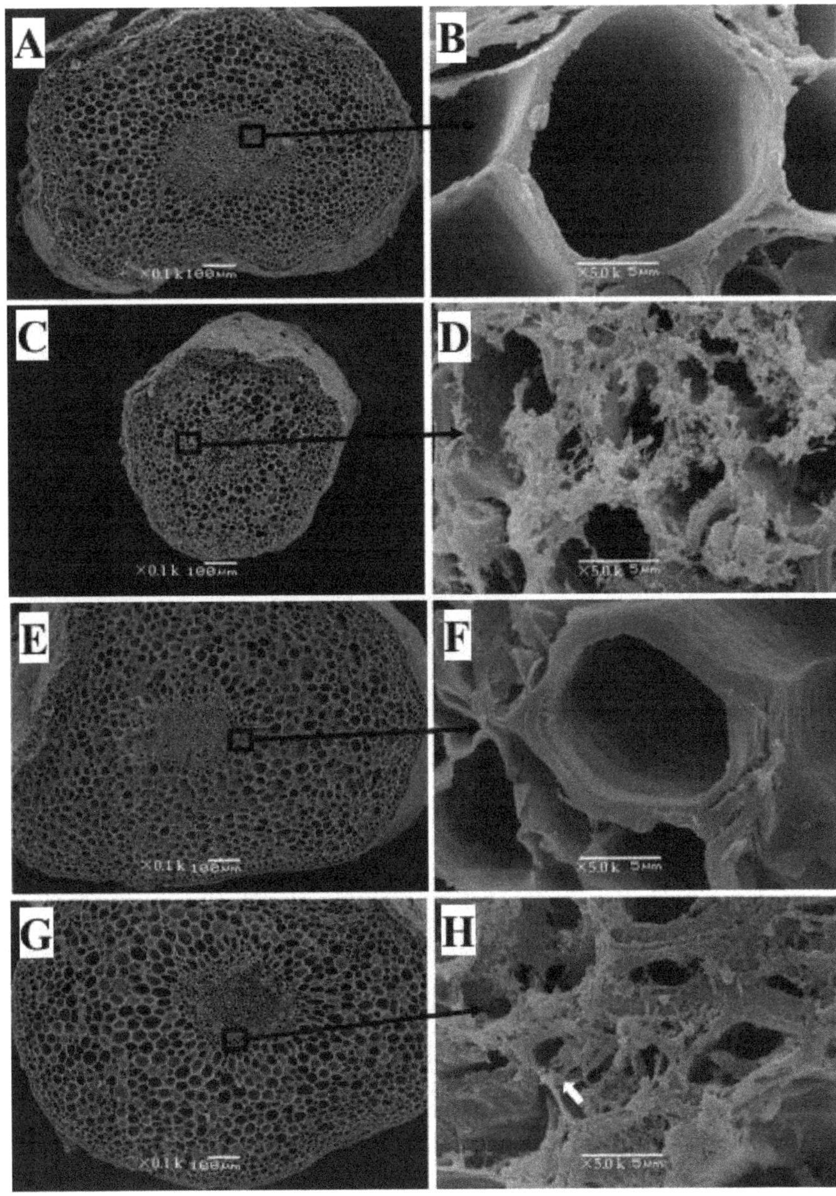

Figure 4. The SEM images of cross-section of 0.5 μM (**A**,**B**,**E**,**F**) and 300 μM Cu-treated (**C**,**D**,**G**,**H**) roots of two citrus species. *C. sinensis*: (**A–D**); *C. grandis*: (**E–H**). The right figures are partial enlargements of the left figures. Images (**A**,**C**,**E**,**G**) show epidermis, exodermis, cortex, endodermis, pericycle, xylem and phloem. Images (**B**,**D**,**F**,**H**) show the xylem of the roots. The arrows represent the blocked xylem of roots by flocculent precipitate in (**D**) and (**H**). The scales in figures (**A**,**C**,**E**,**G**) represent 100 μm. The scales in figures (**B**,**D**,**F**,**H**) represent 5 μm.

2.4. Cu Toxicity Destroyed the Ultrastructure of the Leaves and Roots of C. sinensis *and* C. grandis

Ultrastructural observation under TEM revealed an integral cellular structure of *C. sinensis* leaf under control. As found in Figure 5A, the control leaf cell had an approximately round shape and the membrane structure was clearly visible. The organelles were pushed by a center-located vacuole and closely distributed at the inner side of the plasma membrane. The thylakoids were stacked closely and the lamellar structure was observed inside the chloroplast. Moreover, several starch granules and osmiophilic globules were occasionally found in the chloroplast. The bilayer structure and inner cristae structure of mitochondria were also clearly visible under control (Figure 5B) as was excessive Cu-induced plasmolysis of *C. sinensis* leaf cells (Figure 5C). Moreover, the cellular structure was impaired. The vacuole was broken and replaced by many spherical vesicles. The size of mitochondria decreased, and the bilayer structure was difficult to find under Cu toxicity. The integrity of the chloroplast was destroyed, and the stacked thylakoids were loose (Figure 5D). Strikingly, Cu toxicity increased the number and volume of starch granules in *C. sinensis* leaves.

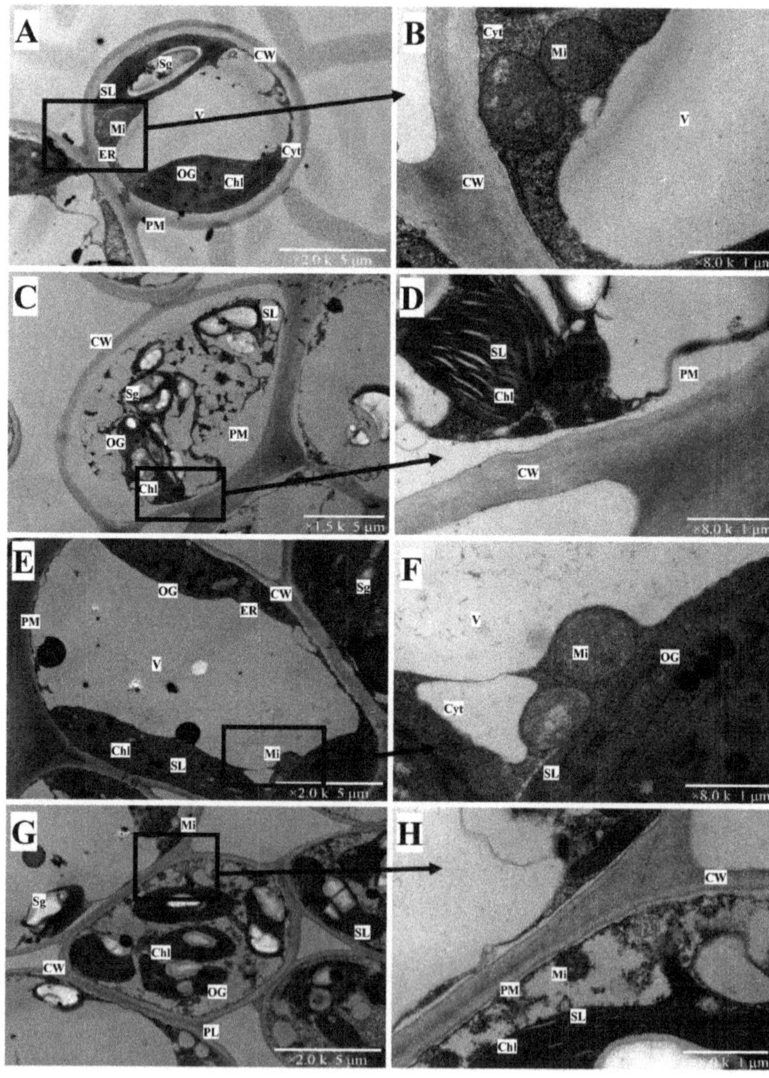

Figure 5. The TEM images of 0.5 μM (**A**,**B**,**E**,**F**) and 300 μM Cu-treated (**C**,**D**,**G**,**H**) leaves of two citrus species. *C. sinensis*: (**A**–**D**); *C. grandis*: (**E**–**H**). The right figures are partial enlargements of the left figures. Chl: chloroplast; CW: cell wall; Cyt: cytoplasm; ER: endoplasmic reticulum; Mi: mitochondrion; OG: osmiophilic globule; PM: plasmalemma; Sg: starch granules; SL: stroma lamella; V: vacuole. The scales in figures (**A**,**C**,**E**,**G**) represent 5 μm. The scales in figures (**B**,**D**,**F**,**H**) represent 1 μm.

The control leaf cells of *C. grandis* were presented in an oval-like shape (Figure 5E). The organelles were well-organized. The stroma lamella of chloroplast and bilayer structure of the mitochondria were apparent in the enlargement in Figure 5F. Under Cu toxicity, the vacuole disappeared. The chloroplasts occupied most of the protoplast. The enlarged starch granules could be found in each chloroplast. The mitochondria of *C. grandis* leaf under Cu toxicity were smaller than in the control (Figure 5F). Compared with *C. sinensis*, *C. grandis* had a relatively integral chloroplast structure.

In the root, the vacuole almost took up the protoplast of the control cell in *C. sinensis* (Figure 6A) and *C. grandis* (Figure 6E). The mitochondria were distributed on the plasma membrane's inner side (Figure 6B,F). Under Cu toxicity, the vacuole disappeared in *C. sinensis* roots. The mitochondrial structure was also difficult to observe (Figure 6C,D). Differentially, *C. grandis* maintained a relatively integral structure of the root cell, supported by an intact structure of mitochondria and a clear edge between the plasma membrane and cell wall (Figure 6G). Interestingly, both citrus cell wall thicknesses increased under Cu toxicity compared with the control.

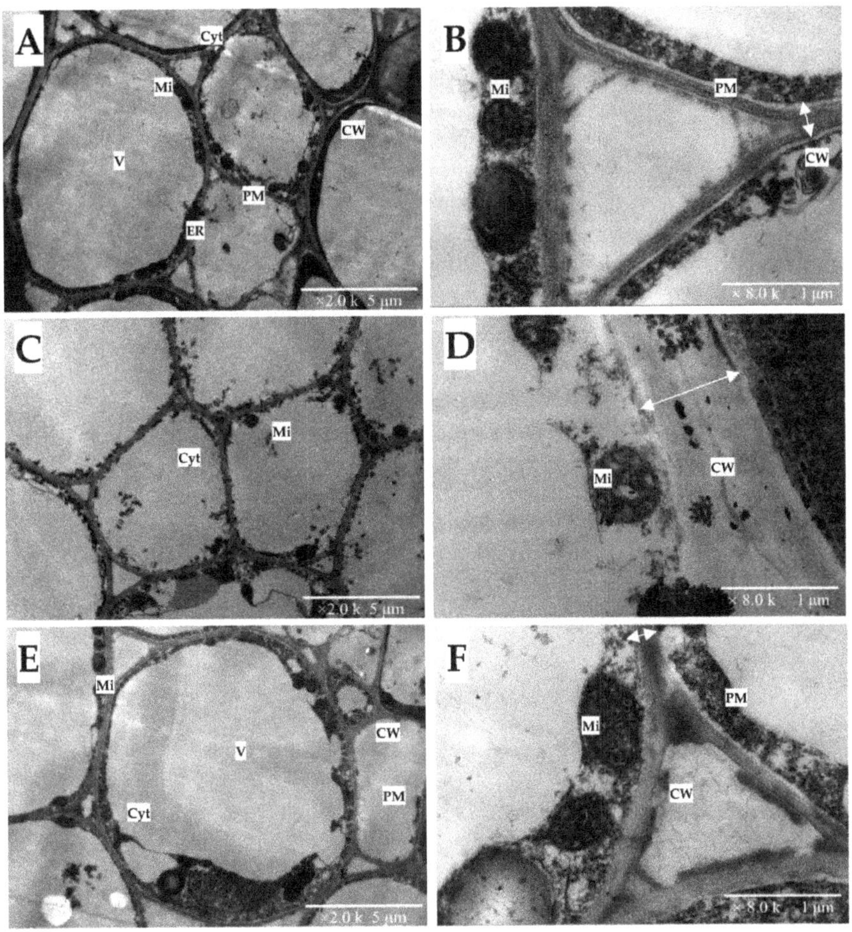

Figure 6. *Cont.*

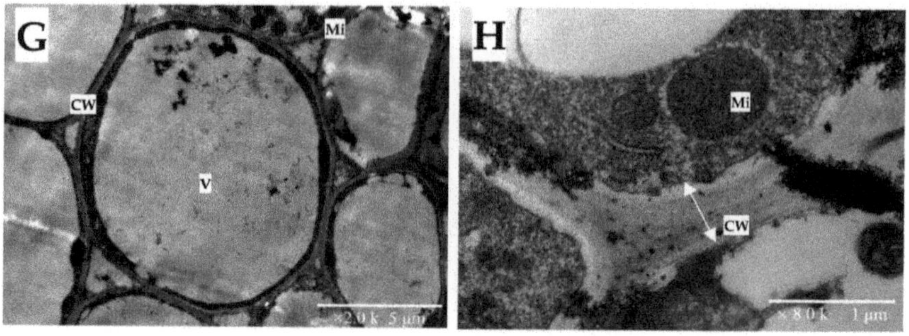

Figure 6. The TEM images of 0.5 μM (**A**,**B**,**E**,**F**) and 300 μM Cu-treated (**C**,**D**,**G**,**H**) roots of two citrus species. *C. sinensis*: (**A–D**); *C. grandis*: (**E–H**). CW: cell wall; Cyt: cytoplasm; ER: endoplasmic reticulum; Mi: mitochondrion; PL: plasmalemma; V: vacuole. The two-way arrow represents the average thickness of the cell wall. The scales in figures (**A**,**C**,**E**,**G**) represent 5 μm. The scales in figures (**B**,**D**,**F**,**H**) represent 1 μm.

3. Discussion

Heavy metal (HM) stress had detrimental effects on seed germination and membrane stability of horticultural plants, which inhibited plant growth and crop yield [25]. Similarly, Cu toxicity decreased the leaf chlorophyll content and depressed the shoot and the root development of citrus species, which is in line with our previous reports [26,27]. Additionally, Cu toxicity significantly downregulated the shoot biomass to dry weight ratio whereas it upregulated the root biomass to dry weight of the two citrus species (Table 1). The changes in biomass distribution might be due, at least in part, to the limited nutrient transport to the shoots. The Cu-toxicity-induced increased ratio of root biomass to shoot biomass agreed with that in excessive-Cu treated *Hymenaea courbaril* L. (Caesalpinioideae), a promising woody species for phytoremediation [28]. Cu toxicity significantly decreased the total chlorophyll content of the two citrus species (Table 1). The decrement was also reported in Cu-stressed falx leaves [29]. Coherently, the total chlorophyll content of soybean leaf decreased with the increasing level of soil Cu [30].

Cu immobilized by the roots contributed to Cu detoxification of horticultural plants [31]. The present results indicate that *C. grandis* had a significantly higher TF than *C. sinensis* under control. However, no significant difference in TF was found between the two citrus species under Cu toxicity (Table 1). This finding suggests that a potential strategy governing Cu mobility might exist in *C. grandis* under Cu toxicity. This hypothesis was evidenced by the upregulation of genes involved in Cu-binding in excessive-Cu-treated roots of *C. grandis* [20]. Beyond the transcriptional evidence, it has also been reported that tiny crystals accumulated in the root xylem of Cu-stressed Moso Bamboo (*Phyllostachys pubescens*), which block the Cu transport in the vascular tissue of the root xylem [32]. In the present study, citrus roots that received Cu excess had similar amorphous material accumulated on the xylem (Figure 4D,H). Interestingly, much more flocculent precipitate was found in *C. grandis* than in *C. sinensis*.

The anatomical investigation provided essential information for understanding the Cu-adaptation of citrus species. Herein, we found Cu toxicity increased starch granules accumulation in the leaf palisade and spongy parenchymatous tissues (Figure 2C,D,G,H) and root cells (Figure 3D,H) of two citrus species compared with control. Li et al. [26] demonstrated the increased starch accumulation in the leaves and roots of two citrus species. It has been proposed that Cu-toxicity-induced starch accumulation in cucumber leaves is related to a decreased phloem loading [33], leading to feedback inhibition of photosynthesis. Meanwhile, the accumulated starch in the root has been considered an osmotic regulation in turgor pressure under stress [34,35]. Those findings, and an increased ratio of root biomass to plant dry weight, imply a crucial role of starch accumulation in citrus roots conferring Cu toxicity. Likewise, the enhanced starch content has also been reported in excessive Cd-treated poplar species [36] and apple seedlings [37]. Additionally, Cu toxicity enhanced the thickness of the leaf epidermis of two citrus species compared with the control (Figure 2C,G). The leaf thickening has also been observed in Cu-stressed oregano leaves [38] and Al-treated sunflower (*Helianthus annuus* L.) leaf [39]. Bouazizi et al. [40] have proposed that the Cu-toxicity-induced cell wall lignification contributes to cell wall thickening, which ultimately strengthens the Cu tolerance of bean leaves. Further studies have focused on phenylalanine ammonia-lyase activity (PAL, EC 4.3.1.5), a rate-limited enzyme in lignin synthesis, and research on the lignin deposited on the cell wall should be carried out to disclose the structural modification of Cu toxicity on citrus leaves and roots.

Based on TEM observation, we believe that the ultrastructural alterations reflect the status of plant organelles under heavy metal toxicity, which are complementary to the plant's physiological parameters [41]. For instance, our previous report has proven that Cu toxicity resulted in oxidative damage of citrus leaves by overproduction of reactive oxygen species [19]. It has also been further reported that oxidative stress accounted for the membrane damage of plant cells exposed to excess Cu [42]. Therefore, the present results, in which Cu-toxicity induced damage to the membrane structure of citrus leaves, verify our previous findings and extend our knowledge of the response to Cu toxicity in citrus roots and leaves. Additionally, it was striking to find a severely disrupted structure of chloroplast and increased starch granules in Cu-stressed *C. sinensis* leaves (Figure 5D). Studies have proposed that heavy metal stress decreased the leaf chlorophyll content by downregulating the activities of enzymes involved in chlorophyll synthesis [43,44]. Similarly, the inhibited alpha-amylase and invertase isoenzymes have been reported for their restricted breakdown of starch granules, leading to starch accumulation under Cu toxicity [45]. The increasing number of starch granules has also been found in excessive-Mn-treated leaves of *Xanthium strumarium* L., a candidate species for Mn-phytoremediation [46]. Compared with Cu-treated *C. sinensis* roots, *C. grandis* had a relatively robust chloroplast structure (Figure 5C,G).

The thickening of the cell wall is significant for plant roots conferring heavy metal toxicity [47]. Primarily composed of polysaccharides (such as pectin, cellulose and lignin), the cell wall had numerous negatively charged groups, such as hydroxyl ($-OH^-$) and carboxyl ($-COOH^-$) groups. Thickening the cell wall increased the negative charge, which restricted the translocation of the toxic cation. In the present study, the observation under stereoscopic microscopy verified that Cu toxicity expanded the fibrous roots of two citrus species (Figure S1). Furthermore, the TEM observation of citrus roots indicated that the root cell wall's increased thickness might account for the root tips' expansion (Figure 6D,H). Our previous finding that Cu toxicity increased the ratio of cell wall biomass to the dry weight of citrus roots is in line with present results [20]. Similar findings have also been reported in excessive-Cu treated roots of *Pinus pinaster* Ait (Arduini et al., 1995), *Arabidopsis thaliana* [48] and *Allium cepa* L. [49]. The comparison between Figure 6C,G implies that *C. grandis* maintained a much more integral vacuole structure.

4. Materials and Methods

4.1. Plant Material and Treatments

Surface-sterilized seeds of *C. sinensis* and *C. grandis* were germinated in moist sand in a greenhouse of Fujian Agriculture and Forestry University, Jinshan campus (26°5′ N, 119°14′ E) in March 2020. Six weeks after germination, the uniformly sized seedlings (about 10 cm) were selected and precultured with 1/4 strength modified Hoagland nutrient solution, which contained 2.5 mM KNO_3, 2.5 mM $Ca(NO_3)_2$, 0.5 mM KH_2PO_4, 1 mM $MgSO_4$, 10 μM H_3BO_3, 2 μM $MnCl_2$, 2 μM $ZnSO_4$, 0.5 μM $CuCl_2$, 0.065 μM$(NH_4)_6Mo_7O_{24}$ and 20 μM Fe-EDTA. Six weeks after preculture, the citrus seedlings were treated with nutrient solution (pH = 4.30–4.50) containing 0.5 μM $CuCl_2$ (as control) or 300 μM $CuCl_2$ (as Cu toxicity) daily for 18 weeks. By the end of the treatments, citrus leaves, stems and roots were sampled separately. For sampling, the citrus roots were first washed with tap water to remove excess sandy particles, followed by soaking in 0.5 M EDTA-Na_2 for 10 min (three times) to remove Cu residue on the root surface. After being washed with distilled water three times, fresh lateral roots of citrus seedlings were fixed in 3% glutaraldehyde–1.5% paraformaldehyde solution (fixing solution) in 0.1 M phosphate buffer solution (PBS, pH = 7.2) to prepare the sample for microscopic observation. There were three replicates for each treatment. After being cleaned with distilled water, the remaining fresh citrus leaves were collected for chlorophyll content measurement. The remaining tissues of citrus seedlings were dried before biomass analyses and Cu quantification. There were four replicates of each treatment for biomass analyses.

4.2. Measurements of Cu Translocation Factor and Leaf Total Chlorophyll Contents of Two Citrus Species

The Cu concentration of plant tissues and the leaf chlorophyll content were measured according to Li et al. [16]. The Cu translocation factor was expressed as the ratio of Cu concentration in the leaves to the lateral roots [50]. The leaf total chlorophyll content was calculated as: $7.15 \times A_{663} + 18.71 \times A_{646}$, in which, A_{663} and A_{646} represented the absorbances of leaf pigment extractant under the wavelength of 663 and 646, respectively. There were four replicates of each treatment.

4.3. Scanning Electron Microscopy (SEM) and Transmission Electron Microscopy (TEM) Analyses of Citrus Leaves and Roots

The pretreatment of citrus leaves and roots was performed according to Huang et al. [51]. Briefly, the citrus leaf and root samples were vacuumed in a syringe filled with fixing solution under 4 °C for 3 h, followed by three rinses in PBS, each lasting 15 min. The samples were then post-fixed with 1% OsO_4–1.5% potassium hexacyanoferrate for 2 h and washed with distilled water three times. Thereafter, the citrus leaf and root samples were dehydrated in an increasing concentration of ethanol (30%, 50%, 70%, 80%, 90%, 95% and 100%) three times.

SEM observation: The dehydrated samples were rinsed in epoxypropane twice and dried by an HCP-2 critical point dryer (HITACHI Ltd., Tokyo, Japan). Further, the samples were mounted on brass stubs coated with gold in Eiko IB-5 ion-coater (Eiko Engineering, Ibaragi, Japan). The images of citrus tissues were obtained using a JEOL JSM-6380LV electron microscope (JEOL, Ltd., Tokyo, Japan). There were three replicates of each treatment.

TEM observation: The dehydrated samples were rinsed with acetone for 10 min twice, and then replaced by a mixture of resin and acetone on a shaker at the ratios of 1:1 and 3:1 for 2 h, respectively. The samples were then dried with filter paper to remove excess acetone on the surface and infiltrated in resin overnight. After being embedded in epoxy resin 618, the individual leaf and root blocks were cut into sections of approximately 80-nm thickness with an ultra-microtome LEICA EM UC6 (Leica, Wetzlar, Germany). The sections were stained with 2% uranyl acetate before observations with a TEM HITACHI HT7700 (HITACHI, Tokyo, Japan) equipped with a digital camera. There were three replicates of each treatment.

4.4. Statistical Analysis

The data in the study represent mean ± SE (n = 4). Four means (two citrus species × two treatments) were analyzed by two ANOVAs followed by Duncan's multiple range tests at $p < 0.05$ using SPSS (Version 22.0, IBM, Armonk, NY, USA).

5. Conclusions

Cu toxicity induced by 300 μM Cu significantly decreased the ratio of shoot biomass to dry weight, the Cu translocation factor and the total chlorophyll of two citrus species. By contrast, the ultrastructural alterations demonstrated that the starch granules in the leaves and roots, the root flocculent precipitate and the thickness of the root cell wall all increased under Cu stress in the two citrus species. The increased root flocculent precipitate and thickened root cell wall might reduce the Cu translocation from roots to the shoots of the two citrus species. Compared with *C. sinensis*, *C. grandis* maintained a relatively integral root cellular structure under Cu toxicity, which provided a structural basis for a higher Cu tolerance than *C. sinensis*. The present results increase our understanding of the morphological alteration of citrus leaves and roots under Cu toxicity. Further study regarding root metabolites under Cu toxicity should be undertaken to reveal the Cu-tolerant mechanisms of citrus species underlying the structural alteration.

Supplementary Materials: The following supporting information can be downloaded at: https://www.mdpi.com/article/10.3390/plants12020351/s1, Figure S1: The root tip morphology of two citrus species under 0.5 μM and 300 μM Cu stress. The image was taken with stereoscopic microscopy SMZ18 (Nikon, Tokyo, Japan).

Author Contributions: Data curation, X.X.; investigation, M.-L.L.; methodology, F.L.; project administration, L.-S.C.; software, X.Z.; writing—original draft, X.-Y.L.; writing—review & editing, Z.-R.H. All authors have read and agreed to the published version of the manuscript.

Funding: This research was funded by the Natural Science Foundation of Fujian Province of China (2022J01603) and the National Natural Science Foundation of China (31801950).

Data Availability Statement: Data are archived in L.-S. Chen's lab and are available upon request.

Conflicts of Interest: The authors declare no conflict of interest.

References

1. Cesco, S.; Pii, Y.; Borruso, L.; Orzes, G.; Lugli, P.; Mazzetto, F.; Genova, G.; Signorini, M.; Brunetto, G.; Terzano, R. A smart and sustainable future for viticulture is rooted in soil: How to face Cu toxicity. *Appl. Sci.* **2021**, *11*, 907. [CrossRef]
2. Rehman, M.; Liu, L.; Wang, Q.; Saleem, M.H.; Bashir, S.; Ullah, S.; Peng, D. Copper environmental toxicology, recent advances, and future outlook: A review. *Environ. Sci. Pollut. Res.* **2019**, *26*, 18003–18016. [CrossRef] [PubMed]
3. Gohel, N.M.; Prajapati, B.K.; Srivastava, J.N. Major diseases of citrus and their management. In *Diseases of Horticultural Crops*; Apple Academic Press: New York, NY, USA, 2022; pp. 155–167.
4. El-Gioushy, S.F.; Sami, R.; Al-Mushhin, A.A.; Abou El-Ghit, H.M.; Gawish, M.S.; Ismail, K.A.; Zewail, R.M. Foliar application of $ZnSO_4$ and $CuSO_4$ affects the growth, productivity, and fruit quality of Washington Navel orange trees (*Citrus sinensis* L.) Osbeck. *Horticulturae* **2021**, *7*, 233. [CrossRef]
5. Triantafyllidis, V.; Zotos, A.; Kosma, C.; Kokkotos, E. Environmental implications from long-term citrus cultivation and wide use of Cu fungicides in Mediterranean soils. *Water Air Soil Pollut.* **2020**, *231*, 218. [CrossRef]
6. He, Z.L.; Yang, X.E.; Stoffella, P.J. Trace elements in agroecosystems and impacts on the environment. *J. Trace Elem. Med. Biol.* **2005**, *19*, 125–140. [CrossRef] [PubMed]
7. Rusjan, D.; Strlič, M.; Pucko, D.; Korošec-Koruza, Z. Copper accumulation regarding the soil characteristics in Sub-Mediterranean vineyards of Slovenia. *Geoderma* **2007**, *141*, 111–118. [CrossRef]
8. Fu, C.; Tu, C.; Zhang, H.; Li, Y.; Li, L.; Zhou, Q.; Scheckel, K.G.; Luo, Y. Soil accumulation and chemical fractions of Cu in a large and long-term coastal apple orchard, North China. *J. Soils Sediments* **2020**, *20*, 3712–3721. [CrossRef]
9. Junior, D.M.; Quaggio, J.A.; Cantarella, H.; Boaretto, R.M.; Zambrosi, F.C.B. Nutrient management for high citrus fruit yield in tropical soils. *Better Crops* **2012**, *96*, 4–7.
10. Morgan, K.T.; Kadyampakeni, D.M. *Nutrition of Florida Citrus Trees*; University of Florida Institute of Food and Agricultural Sciences: Gainesville, FL, USA, 2020.
11. Khattak, R.A.; Hussain, Z. Evaluation of soil fertility status and nutrition of orchards. *Soil Environ.* **2007**, *26*, 22–32.
12. Li, Y.; Han, M.; Lin, F.; Ten, Y.; Lin, J.; Zhu, D.; Guo, P.; Weng, Y.B.; Chen, L. Soil chemical properties, 'Guanximiyou' pummelo leaf mineral nutrient status and fruit quality in the southern region of Fujian province, China. *J. Soil Sci. Plant Nutr.* **2015**, *15*, 615–628. [CrossRef]
13. Hippler, F.W.R.; Boaretto, R.M.; Dovis, V.L.; Quaggio, J.A.; Azevedo, R.A.; Mattos-Jr, D. Oxidative stress induced by Cu nutritional disorders in Citrus depends on nitrogen and calcium availability. *Sci. Rep.* **2018**, *8*, 1641. [CrossRef] [PubMed]
14. Chen, X.; Hua, D.; Zheng, Z.; Zhang, J.; Huang, W.; Chen, H.; Huang, Z.; Yang, L.; Ye, X.; Chen, L. Boron-mediated amelioration of copper-toxicity in sweet orange [*Citrus sinensis* (L.) Osbeck cv. Xuegan] seedlings involved reduced damage to roots and improved nutrition and water status. *Ecotoxicol. Environ. Saf.* **2022**, *234*, 113423. [CrossRef] [PubMed]
15. Smith, P.F. Effects of high levels of copper, zinc, and manganese on tree growth and fruiting of Valencia orange in sand culture. *Proc. Am. Soc. Hortic. Sci.* **1955**, *67*, 202–209.
16. Li, Q.; Chen, H.; Qi, Y.; Ye, X.; Yang, L.; Huang, Z.; Chen, L. Excess copper effects on growth, uptake of water and nutrients, carbohydrates, and PSII photochemistry revealed by OJIP transients in Citrus seedlings. *Environ. Sci. Pollut. Res.* **2019**, *26*, 30188–30205. [CrossRef]

17. Wu, F.; Huang, H.; Peng, M.; Lai, Y.; Ren, Q.; Zhang, J.; Huang, Z.; Yang, L.; Rensing, C.; Chen, L. Adaptive responses of *Citrus grandis* leaves to copper toxicity revealed by RNA-Seq and physiology. *Int. J. Mol. Sci.* **2021**, *22*, 12023. [CrossRef]
18. Huang, W.; Wu, F.; Huang, H.; Huang, W.; Deng, C.; Yang, L.; Huang, Z.; Chen, L. Excess copper-induced alterations of protein profiles and related physiological parameters in citrus leaves. *Plants* **2020**, *9*, 291. [CrossRef]
19. Huang, H.; Ren, Q.; Lai, Y.; Peng, M.; Zhang, J.; Yang, L.; Huang, Z.; Chen, L. Metabolomics combined with physiology and transcriptomics reveals how *Citrus grandis* leaves cope with copper-toxicity. *Ecotoxicol. Environ. Saf.* **2021**, *223*, 112579. [CrossRef]
20. Ren, Q.; Huang, Z.; Huang, W.; Huang, W.; Chen, H.; Yang, L.; Ye, X.; Chen, L. Physiological and molecular adaptations of *Citrus grandis* roots to long-term copper excess revealed by physiology, metabolome and transcriptome. *Environ. Exp. Bot.* **2022**, *203*, 105049. [CrossRef]
21. Chen, H.; Chen, X.; Zheng, Z.; Huang, W.; Guo, J.; Yang, L.; Chen, L. Characterization of copper-induced-release of exudates by *Citrus sinensis* roots and their possible roles in copper-tolerance. *Chemosphere* **2022**, *308*, 136348. [CrossRef]
22. Sánchez-Pardo, B.; Fernández-Pascual, M.; Zornoza, P. Copper microlocalisation and changes in leaf morphology, chloroplast ultrastructure and antioxidative response in white lupin and soybean grown in copper excess. *J. Plant Res.* **2014**, *127*, 119–129. [CrossRef]
23. de Freitas, T.A.; França, M.G.C.; de Almeida, A.F.; de Oliveira, S.J.R.; de Jesus, R.M.; Souza, V.L.; Dos Santos Silva, J.V.; Mangabeira, P.A. Morphology, ultrastructure and mineral uptake is affected by copper toxicity in young plants of *Inga subnuda* subs. *luschnathiana (Benth.) TD Penn. Environ. Sci. Pollut. Res.* **2015**, *22*, 15479–15494. [CrossRef] [PubMed]
24. Minkina, T.; Rajput, V.; Fedorenko, G.; Fedorenko, A.; Mandzhieva, S.; Sushkova, S.; Morin, T.; Yao, J. Anatomical and ultrastructural responses of *Hordeum sativum* to the soil spiked by copper. *Environ. Geochem. Health* **2020**, *42*, 45–58. [CrossRef] [PubMed]
25. Noor, I.; Sohail, H.; Sun, J.; Nawaz, M.A.; Li, G.; Hasanuzzaman, M.; Liu, J. Heavy metal and metalloid toxicity in horticultural plants: Tolerance mechanism and remediation strategies. *Chemosphere* **2022**, *303*, 135196. [CrossRef] [PubMed]
26. Cai, L.; Zhang, J.; Ren, Q.; Lai, Y.; Peng, M.; Deng, C.; Ye, X.; Yang, L.; Huang, Z.; Chen, L. Increased pH-mediated alleviation of copper-toxicity and growth response function in *Citrus sinensis* seedlings. *Sci. Hortic.* **2021**, *288*, 110310. [CrossRef]
27. Li, X.; Lin, M.; Hu, P.; Lai, N.; Huang, Z.; Chen, L. Copper toxicity differentially regulates the seedling growth, copper distribution, and photosynthetic performance of *Citrus sinensis* and *Citrus grandis*. *J. Plant Growth Regul.* **2021**, *42*, 3333–3344. [CrossRef]
28. Marques, D.M.; Veroneze Júnior, V.; Da Silva, A.B.; Mantovani, J.R.; Magalhães, P.C.; de Souza, T.C. Copper toxicity on photosynthetic responses and root morphology of *Hymenaea courbaril* L. (Caesalpinioideae). *Water Air Soil Pollut.* **2018**, *229*, 138. [CrossRef]
29. Saleem, M.H.; Kamran, M.; Zhou, Y.; Parveen, A.; Rehman, M.; Ahmar, S.; Malik, Z.; Mustafa, A.; Anjum, R.M.A.; Wang, B.; et al. Appraising growth, oxidative stress and copper phytoextraction potential of flax (*Linum usitatissimum* L.) grown in soil differentially spiked with copper. *J. Environ. Manag.* **2020**, *257*, 109994. [CrossRef]
30. Gomes, D.G.; Lopes-Oliveira, P.J.; Debiasi, T.V.; da Cunha, L.S.; Oliveira, H.C. Regression models to stratify the copper toxicity responses and tolerance mechanisms of *Glycine max* (L.) Merr. plants. *Planta* **2021**, *253*, 43. [CrossRef]
31. Wan, H.; Yang, F.; Zhuang, X.; Cao, Y.; He, J.; Li, H.; Qin, S.; Lyu, D. Malus rootstocks affect copper accumulation and tolerance in trees by regulating copper mobility, physiological responses, and gene expression patterns. *Environ. Pollut.* **2021**, *287*, 117610. [CrossRef]
32. Chen, J.; Peng, D.; Shafi, M.; Li, S.; Wu, J.; Ye, Z.; Yan, W.; Lu, K.; Liu, D. Effect of copper toxicity on root morphology, ultrastructure, and copper accumulation in Moso bamboo (*Phyllostachys pubescens*). *Z. Nat. C* **2014**, *69*, 399–406. [CrossRef]
33. Alaoui-Sossé, B.; Genet, P.; Vinit-Dunand, F.; Toussaint, M.; Epron, D.; Badot, P. Effect of copper on growth in cucumber plants (*Cucumis sativus*) and its relationships with carbohydrate accumulation and changes in ion contents. *Plant Sci.* **2004**, *166*, 1213–1218. [CrossRef]
34. Velázquez-Márquez, S.; Conde-Martínez, V.; Trejo, C.; Delgado-Alvarado, A.; Carballo, A.; Suárez, R.; Mascorro, J.O.; Trujillo, A.R. Effects of water deficit on radicle apex elongation and solute accumulation in *Zea mays* L. *Plant Physiol. Biochem.* **2015**, *96*, 29–37. [CrossRef] [PubMed]
35. Anjum, S.A.; Tanveer, M.; Hussain, S.; Shahzad, B.; Ashraf, U.; Fahad, S.; Hassan, W.; Jan, S.; Khan, I.; Saleem, M.F. Osmoregulation and antioxidant production in maize under combined cadmium and arsenic stress. *Environ. Sci. Pollut. Res.* **2016**, *23*, 11864–11875. [CrossRef] [PubMed]
36. He, J.; Ma, C.; Ma, Y.; Li, H.; Kang, J.; Liu, T.; Polle, A.; Peng, C.; Luo, Z. Cadmium tolerance in six poplar species. *Environ. Sci. Pollut. Res.* **2013**, *20*, 163–174. [CrossRef]
37. Zhou, J.; Wan, H.; He, J.; Lyu, D.; Li, H. Integration of cadmium accumulation, subcellular distribution, and physiological responses to understand cadmium tolerance in apple rootstocks. *Front. Plant Sci.* **2017**, *8*, 966. [CrossRef]
38. Panou-Filotheou, H.; Bosabalidis, A.M.; Karataglis, S. Effects of copper toxicity on leaves of oregano (*Origanum vulgare* subsp. hirtum). *Ann. Bot.* **2001**, *88*, 207–214. [CrossRef]
39. Jesus, D.D.S.D.; Martins, F.M.; Azevedo Neto, A.D.D. Structural changes in leaves and roots are anatomical markers of aluminum sensitivity in sunflower. *Pesqui. Agropecu. Trop.* **2016**, *46*, 383–390. [CrossRef]
40. Bouazizi, H.; Jouili, H.; Geitmann, A.; El Ferjani, E. Cell wall accumulation of Cu ions and modulation of lignifying enzymes in primary leaves of bean seedlings exposed to excess copper. *Biol. Trace Element Res.* **2011**, *139*, 97–107. [CrossRef]

41. Giannakoula, A.; Therios, I.; Chatzissavvidis, C. Effect of lead and copper on photosynthetic apparatus in citrus (*Citrus aurantium* L.) plants. The role of antioxidants in oxidative damage as a response to heavy metal stress. *Plants* **2021**, *10*, 155. [CrossRef]
42. Saleem, M.H.; Fahad, S.; Khan, S.U.; Din, M.; Ullah, A.; Sabagh, A.E.; Hossain, A.; Llanes, A.; Liu, L. Copper-induced oxidative stress, initiation of antioxidants and phytoremediation potential of flax (*Linum usitatissimum* L.) seedlings grown under the mixing of two different soils of China. *Environ. Sci. Pollut. Res.* **2020**, *27*, 5211–5221. [CrossRef]
43. Ulhassan, Z.; Gill, R.A.; Huang, H.; Ali, S.; Mwamba, T.M.; Ali, B.; Huang, Q.; Hamid, Y.; Khan, A.R.; Wang, J. Selenium mitigates the chromium toxicity in *Brassicca napus* L. by ameliorating nutrients uptake, amino acids metabolism and antioxidant defense system. *Plant Physiol. Biochem.* **2019**, *145*, 142–152. [CrossRef] [PubMed]
44. Handa, N.; Kohli, S.K.; Sharma, A.; Thukral, A.K.; Bhardwaj, R.; Abd Allah, E.F.; Alqarawi, A.A.; Ahmad, P. Selenium modulates dynamics of antioxidative defence expression, photosynthetic attributes and secondary metabolites to mitigate chromium toxicity in *Brassica juncea* L. plants. *Environ. Exp. Bot.* **2019**, *161*, 180–192. [CrossRef]
45. Mir, A.R.; Pichtel, J.; Hayat, S. Copper: Uptake, toxicity and tolerance in plants and management of Cu-contaminated soil. *Biometals* **2021**, *34*, 737–759. [CrossRef] [PubMed]
46. Pan, G.; Yan, W.; Zhang, H.; Xiao, Z.; Li, X.; Liu, W.; Zheng, L. Subcellular distribution and chemical forms involved in manganese accumulation and detoxification for *Xanthium strumarium* L. *Chemosphere* **2019**, *237*, 124531. [CrossRef] [PubMed]
47. Menna, A.; Dora, S.; Sancho-Andrés, G.; Kashyap, A.; Meena, M.K.; Sklodowski, K.; Gasperini, D.; Coll, N.S.; Sánchez-Rodríguez, C. A primary cell wall cellulose-dependent defense mechanism against vascular pathogens revealed by time-resolved dual transcriptomics. *BMC Biol.* **2021**, *19*, 161. [CrossRef] [PubMed]
48. Lequeux, H.; Hermans, C.; Lutts, S.; Verbruggen, N. Response to copper excess in Arabidopsis thaliana: Impact on the root system architecture, hormone distribution, lignin accumulation and mineral profile. *Plant Physiol. Biochem.* **2010**, *48*, 673–682. [CrossRef]
49. Macar, T.K.; Macar, O.; Yalçın, E.; Çavuşoğlu, K. Resveratrol ameliorates the physiological, biochemical, cytogenetic, and anatomical toxicities induced by copper (II) chloride exposure in Allium cepa L. *Environ. Sci. Pollut. Res.* **2020**, *27*, 657–667. [CrossRef]
50. Zhou, H.; Zeng, M.; Zhou, X.; Liao, B.; Peng, P.; Hu, M.; Zhu, W.; Wu, Y.; Zou, Z. Heavy metal translocation and accumulation in iron plaques and plant tissues for 32 hybrid rice (*Oryza sativa* L.) cultivars. *Plant Soil* **2015**, *386*, 317–329. [CrossRef]
51. Huang, J.; Cai, Z.; Wen, S.; Guo, P.; Ye, X.; Lin, G.; Chen, L. Effects of boron toxicity on root and leaf anatomy in two Citrus species differing in boron tolerance. *Trees* **2014**, *28*, 1653–1666. [CrossRef]

plants

MDPI

Article

The Variations of Leaf $\delta^{13}C$ and Its Response to Environmental Changes of Arbuscular and Ectomycorrhizal Plants Depend on Life Forms

Shan Zhang [1,2,3], Mingli Yuan [4], Zhaoyong Shi [1,2,3,*], Shuang Yang [1,2,3], Mengge Zhang [1,2,3], Lirong Sun [1,2,3], Jiakai Gao [1,2,3] and Xugang Wang [1,2,3]

1 College of Agriculture, Henan University of Science and Technology, Luoyang 471023, China
2 Luoyang Key Laboratory of Symbiotic Microorganism and Green Development, Luoyang 471023, China
3 Henan Engineering Research Center of Human Settlements, Luoyang 471023, China
4 School of Agriculture and Animal Husbandry Engineering, Zhoukou Vocational and Technical College, Zhoukou 466000, China
* Correspondence: shizy1116@126.com

Abstract: Arbuscular mycorrhiza (AM) and ectomycorrhiza (ECM) are the two most common mycorrhizal types and are paid the most attention to, playing a vital common but differentiated function in terrestrial ecosystems. The leaf carbon isotope ratio ($\delta^{13}C$) is an important factor in understanding the relationship between plants and the environment. In this study, a new database was established on leaf $\delta^{13}C$ between AM and ECM plants based on the published data set of leaf $\delta^{13}C$ in China's C_3 terrestrial plants, which involved 1163 observations. The results showed that the differences in leaf $\delta^{13}C$ between AM and ECM plants related closely to life forms. Leaf $\delta^{13}C$ of ECM plants was higher than that of AM plants in trees, which was mainly led by the group of evergreen trees. The responses of leaf $\delta^{13}C$ to environmental changes were varied between AM and ECM plants. Among the four life forms, leaf $\delta^{13}C$ of ECM plants decreased more rapidly than that of AM plants, with an increase of longitude, except for deciduous trees. In terms of the sensitivity of leaf $\delta^{13}C$ to temperature changes, AM plants were higher than ECM plants in the other three life forms, although there was no significant difference in evergreen trees. For the response to water conditions, the leaf $\delta^{13}C$ of ECM plants was more sensitive than that of AM plants in all life forms, except evergreen and deciduous trees. This study laid a foundation for further understanding the role of mycorrhiza in the relationship between plants and the environment.

Keywords: leaf $\delta^{13}C$; arbuscular mycorrhiza (AM); ectomycorrhiza (ECM); environment factors; life forms

Citation: Zhang, S.; Yuan, M.; Shi, Z.; Yang, S.; Zhang, M.; Sun, L.; Gao, J.; Wang, X. The Variations of Leaf $\delta^{13}C$ and Its Response to Environmental Changes of Arbuscular and Ectomycorrhizal Plants Depend on Life Forms. *Plants* **2022**, *11*, 3236. https://doi.org/10.3390/plants11233236

Academic Editor: Lorenzo Rossi

Received: 28 September 2022
Accepted: 22 November 2022
Published: 25 November 2022

1. Introduction

Leaf carbon isotope discrimination ($\delta^{13}C$) plays an important role in our understanding of the relationship between plants and environment. It not only provides a series of climate and environmental information related to plant growth processes, but it can also indicate how plants interact with the environment, and respond [1–4]. Thus, the relationship between leaf $\delta^{13}C$ and the environment has aroused interest [5–8].

There are many studies which have demonstrated that the characteristics of the plant itself can affect the leaf $\delta^{13}C$ [9,10]. Study found significant differences in $\delta^{13}C$ among different species of *R. natans*, which may be caused by stomatal limitation [4]. Li et al. [7] found the order of the averaged $\delta^{13}C$ for plant life forms from most positive to most negative was subshrubs > herbs = shrubs > trees > subtrees after studying 2538 plants in China. He et al. [11] found that tree height affected the leaf $\delta^{13}C$ of plants, and leaf $\delta^{13}C$ increased with the increase in tree height.

Most studies show that climatic variables have important roles in shaping the patterns of leaf $\delta^{13}C$, in addition to the physiological characteristics of plants [12–14]. Zhou et al. [15]

studied the relationship between δ^{13}C and temperature and precipitation in temperate grassland plants in Inner Mongolia and found that the δ^{13}C of all plants increased with increasing temperature, and decreased with increasing precipitation. Ma et al. [16] studied the spatial variation of stable carbon isotope composition in leaves of three species of Caragana in Northern China. The results showed that the δ^{13}C in leaves of three species decreased significantly with the increase of MAP and RH, and increased with the increase of altitude and MAT. Liu et al. [17] studied the changes of δ^{13}C in desert plant leaves in Northern China along climatic gradients and found that δ^{13}C in plant leaves decreased with increasing precipitation. These studies are mainly about how environmental conditions affect plant leaf δ^{13}C and how plants adapt to the dynamic changes of habitat by adjusting leaf δ^{13}C. However, the effect of mycorrhizal type on plant leaf δ^{13}C, especially the relationship between mycorrhizal type and plant leaf δ^{13}C on a large scale, is still unknown.

The response of leaf δ^{13}C traits to biological factors, such as mycorrhizal characteristics and their interrelationships, are of great significance to plants. Mycorrhizal fungi play a crucial role in the regulation of leaf δ^{13}C [18,19]. These fungi are obligate symbionts that form a mutualistic relationship with plant roots, known as mycorrhiza. Previous studies have shown that mycorrhizal fungi affect the leaf δ^{13}C of host plants by influencing the gas exchange parameters of plant leaves, such as photosynthetic rate, cellular CO_2 concentration, and water use efficiency [20–22]. Although there are serval distinct types of mycorrhiza status in nature, we focused on arbuscular mycorrhizal (AM) and ectomycorrhizal (ECM) associations because they are the two most universal and best-studied types in terrestrial ecosystems. AM fungi are widespread, forming symbiotic associations with 85% of all terrestrial plants [23]. By contrast, ECM fungi are restricted to a smaller number of host plant species [24]. Vargas et al. [25] found that ecosystem CO_2 fluxes were differently influenced by arbuscular and ectomycorrhizal-dominated vegetation types. Terrer et al. [26] suggested that mycorrhizal types can affect the plant nutrient acquisition strategies thus affecting the trade-off between plant and soil carbon storage under elevated CO_2. The leaf δ^{13}C and terrestrial ecosystem carbon cycle are closely linked.

However, the effects of different mycorrhizal fungi on the content of δ^{13}C in the leaves of different life forms, and the relationship between δ^{13}C and environmental factors, are still unclear.

Hence, we researched the difference in leaf δ^{13}C between AM plants and ECM plants across life forms and further analyzed the relationship between environmental factors and leaf δ^{13}C in AM plants and ECM plants across life forms. We considered that different types of mycorrhizal fungi have different effects on the physiological processes of plant leaves, especially on the photosynthetic processes of plants, and the geographical distribution patterns of different types of mycorrhizal fungi. We also propose two hypotheses: (1) the leaf δ^{13}C value varies with mycorrhizal types; and (2) the leaf δ^{13}C of different mycorrhizal types varies with environmental factors.

2. Materials and Methods

2.1. Assembly of Database

Most C_3 plants are positive among land plants, and the δ^{13}C of C_3 plants can better help us understand the relationship between land plants and their environment [27,28]. In this study, leaf 1δ^{13}C data of Chinese C_3 terrestrial plants were obtained from the database constructed by Li et al. [7] along with the life form of each plant, photosynthesis type and environmental data such as mean annual precipitation (MAP, mm), relative humidity (RH, %), mean annual temperature (MAT, °C) and solar hours (SH, hours) for each sampling site. We established a new database of leaf δ^{13}C of different mycorrhizal types of C_3 plants in China based on the database constructed by Li et al. [7].

Based on the reference information provided by Li et al., we identified the specific plant species corresponding to each observation. The mycorrhizal type of each plant species was ascertained according to the published literature, especially by Wang and Qiu [29],

Averill et al. [24], and Shi et al. [30]. We classified all the plants with typical AM and ECM structures as AM type and ECM type.

In order to compare the differences between AM and ECM in different plant life types, plants were subdivided into two subgroups based on their growth forms, i.e., woody species and herbaceous species. In this study, we refer to those woody plants with independent trunks as trees that occur from roots and have a distinct trunk and crown, usually higher than 6 m. Woody plants without a distinct trunk and in a clumped state are dwarf, usually less than 6 m, and are referred to as shrubs. Plants with herbaceous or fleshy stems with less developed woody parts, whose above-ground parts mostly die in the same year, are called herbaceous plants. The woody species were divided into two sub-sub groups, i.e., trees and shrubs; whereas the herbaceous species were divided into the annual herb and perennial herb. The herbaceous species included 53 annual herbs and 337 perennial herbs. Of these, 53 annual herbaceous plants and 336 perennial herbaceous plants belong to AM plants and only 1 perennial plant belongs to ECM plants. The trees were further divided into deciduous trees and evergreen trees; According to statistics, among the plants in the database, 817 species belong to the AM group, including 153 tree species, 275 shrub species, and 389 herb species; and 167 species belong to the ECM group, including 160 tree species, 6 shrub species, and 1 herb species. Since the number of shrub species and herb species in ECM plants was too little to compare, for the accuracy of the conclusion, we only compare the leaf δ13C of AM plants and ECM plants in four groups (i.e., total plants, trees, evergreen trees, deciduous trees).

2.2. Data Analysis

Carbon isotopic value is expressed as the standard notation relative to the Vienna Pee Dee Belemnite standard using the following equation: $\delta^{13}C = (R_{sample}/R_{standard} - 1) \times 1000$ (‰), where R_{sample} and $R_{standard}$ are the $^{13}C/^{12}C$ ratios of the sample and the standard, respectively [31]. In our study, leaf $\delta^{13}C$ values of C_3 plants were obtained from the database established by Li et al. [7]. To compare the differences in leaf $\delta^{13}C$ between AM plants and ECM plants, we calculated the means of leaf $\delta^{13}C$ of AM plants and ECM plants for each group. The $\delta^{13}C$ of AM plants and ECM plants variation in four groups was examined by one-way analysis of variance. Then, after having diagnosed the covariance of environmental factors, we attempted to establish the relationship between environmental factors and plant leaf $\delta^{13}C$ using multiple regression analysis, comparing the adjusted R^2 of the best multiple regression model to assess the relative importance of different environmental factors in determining plant leaf $\delta^{13}C$. All statistical analyses were conducted using the SPSS software (2012, ver. 22.0; SPSS Inc., Chicago, IL, USA).

We tested the relationship between environmental factors and leaf $\delta^{13}C$ of two mycorrhizal-type plants individually in each group. The slopes of regressions with 95% confidence intervals were displayed with scatter plots.

3. Results

3.1. The Differences in Leaf $\delta^{13}C$ between AM and ECM Plants

The results of the one-way analysis of variance revealed that the differences in leaf $\delta^{13}C$ between AM and ECM plants varied with life forms. When all vegetation types were considered, the leaf $\delta^{13}C$ did not vary between AM and ECM plants (Figure 1a).

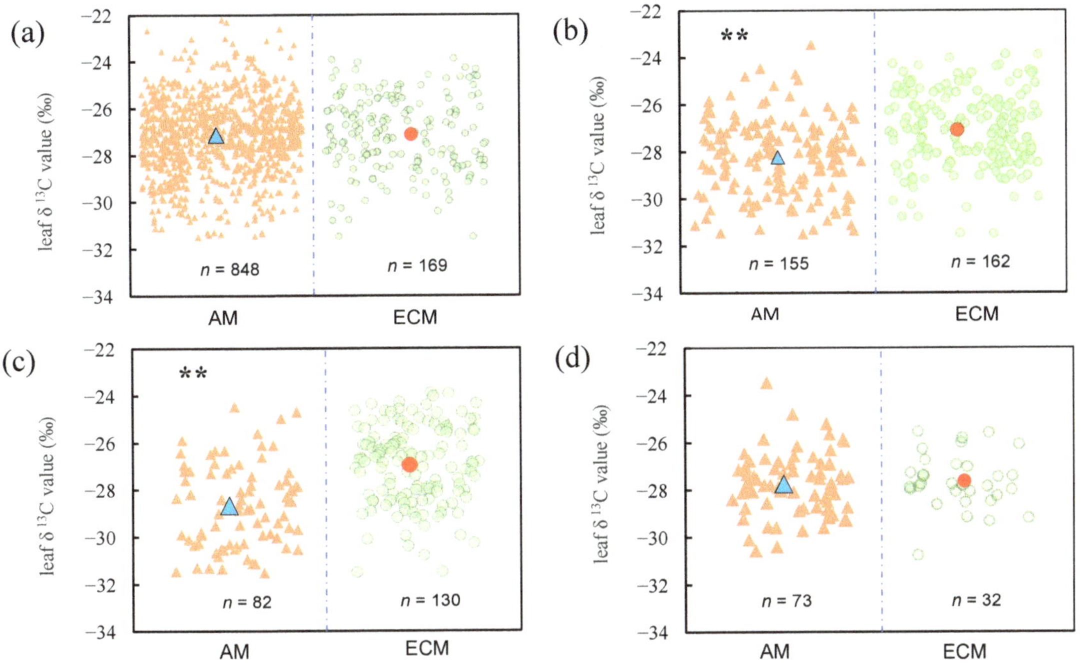

Figure 1. Leaf $\delta^{13}C$ values of AM and ECM plants with different life forms. (**a**) Total plants, include all plants in the database without distinguishing their life forms (**b**) Trees, only contain all the tree species in the database (**c**) Evergreen trees, only contain all evergreen trees of the tree species (**d**) Deciduous trees, only contain all deciduous trees of the tree species. All yellow triangles represent AM plants and all green rectangles represent ECM plants. The blue triangle represents the average of leaf $\delta^{13}C$ values of AM plants and the red circle represents the average of leaf $\delta^{13}C$ in ECM plants. The asterisks above the line bars show the results of one-way ANOVA at the level of $p < 0.05$. Two asterisks indicate extremely significant differences.

For AM plants, the mean of leaf $\delta^{13}C$ is −27.01‰ and the value in ECM plants is −27.12‰. But the leaf $\delta^{13}C$ of AM and ECM plants varied significantly among trees (Figure 1b, $p < 0.001$). For AM plants, the mean of leaf $\delta^{13}C$ is −28.23‰ and the value in ECM plants is −27.09‰. When only evergreen trees were considered, a one-way analysis of variance results revealed that there were significant variations of leaf $\delta^{13}C$ between AM and ECM plants, leaf $\delta^{13}C$ of ECM plants was significantly higher than that of AM plants (Figure 1c, $p < 0.001$). By contrast, there was no significant difference in $\delta^{13}C$ between AM plants and ECM plants.

3.2. Variation of Leaf $\delta^{13}C$ across Longitude and Latitude in AM and ECM Plants

For all plants, the leaf $\delta^{13}C$ of AM and ECM both decreased significantly with the increase in longitude and there is a visible difference in leaf $\delta^{13}C$ in AM and ECM plants ($P_{AM} < 0.05$; $P_{ECM} < 0.001$; Figure 2a). With the increase of longitude, the decreasing rate of $\delta^{13}C$ value of ECM plant leaves was significantly higher than that of AM plant leaves ($P_{AM\&ECM} < 0.001$). For all trees, the leaf $\delta^{13}C$ of ECM plants decreased significantly with the increase in longitude ($P_{ECM} < 0.001$), and there were no significant interactions between the leaf $\delta^{13}C$ of AM plants and the longitude (Figure 2b). For evergreen trees, the leaf δ13C in AM plants decreased significantly with the increase of longitude ($P_{ECM} < 0.001$; Figure 2c), yet no notable linear regression relation was found between leaf $\delta^{13}C$ of AM plants and longitude. For deciduous trees, both the leaf $\delta^{13}C$ in AM plants and ECM plants do not correlate with the longitude (Figure 2d).

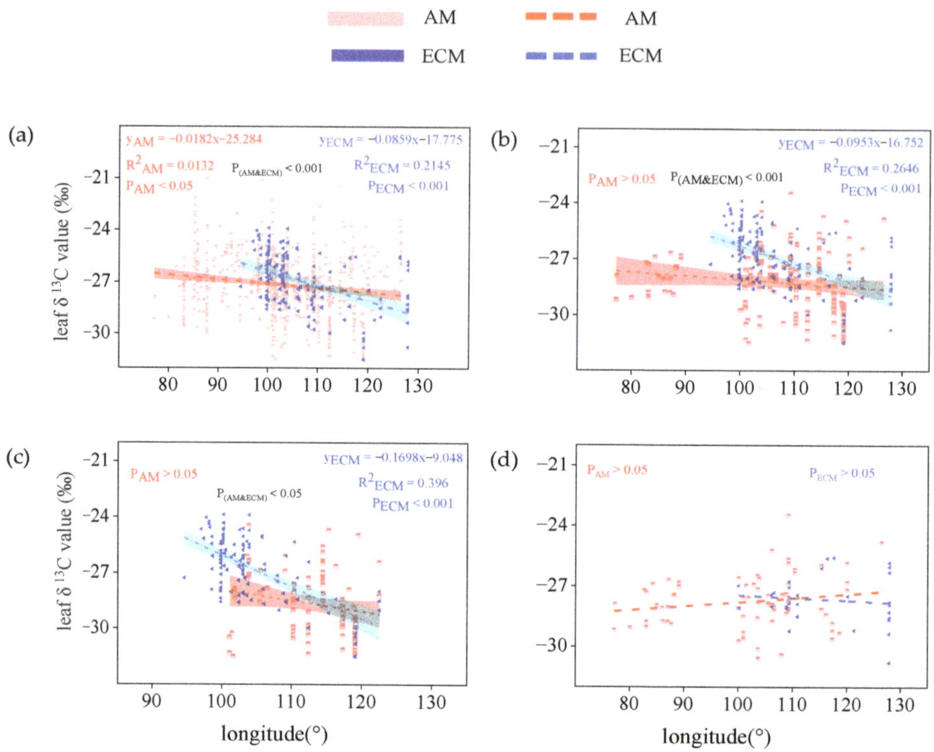

Figure 2. Variation of leaf $\delta^{13}C$ in AM plants and ECM plants with different life forms along longitude. (**a**) Total plants, include all plants in the database without distinguishing their life forms (**b**) trees, only contain all the tree species in the database (**c**) evergreen trees, only contain all evergreen trees of the tree species (**d**) deciduous trees, only contain all deciduous trees of the tree species. The red broken line means AM, the blue broken line means ECM. The red and cyan bands represent, respectively, the prediction intervals of AM and ECM plants. P(AM and ECM) represents the significant difference in the slope of the two regression lines.

When all vegetation types were considered, the leaf $\delta^{13}C$ of AM and ECM both first increased and then decreased significantly with the increase of latitude ($P_{AM} < 0.001$; $P_{ECM} < 0.001$; Figure 3a). It is noteworthy that even though the two curves follow the same trend, the maximum value of leaf $\delta^{13}C$ is different in latitude. In AM plants, the leaf $\delta^{13}C$ had the highest value at the latitude of 42.10°. In ECM plants, the leaf $\delta^{13}C$ had the highest value at the latitude of 37.56°. In total plants, the maximum value of $\delta^{13}C$ in the leaves of AM plants was −21.05‰, which was higher than the maximum value of $\delta^{13}C$ in the leaves of ECM plants (−23.88‰). When only trees were considered, the relationship between leaf $\delta^{13}C$ of two kinds of mycorrhizal type plants and latitude in trees corresponded to that in total plants ($P_{AM} < 0.001$; $P_{ECM} < 0.001$; Figure 3b). In AM plants, the leaf $\delta^{13}C$ had the highest value at the latitude of 45.08°. In ECM plants, the leaf $\delta^{13}C$ had the highest value at the latitude of 36.71°. In trees, however, the difference between the leaf $\delta^{13}C$ maxima of AM plants and ECM plants decreased significantly, and the leaf $\delta^{13}C$ maxima of AM plants (−23.26‰) were higher slightly than those of ECM plants (−23.88‰). When only evergreen trees were considered, even though the two parabolas open in opposite directions, leaf $\delta^{13}C$ of two kinds of mycorrhizal types have been rising as latitude within the scope of the study. In evergreen plants, the maximum value of leaf $\delta^{13}C$ was −23.88‰ for ECM plants. In deciduous plants, the maximum value of leaf $\delta^{13}C$ was −23.46‰ for AM plants.

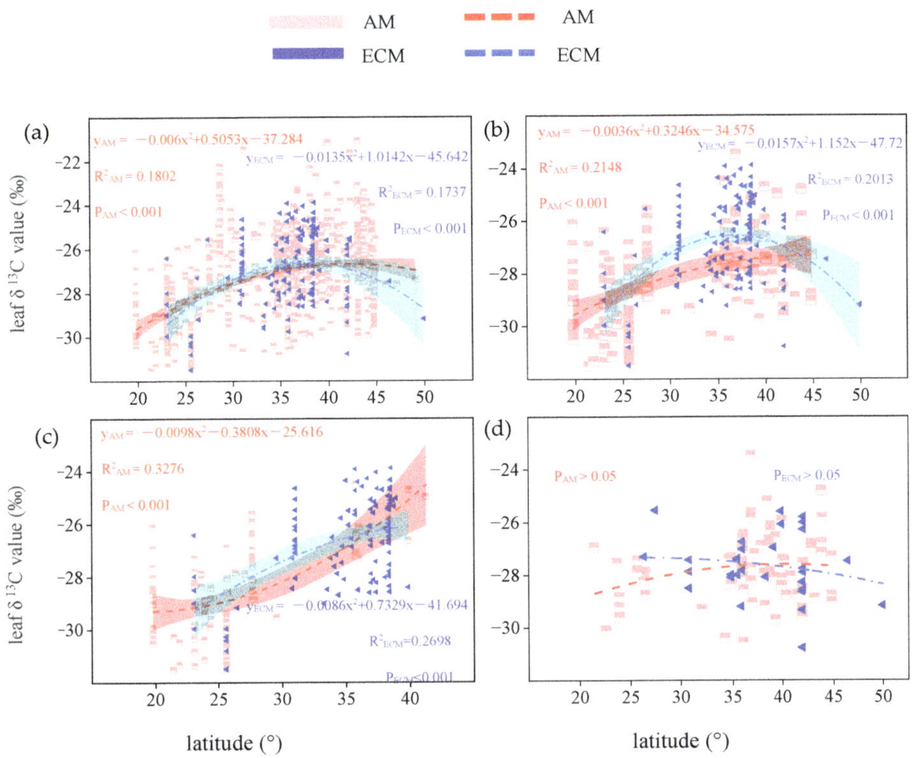

Figure 3. Variation of leaf $\delta^{13}C$ in AM plants and ECM plants with different life forms along latitude. (**a**) Total plants, include all plants in the database without distinguishing their life forms (**b**) trees, only contain all the tree species in the database (**c**) evergreen trees, only contain all evergreen trees of the tree species. (**d**) deciduous trees, only contain all deciduous trees of the tree species. The red broken line means AM, the blue broken line means ECM. The red and cyan bands represent, respectively, the prediction intervals of AM and ECM plants.

3.3. Variation of Leaf $\delta^{13}C$ with Environmental Factors in AM and ECM Plants

For all species, the leaf $\delta^{13}C$ in AM and ECM plants both decreased significantly with the increase of MAP ($P_{AM} < 0.001$; $P_{ECM} < 0.001$), with a similar slope (Figure 4a). For all trees, the change of leaf $\delta^{13}C$ in AM and ECM plants still had the same response to the change of MAP. However, the effect of MAP on leaf 13C of ECM plants was greater than that of AM plants. With the increase in rainfall, the decrease rate of leaf 13C of ECM plants was twice that of AM plants (Figure 4b). For evergreen trees, leaf $\delta^{13}C$ showed similar relationships with MAP across mycorrhizal-type plants, with the slopes being −0.0018 ($p < 0.001$) in AM plants, and −0.0021 ($p < 0.001$) in ECM plants, respectively. But the effect of MAP on ECM plants ($R^2 = 0.27$) explained 10% more of the variation in leaf $\delta^{13}C$ than the effect of MAP on AM plants ($R^2 = 0.17$, Figure 4c). For deciduous trees, the leaf $\delta^{13}C$ is uncorrelated with MAP, whether in AM types of plants or ECM types of plants (Figure 4d).

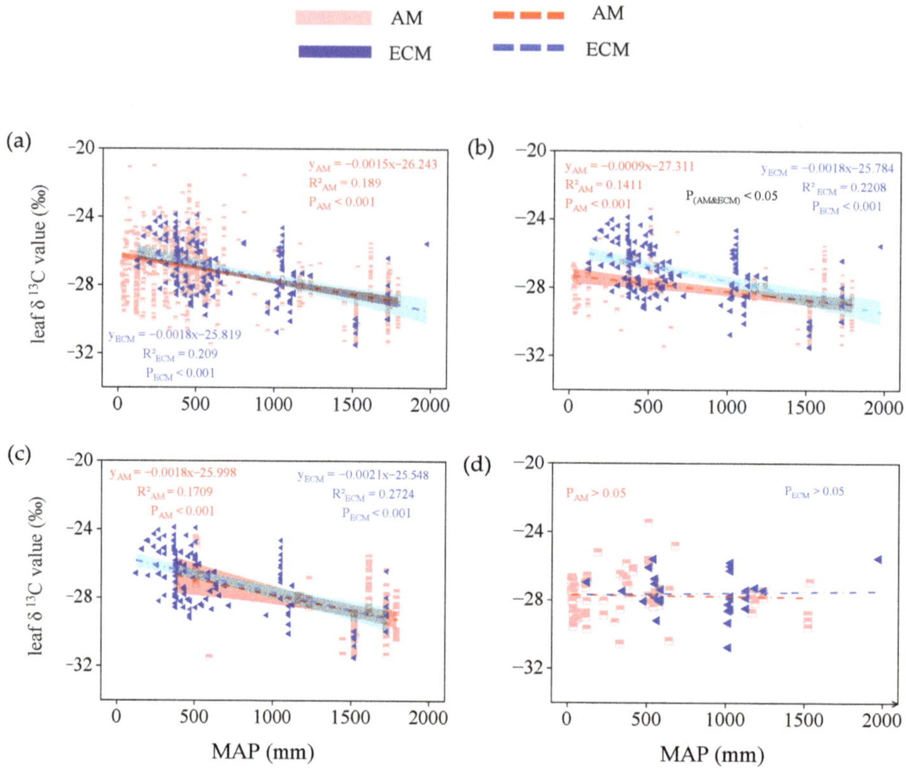

Figure 4. The relationship between leaf δ13C in AM plants and ECM plants with different life forms and mean annual precipitation (MAP, mm). (**a**) Total plants, include all plants in the database without distinguishing their life forms. (**b**) Trees, only contain all the tree species in the database. (**c**) Evergreen trees, only contain all evergreen trees of the tree species. (**d**) Deciduous trees, only contain all deciduous trees of the tree species. The red broken line means AM, the blue broken line means ECM. The red and cyan bands represent, respectively, the prediction intervals of AM and ECM plants. P(AM and ECM) represents the significant difference in the slope of the two regression lines.

When all vegetation types were considered, leaf $\delta^{13}C$ in AM plants and ECM plants both decreased significantly with increasing MAT ($P_{AM} < 0.001$; $P_{ECM} < 0.05$). The slope of leaf $\delta^{13}C$ in AM plants across MAT is twice as large as it is in ECM plants (Figure 5a). When only evergreen trees were considered, leaf $\delta^{13}C$ in AM plants and ECM plants was also positively related to MAT, while the response of leaf $\delta^{13}C$ to MAT tended to be more sensitive in AM plants (slope = 0.10) in comparison to ECM plants (slope = 0.06; Figure 5c). By contrast, leaf $\delta^{13}C$ in AM plants and ECM plants had no correlation with MAT, in deciduous trees (Figure 5d).

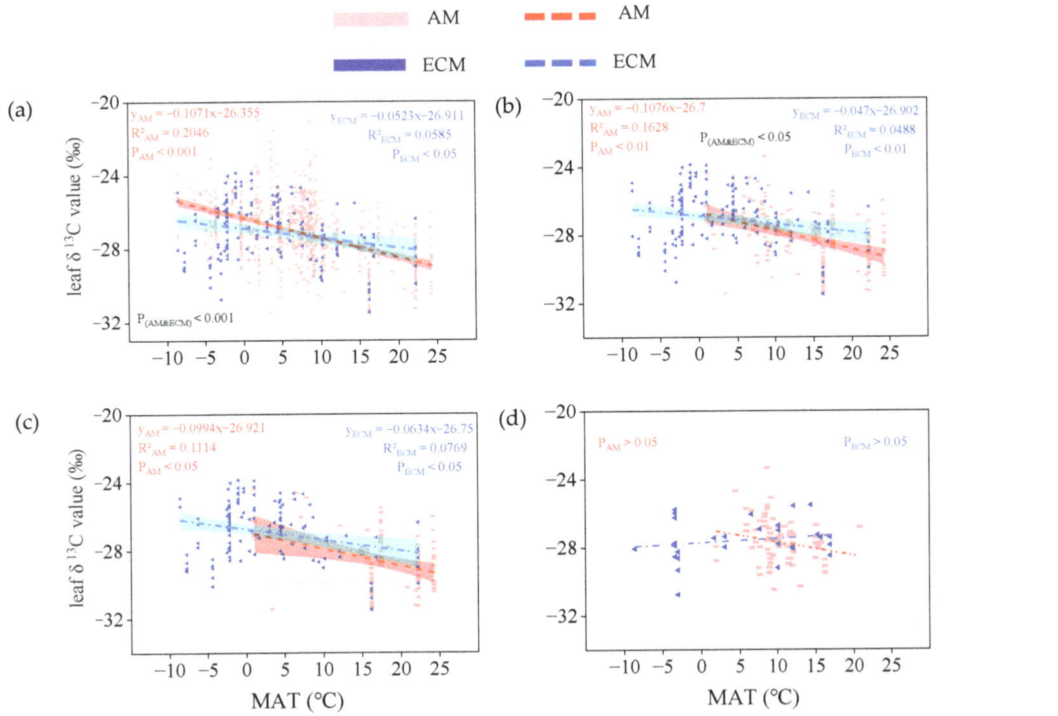

Figure 5. The relationship between leaf $\delta^{13}C$ in AM plants and ECM plants with different life forms and mean annual temperature (MAT, °C). (**a**) Total plants, include all plants in the database without distinguishing their life forms. (**b**) Trees, only contain all the tree species in the database. (**c**) Evergreen trees, only contain all evergreen trees of the tree species. (**d**) Deciduous trees, only contain all deciduous trees of the tree species. The red broken line means AM, the blue broken line means ECM. The red and cyan bands represent, respectively, the prediction intervals of AM and ECM plants. $P_{(AM\ and\ ECM)}$ represents the significant difference in the slope of the two regression lines.

In all vegetation type group, leaf $\delta^{13}C$ tended to decrease with the increasing relative humidity, which is both significant in AM plants ($p < 0.001$) and in ECM plants ($p < 0.001$), but, the effect of relative humidity change on $\delta^{13}C$ in ECM plant leaves was more obvious, and the value of R^2 was higher (Figure 6a). In the trees group, leaf $\delta^{13}C$ in AM plants and ECM plants both decreased significantly with increasing relative humidity ($P_{AM} < 0.001$, $P_{ECM} < 0.001$). The slope of leaf $\delta^{13}C$ in AM plants across relative humidity is half as large as it is in ECM plants (Figure 6b). In the evergreen trees group, the leaf $\delta^{13}C$ in AM plants and ECM plants both were negatively correlated with relative humidity ($P_{AM} < 0.001$, $P_{ECM} < 0.001$). Although the $\delta^{13}C$ of AM plants is more sensitive to changes in RH than the $\delta^{13}C$ of ECM plants, the effect of RH on ECM plants ($R^2 = 0.32$) explained 5% more of the variation in leaf $\delta^{13}C$ than the effect of MAP on AM plants ($R^2 = 0.27$; Figure 6c). In the deciduous trees group, leaf $\delta^{13}C$ in AM and ECM plants slightly increased with higher relative humidity, but the relationship was not significant (Figure 6d).

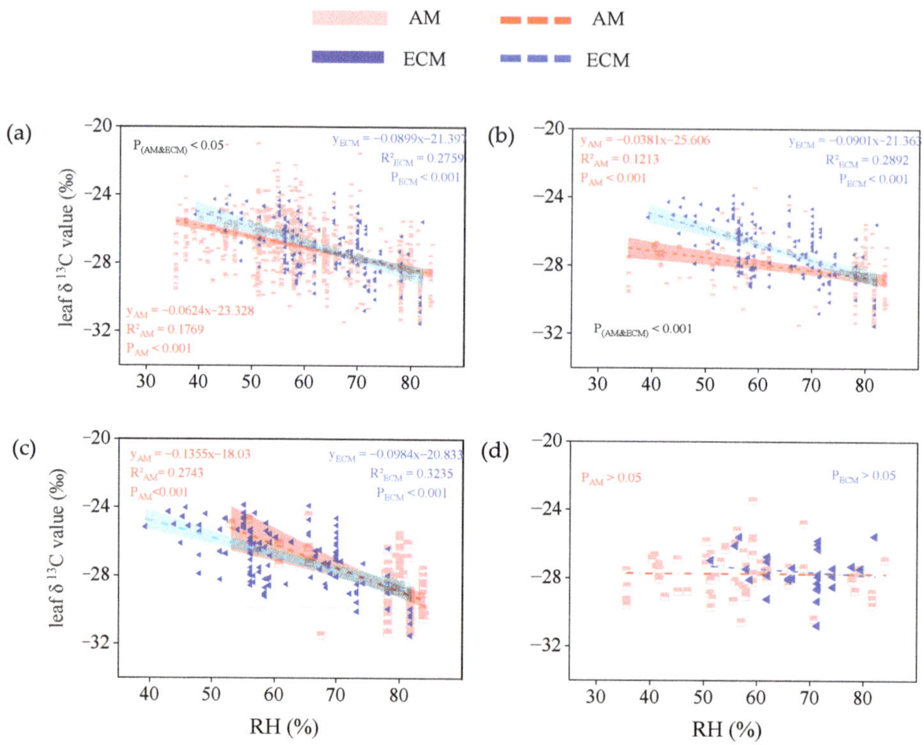

Figure 6. The relationship between leaf $\delta^{13}C$ in AM plants and ECM plants with different life forms and relative humidity (RH, %). (**a**) Total plants, include all plants in the database without distinguishing their life forms. (**b**) Trees, only contain all the tree species in the database. (**c**) Evergreen trees, only contain all evergreen trees of the tree species. (**d**) Deciduous trees, only contain all deciduous trees of the tree species. The red broken line means AM, the blue broken line means ECM. The red and cyan bands represent, respectively, the prediction intervals of AM and ECM plants. P(AM and ECM) represents the significant difference in the slope of the two regression lines.

For all vegetation types, for both AM plants and ECM plants, the leaf $\delta^{13}C$ increased significantly with altitude ($P_{AM} < 0.001$; $P_{ECM} < 0.05$) and the change in altitude had the same effect on both species (Figure 7a). This correlation is also suited to that for trees, leaf $\delta^{13}C$ in AM plants and ECM plants both related positively to altitude ($P_{AM} < 0.001$; $P_{ECM} < 0.001$). Besides, the effect of altitude on leaf $\delta^{13}C$ of AM plants was similar to the effect of altitude on leaf $\delta^{13}C$ of ECM plants (Figure 7b). For evergreen trees, leaf $\delta^{13}C$ in AM plants and ECM plants both slightly increased with higher altitude but the relationship between $\delta^{13}C$ in ECM plants and altitude was not significant (Figure 7c). By contrast, for deciduous trees, leaf $\delta^{13}C$ tended to decrease with altitude and there was no correlation with altitude, whether the leaf $\delta^{13}C$ in AM plants or ECM plants (Figure 7d).

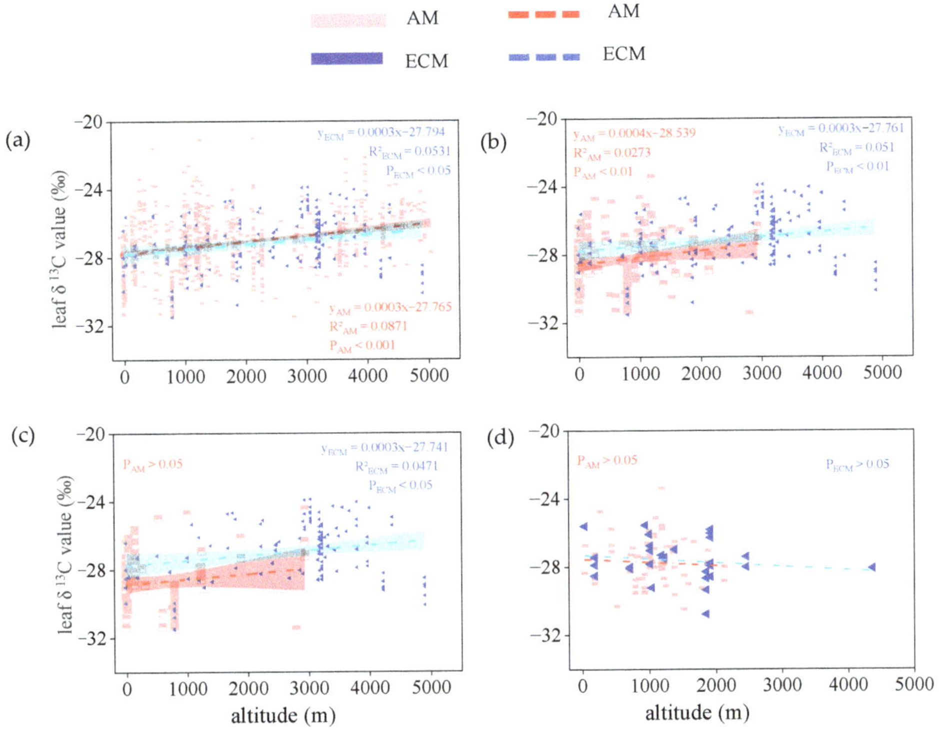

Figure 7. The relationship between leaf $\delta^{13}C$ in AM plants and ECM plants with different life forms and altitudes (m). (**a**) Total plants, include all plants in the database without distinguishing their life forms. (**b**) Trees, only contain all the tree species in the database. (**c**) Evergreen trees, only contain all evergreen trees of the tree species. (**d**) Deciduous trees, only contain all deciduous trees of the tree species. The red broken line means AM, the blue broken line means ECM. The red and cyan bands represent, respectively, the prediction intervals of AM and ECM plants. $P_{(AM\ and\ ECM)}$ represents the significant difference in the slope of the two regression lines.

When all vegetation types were considered, the leaf $\delta^{13}C$ in AM plants and ECM plants both increased significantly as sunshine hours (P_{AM} < 0.001; P_{ECM} < 0.001) and were close in their sensitivity to sunshine hours responses (Figure 8a). When only trees were considered, leaf $\delta^{13}C$ in AM plants and ECM plants were both positively correlated with sunshine hours (P_{AM} < 0.001; P_{ECM} < 0.001). The response of leaf $\delta^{13}C$ to sunshine hours tended to be more sensitive in ECM plants (slope = 0.0014) in comparison to AM plants (slope = 0.0008). We further analyzed the linear relationship between leaf $\delta^{13}C$ and sunshine hours of AM plants and ECM plants in evergreen trees and deciduous trees, respectively. The results showed that the effect of sunshine hours on leaf $\delta^{13}C$ of AM plants and ECM plants in evergreen trees was significantly (P_{AM} < 0.05; P_{ECM} < 0.01) greater than the effect of sunshine hours on leaf $\delta^{13}C$ of AM plants and ECM plants in deciduous trees (Figure 8c,d). In evergreen trees, the effect of sunshine hours on ECM plants (R^2 = 0.1867) explained 10.6% more of the variation in leaf $\delta^{13}C$ than AM plants, but with a similar slope (Figure 8c).

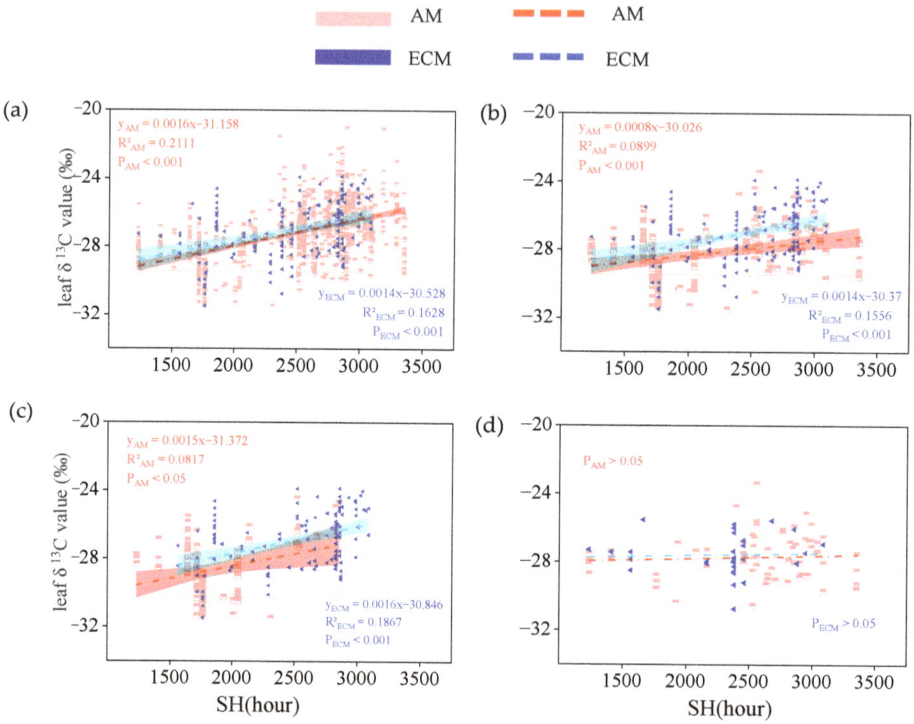

Figure 8. The relationship between leaf $\delta^{13}C$ in AM plants and ECM plants with different life forms and sunshine hours (SH, hours). (**a**) Total plants, include all plants in the database without distinguishing their life forms. (**b**) Trees, only contain all the tree species in the database. (**c**) Evergreen trees, only contain all evergreen trees of the tree species. (**d**) Deciduous trees, only contain all deciduous trees of the tree species. The red broken line means AM, the blue broken line means ECM. The red and cyan bands represent, respectively, the prediction intervals of AM and ECM plants. $P_{(\text{AM and ECM})}$ represents the significant difference in the slope of the two regression lines.

3.4. Stepwise Regression Analysis of Leaf $\delta^{13}C$ and Environmental Factors in AM and ECM Plants

The model summary was shown in Table 1. When all vegetation types were considered, in AM plants, the value of R^2 was 0.266, which indicated that there were 26.6% changes in the response variable (leaf $\delta^{13}C$) because of changes in the combination of four controlled variables including Lat, Lon, MAT, and SH. Among four controlled variables that affected $\delta^{13}C$ values, MAT was the most profound environment factor ($\beta = -0.265$, $p < 0.001$), while SH was the secondary environment factor ($\beta = 0.245$, $p < 0.001$). In ECM plants, the R^2 value of the stepwise regression equation was higher than that of AM plants, which was 0.285. According to the results, the variation of leaf $\delta^{13}C$ was mainly attributed to four variables, these variables were: RH ($\beta = -0.449$, $p < 0.001$), LAT ($\beta = -0.413$, $p < 0.001$), Altitude ($\beta = 0.263$, $p < 0.001$). When only trees were considered, the multivariate stepwise regression equation of leaf $\delta^{13}C$ in AM plants only screened out LAT as a significant influencing factor ($\beta = 0.448$, $p < 0.001$). It explained 19.6 % of the total variation of leaf $\delta^{13}C$ in AM plants. The multivariate stepwise regression equation of $\delta^{13}C$ in leaves of ECM plants screened three significant influencing factors, namely Lat ($\beta = -0.445$, $p < 0.001$), RH ($\beta = -0.459$, $p < 0.001$), Altitude ($\beta = 0.279$, $p < 0.001$), which together explained 41.5% of the total variation of $\delta^{13}C$ in leaves. When only evergreen trees were considered, in AM groups, the best leaf $\delta^{13}C$ model showed that Lat ($\beta = 0.789$, $p < 0.001$), Lon ($\beta = -0.417$, $p < 0.001$), and Altitude ($\beta = -0.465$, $p < 0.001$) in combination explained 43.2% of the total variation. Compared to this model, the leaf $\delta^{13}C$ of ECM plants was explained by a set of three environmental factors including Lat ($\beta = -1.773$, $p < 0.001$), Lon ($\beta = -1.483$, $p < 0.001$), and RH ($\beta = -0.583$, $p < 0.001$), the explanation rate of the model is 49.4%.

Table 1. Results of the multiple regression analyses for predicting the leaf $\delta^{13}C$ of AM and ECM plants with the combination of all traits.

Groups	Mycorrhizal Types	Standardized Coefficients							R^2	Sig
		LAT	LON	MAP	MAT	RH	Altitude	SH		
Total plants	AM	0.182	0.120	-	−0.265	-	-	0.245	0.266	***
	ECM	−0.413	-	-	-	−0.449	0.263	-	0.285	***
Trees	AM	0.448	-	-	-	-	-	-	0.196	***
	ECM	−0.445	-	-	-	−0.459	0.279	-	0.415	***
Evergreen trees	AM	0.789	−0.417	-	-	-	−0.465	-	0.432	***
	ECM	−1.773	−1.483	-	-	−0.583	-	-	0.494	***

AM: arbuscular mycorrhiza, ECM: ectomycorrhiza, LAT: latitude, LON: longitude, MAP: mean annual precipitation, MAT: mean annual temperature, RH: relative humidity, SH: sun hours. - means that the environmental factors were not included in the best model. *** $p < 0.001$.

4. Discussion

Here, we evaluated firstly the leaf $\delta^{13}C$ and their relationship with environmental factors of C_3 plants between AM and ECM types. The influence of mycorrhiza on plant development and its response to climate change vary with different mycorrhizal types, both at the individual level and at the ecological level [25,32]. Vargas et al. [25] suggested that ecosystem CO2 fluxes of ECM symbiosis tend to become dominant in woody vegetation types where interannual variation in ecosystem CO_2 fluxes is primarily controlled by changes in temperature, whereas the AM symbiosis dominates grassland or woody vegetation types where interannual variation in CO_2 fluxes was largely controlled by changes in precipitation. Shi et al. [30] presented that the world leaf economic spectrum traits were greatly linked with mycorrhizal traits and are woody plants, especially trees, have shorter leaf lifespans, lower leaf mass per area, and higher leaf nitrogen concentration, photosynthetic capacity, and dark respiration rate than nonwoody ones.

Our results further support this conclusion and point out that plant leaf $\delta^{13}C$ is closely related to plant mycorrhizal types. The data presented here clearly show that variations in mean annual temperature (MAT), mean annual precipitation (MAP), and relative humidity (RH) are important environmental drivers for leaf $\delta^{13}C$, but influence AM and ECM-dominated vegetation types differently.

4.1. *Overall Differences in Leaf $\delta^{13}C$ between AM Plants and ECM Plants*

Our analysis indicated that the arithmetic means of leaf $\delta^{13}C$ in AM plants is −27.01‰ and ECM plants is −27.12‰, which were both nearly identical to the global average of leaf $\delta^{13}C$, −27.0‰. The latter was reported by Kohn [33], who collected leaf $\delta^{13}C$ values from approximately 570 sites on a global scale. Moreover, our results were slightly higher than the global average (−27.25‰) reported by O'Leary [34].

Our results suggested that there were no differences in leaf $\delta^{13}C$ between AM plants and ECM plants in total plants while the leaf $\delta^{13}C$ of AM plants was less than ECM plants in trees (Figure 1a,b). Compared to the total plant group, we found leaf $\delta^{13}C$ of AM plants decreased significantly but the leaf $\delta^{13}C$ of ECM plants hardly change in the tree group. The analysis results show that this phenomenon was mainly caused by shrubs and herbs, because shrubs and herbs constitute the majority of AM plants, and the leaves of shrubs and herbs $\delta^{13}C$ was greater than the leaves of trees $\delta^{13}C$. Previous studies presented that there were significant differences in leaf $\delta^{13}C$ among different life forms [35].

After further analysis, we found that the differences in leaf $\delta^{13}C$ between two mycorrhizal types of plants in evergreen trees were the main reason for the differences in trees (Figure 1c). The average value of leaf $\delta^{13}C$ of AM plants was less than ECM plants in evergreen trees. Numerous studies have demonstrated that different mycorrhizal types differ in their use to improve nutrient uptake by plants; for example, AM mainly improves P nutrition of plants, while ECM facilitates N uptake by plants. Therefore, different mycorrhiza fungi also have different effects on the physiological processes of plants. Zhao et al. [36] found that the hydraulic conductivity of ECM trees was significantly higher than that of AM species, indicating that ECM species have higher photosynthetic rates. In summary, there are significant differences between mycorrhizal types in terms of their effects on plant physiological processes and nutrient cycling in the ecosystem. As an important characteristic of plant leaves, leaf $\delta^{13}C$ is an important indicator to study the relationship between plants and their environment, and our results show that there are significant differences in leaf $\delta^{13}C$ among different mycorrhizal types. This may be related to the different effects of different mycorrhizal types on photosynthetic rates as well as the stomatal conductance of their host plants. The exact reasons for this are to be further verified.

4.2. Differences in the Spatial Distribution of Leaf $\delta^{13}C$ in AM Pants and ECM Plants

As an important characteristic of leaves, most studies demonstrated that leaf $\delta^{13}C$ and geolocation information were related closely [37,38]. Li et al. [7] invested the spatial pattern of leaf $\delta^{13}C$ values in China and found that leaf $\delta^{13}C$ slightly decreased as longitude increased, but first increased, and then decreased as the latitude increased. Based on the study researched by Li et al. [7], we analyzed the relationship between leaf $\delta^{13}C$ and longitude, and latitude, respectively, after classifying vegetation types as AM or ECM dominant. In our study, the leaf $\delta^{13}C$ of AM plants and ECM plants both decreased significantly as longitude increased (except leaf $\delta^{13}C$ of AM plants in trees) while the leaf $\delta^{13}C$ of ECM plants was more sensitive to longitude change than AM plants (Figure 2a,b). The leaf $\delta^{13}C$ of AM and ECM plants both first increased and then decreased as the latitude increased in total plants and tree groups. However, the maximum leaf $\delta^{13}C$ of AM plants and ECM plants appeared at different latitudes. The latitude at which the maximum of leaf $\delta^{13}C$ in AM plants occurs was greater than ECM plants in total plants and tree plants (Figure 3a,b). Our results also show that in different groups, latitude always has opposite effects on leaf $\delta^{13}C$ of AM plants and ECM plants. It may be related to the global distribution of mycorrhizal fungi [39]. Due to the different selectivity of mycorrhizal fungi to host plants and adaptability to environmental conditions or the historical reasons in the evolutionary process, the distribution of mycorrhizal fungi in the natural ecosystem is different [40]. Numerous studies showed that mycorrhizal status significantly affects the responses of leaf characteristics to geographical distribution [28,40], our results further confirm this conclusion. Lu et al. [41] found that leaf ash concentration in arbuscular mycorrhizal plants was significant to be controlled by latitude and temperature factors, while ECM plants were not. Our results suggested that the spatial distribution of $\delta^{13}C$ in plant leaves will be affected by plant mycorrhizal types consistent with previous studies. Due to the lack of relevant studies, the specific reasons need to be further explored.

4.3. Variations in Response of Leaf $\delta^{13}C$ to Environmental Factors between AM Plants and ECM Plants

Water change has a lasting impact on spatial variation and the characteristics of vegetation, and maybe alter leaf $\delta^{13}C$ of plants [8,42–45]. In our study, we analyzed and quantified the relationship between water availability (including MAP and RH) and leaf $\delta^{13}C$ in AM plants and ECM plants based on our database (Figures 4 and 6). Our results showed that leaf $\delta^{13}C$ in AM and ECM plants both decreased as water availability increased (including MAP and RH), which indicated that the higher precipitation, the lower $\delta^{13}C$. Meanwhile, the results of step-by-step analysis showed that the relative humidity only affected the leaf $\delta^{13}C$ of ECM plants, which further explained that the leaf $\delta^{13}C$ of different mycorrhizal plants had different responses to water conditions (Table 1). The leaf $\delta^{13}C$ of ECM plants has a higher change and explain rate than AM plants under changing water availability, both in total plants and trees (Figure 6a,b). This result indicated that mycorrhizal fungi can affect the response of leaf $\delta^{13}C$ to changes in water availability. This may be due to the different sensitivity of the two kinds of mycorrhizal fungi to water changes [5,46]. Most studies suggested that ECM plants were more sensitive to water availability than AM plants [37,47]. Shi et al. [47] compared separately the net primary productivity of AM and ECM type forests and found that ECM type forests were more sensitive to precipitation changes than AM type forests. Zhao et al. [36] found that ECM trees have stronger drought resistance ability and higher water use efficiency compared with AM trees under the background of increasing drought in subtropical forest. Studies have shown that mycorrhiza can affect the stomatal conductance of plants [48,49]. The response of leaf characteristics to changes in water availability was affected by mycorrhizal types [30,41]. Lu et al. [41] studied the responses of leaf ash concentration of AM plants and ECM plants to climate change and found that the response of AM plants was more susceptible to mean annual precipitation with a 1.61 times response amplitude compared to ECM plants. All the above studies showed that leaf characteristics of different mycorrhizal types of plants were different in response to climate change. As one of the characteristics of plant leaves, the leaf $\delta^{13}C$ of different mycorrhizal plants has different responses to the change in water availability.

In our study, we found a negative liner for the relationship of leaf $\delta^{13}C$ in AM and ECM plants with MAT, but in the tree leaf $\delta^{13}C$ of AM, plants were more sensitive to the changes in MAT than ECM plants trees (Figure 5a,b). Temperature affects the $\delta^{13}C$ of plants mainly by affecting enzymes involved in photosynthesis [50]; AM and ECM both can promote photosynthesis in plants [51,52]. Wang et al. [53] found that AM inoculation had a significant effect on the photosynthetic of Sinocalycanthus Chinensis under simulated warming conditions. Compared to ECM, AM was more closely related to the photosynthesis of plants. This may be related to the different metrological characteristics of plant leaves of different mycorrhizal types. Studies have shown that AM plays the most significant role in improving the P nutrient status of plants, while ECM plays a greater role in the N nutrient absorption of plants [24]. As an important component of Rubisco and energy substance ATP, phosphorus plays an important role in plant photosynthesis. As this study is the first time to explore, the specific reasons need to be studied in the future.

Although the $\delta^{13}C$ in AM and ECM plants have different responses to changes in temperature and water availability, they have the same response to changes in other environmental factors (e.g., altitude, sunshine hours). In our study, we found that the leaf $\delta^{13}C$ of AM plants and ECM plants both increased weakly as altitude and sunshine hours increased (Figures 7 and 8). This result indicated that certain environmental factors affect mycorrhizal fungi in the same way.

Overall, although plant life forms and environmental conditions are inherently variable, the variation of plant leaf $\delta^{13}C$ along environmental gradients offers one way to evaluate potential plant responses to climate change. The different responses of the leaf $\delta^{13}C$ of mycorrhizal types to climate change might also provide a reference for future studies, simulating the response of vegetation distribution to climate change. As this study is the first to explore this, the specific reasons need to be studied in the future.

5. Conclusions

All life forms will affect the leaf $\delta^{13}C$ content of plants with different mycorrhizal types. The effect of arbuscular and ectomycorrhizal on the responses of leaf $\delta^{13}C$ to changes in environmental factors was different. Further analysis showed that the results showed that the response of AM plant leaf $\delta^{13}C$ and ECM plant leaf $\delta^{13}C$ to climate variations depended on plant life forms. This study initially explored the effect of mycorrhizal on leaf $\delta^{13}C$. It also provides data in support of future exploration of the response of leaf $\delta^{13}C$ in AM and ECM plants to changes in climate and environmental factors. Findings showed that the responses of leaf $\delta^{13}C$ to changes in environmental factors are differentially affected by arbuscular and ectomycorrhizal types; the leaf $\delta^{13}C$ of AM plants was mainly affected by temperature; while the leaf $\delta^{13}C$ of ECM plants was more sensitive to moisture content.

Our results have important implications for understanding the relationship between leaf $\delta^{13}C$ and climatic factors, and the climatic and environmental significance indicated by plant leaf $\delta^{13}C$. This is a previously unrecognized study that has important implications for our understanding of the impacts of arbuscular and ectomycorrhizal on plants' adaptation to climate change.

Author Contributions: S.Z.: conceptualization, methodology, writing—original draft; M.Y. and Z.S.: conceptualization, methodology, writing—original draft and review; S.Y.: data curation, writing—review and editing; M.Z.: data curation, writing—editing; J.G.: data curation, writing—review; L.S. and X.W.: data curation,—review. All authors have read and agreed to the published version of the manuscript.

Funding: This work was funded by NSFC (32171620, 31670499), Scientific and technological research projects in Henan province (192102110128), Program for Science & Technology Innovation Talents in Universities of Henan Province (18HASTIT013), Key Laboratory of Mountain Surface Processes and Ecological Regulation, CAS (20160618).

Data Availability Statement: The raw data for this study available via https://figshare.com/s/fcfdd37541c8ba1b8278 (accessed on 3 April 2022).

Conflicts of Interest: The authors declare that the research was conducted in the absence of any commercial or financial relationships that could be constructed as a potential conflict of interest.

References

1. Wang, N.; Xu, S.S.; Jia, X.; Gao, J.; Zhang, W.P.; Qiu, Y.P.; Wang, G.X. Variations in foliar stable carbon isotopes among functional groups and along environmental gradients in China—A meta-analysis. *Plant Biol.* **2013**, *15*, 144–151. [CrossRef] [PubMed]
2. Zhou, Y.C.; Li, H.B.; Xu, X.Y.; Li, Y.H. Responses of carbon isotope composition of common C_3 and C_4 plants to climatic factors in temperate grasslands. *Sustainability* **2022**, *14*, 7311. [CrossRef]
3. Mueller, K.E.; Blumenthal, D.M.; Pendall, E.; Carrillo, Y.; Dijkstra, F.A.; Williams, D.G.; Follett, R.F.; Morgan, J.A. Impacts of warming and elevated CO_2 on a semiarid grassland are non-additive, shift with precipitation, and reverse over time. *Ecol. Lett.* **2016**, *19*, 956–966. [CrossRef] [PubMed]
4. Li, Z.Q.; Yang, L.; Lu, W.; Guo, X.S.; Xu, J.; Yu, D. Spatial patterns of leaf carbon, nitrogen stoichiometry and stable carbon isotope composition of *Ranunculus natans* C.A. Mey. (Ranunculaceae) in the arid zone of northwest China. *Ecol. Eng.* **2015**, *77*, 9–17. [CrossRef]
5. Chen, Z.X.; Wang, G.A.; Jia, Y.F. Foliar $\delta^{13}C$ showed no altitudinal trend in an arid region and atmospheric pressure exerted a negative effect on plant $\delta^{13}C$. *Front. Plant Sci.* **2017**, *8*, 1070. [CrossRef]
6. Yoneyama, T.; Okada, H.; Ando, S. Seasonal variations in natural $\delta^{13}C$ abundances in C_3 and C_4 plants collected in Thailand and the Philippines. *Soil Sci. Plant Nutr.* **2010**, *56*, 422–426. [CrossRef]
7. Li, M.X.; Peng, C.H.; Wang, M.; Yang, Y.Z.; Zhang, K.R.; Li, P.; Yang, Y.Z.; Ni, J.; Zhu, Q.A. Spatial patterns of leaf $\delta^{13}C$ and its relationship with plant functional groups and environmental factors in China. *J. Geophys. Res. Biogeosci.* **2017**, *122*, 1564–1575. [CrossRef]
8. Adams, M.A.; Buckley, T.N.; Turnbull, T.L. Diminishing CO_2-driven gains in water-use efficiency of global forests. *Nat. Clim. Chang.* **2020**, *10*, 466–471. [CrossRef]
9. Stein, R.A.; Sheldon, N.D.; Smith, S.Y. C_3 plant carbon isotope discrimination does not respond to CO_2 concentration on decadal to centennial timescales. *New Phytol.* **2020**, *229*, 2576–2585. [CrossRef]
10. Brooks, J.R.; Flanagan, L.B.; Buchmann, N.; Ehleringer, J.R. Carbon isotope composition of boreal plants: Functional grouping of life. *Oecologia* **1997**, *110*, 301–311. [CrossRef]
11. He, C.X.; Li, J.Y.; Zhou, P.; Guo, M.; Zheng, Q.S. Changes of leaf morphological, anatomical structure and carbon isotope ratio with the height of the Wangtian tree (*Parashorea chinensis*) in Xishuangbanna, China. *J. Integr. Plant Biol.* **2008**, *50*, 168–173. [CrossRef] [PubMed]
12. Ale, R.; Zhang, L.; Li, X.; Raskoti, B.B.; Pugnaire, F.I.; Luo, T.X. Leaf delta $\delta^{13}C$ as an indicator of water availability along elevation gradients in the dry Himalayas. *Ecol. Indic.* **2018**, *94*, 266–273. [CrossRef]
13. Cermusak, L.A.; Ubierna, N.; Winter, K.; Holtum, J.A.M.; Marshall, J.D.; Farquhar, G.D. Environmental and physiological determinants of carbon isotope discrimination in terrestrial plants. *New Phytol.* **2013**, *200*, 950–965. [CrossRef] [PubMed]
14. Shtangeeva, I.; Busa, L.; Viksna, A. Carbon and nitrogen stable isotope ratios of soils and grasses as indicators of soil characteristics and biological taxa. *Appl. Geochem.* **2019**, *104*, 19–24. [CrossRef]
15. Zhou, Y.C.; Zhang, W.B.; Cheng, X.L.; Harris, W.; Schaeffer, S.M.; Xu, X.Y.; Zhao, B. Factors affecting $\delta^{13}C$ enrichment of vegetation and soil in temperate grasslands in Inner Mongolia, China. *J. Soils Sediments* **2019**, *19*, 2190–2199. [CrossRef]
16. Ma, F.; Liang, W.Y.; Zhou, Z.N.; Xiao, G.J.; Liu, J.L.; He, J.; Jiao, B.Z.; Xu, T.T. Spatial variation in leaf stable carbon isotope composition of three caragana species in northern China. *Forests* **2018**, *9*, 297. [CrossRef]

17. Liu, J.; Su, Y.G.; Li, Y.; Huang, G. Shrub colonization regulates $\delta^{13}C$ enrichment between soil and vegetation in deserts by affecting edaphic variables. *Catena* **2021**, *203*, 105365. [CrossRef]
18. Verbruggen, E.; Jansa, J.; Hammer, E.C.; Rillig, M.C. Do arbuscular mycorrhizal fungi stabilize litter-derived carbon in soil? *J. Ecol.* **2016**, *104*, 261–269. [CrossRef]
19. Tian, Y.; Yan, C.; Wang, Q.; Ma, W.; Lu, H. Glomalin-related soil protein enriched in $\delta^{13}C$ and $\delta^{15}N$ excels at storing blue carbon in mangrove wetlands. *Sci. Total Environ.* **2020**, *732*, 138327. [CrossRef]
20. Auge, R.M.; Toler, H.D.; Saxton, A.M. Mycorrhizal stimulation of leaf gas exchange about root colonization, shoot size, leaf phosphorus, and nitrogen: A quantitative analysis of the literature using meta-regression. *Front. Plant Sci.* **2016**, *7*, 1084. [CrossRef]
21. Zhang, H.Y.; Lu, X.T.; Hartmann, H.; Keller, A.; Han, X.G.; Trumbore, S.; Phillips, R.P. Foliar nutrient resorption differs between arbuscular mycorrhizal and ectomycorrhizal trees at local and global scales. *Global Ecol. Biogeogr.* **2018**, *27*, 875–885. [CrossRef]
22. El-Nashar, Y.I.; Hassan, B.A.; Aboelsaadat, E.M. Response of *Nemesia* (*Nemesia* × *hybridus*) plants to different irrigation water sources and arbuscular mycorrhizal fungi inoculation. *Agr. Water Manag.* **2021**, *243*, 106416. [CrossRef]
23. Smith, S.E.; Read, D.J. *Mycorrhizal Symbiosis*, 3rd ed.; Academic Press: London, UK, 2008.
24. Averill, C.; Bhatnagar, J.M.; Dietze, M.C.; Pearse, W.D.; Kivlin, S.N. Global imprint of mycorrhizal fungi on whole-plant nutrient economics. *Proc. Natl. Acad. Sci. USA* **2019**, *116*, 23163–23168. [CrossRef] [PubMed]
25. Vargas, R.; Baldocchi, D.D.; Querejeta, J.I.; Curtis, P.S.; Hasselquist, N.J.; Janssens, I.A.; Allen, M.F.; Montagnani, L. Ecosystem CO_2 fluxes of arbuscular and ectomycorrhizal dominated vegetation types are differentially influenced by precipitation and temperature. *New Phytol.* **2010**, *185*, 226–236. [CrossRef] [PubMed]
26. Terrer, C.; Phillips, R.P.; Hungate, B.A.; Rosende, J.; PettRidge, J.; Craig, M.E.; van Groenigen, K.J.; Keenan, T.F.; Sulman, B.N.; Stocker, B.D.; et al. A trade-off between plant and soil carbon storage under elevated CO_2. *Nature* **2021**, *591*, 599–603. [CrossRef] [PubMed]
27. Rao, Z.G.; Guo, W.K.; Cao, J.T.; Shi, F.X.; Jiang, H.; Li, C.Z. Relationship between the stable carbon isotopic composition of modern plants and surface soils and climate: A global review. *Earth-Sci. Rev.* **2017**, *165*, 110–119. [CrossRef]
28. Li, J.Z.; Wang, G.A.; Liu, X.Z.; Han, J.M.; Liu, M.; Liu, X.J. Variations in carbon isotope ratios of C_3 plants and distribution of C_4 plants along an altitudinal transect on the eastern slope of Mount Gongga. *Sci. China Ser. D Earth Sci.* **2009**, *52*, 1714–1723. [CrossRef]
29. Wang, B.; Qiu, Y.L. Phylogenetic distribution and evolution of mycorrhizas in land plants. *Mycorrhiza* **2006**, *16*, 299–363. [CrossRef]
30. Shi, Z.Y.; Li, K.; Zhu, X.Y.; Wang, F.Y. The worldwide leaf economic spectrum traits are closely linked with mycorrhizal traits. *Fungal Ecol.* **2020**, *43*, 100877. [CrossRef]
31. Gong, X.S.; Xu, Z.Y.; Peng, Q.T.; Tian, Y.Q.; Hu, Y.; Li, Z.Q.; Hao, T. Spatial patterns of leaf $\delta^{13}C$ and $\delta^{15}N$ of aquatic macrophytes in the arid zone of northwestern China. *Ecol. Evol.* **2021**, *11*, 3110–3119. [CrossRef]
32. Bahram, M.; Netherway, T.; Hildebrand, F.; Pritsch, K.; Drenkhan, R.; Loit, K.; Anslan, S.; Brok, P.; Tedersoo, L. Plant nutrient-acquisition strategies drive topsoil microbiome structure and function. *New Phytol.* **2020**, *227*, 1189–1199. [CrossRef] [PubMed]
33. Kohn, M.J. Carbon isotope compositions of terrestrial C_3 plants as indicators of (paleo)ecology and (paleo)climate. *Proc. Natl. Acad. Sci. USA* **2010**, *107*, 19691–19695. [CrossRef] [PubMed]
34. O'Leary, M.H. Carbon isotopes in photosynthesis. *Bioscience* **1988**, *38*, 328–336. [CrossRef]
35. Zhang, H.; Zeng, Z.X.; Zou, Z.G.; Zeng, F.P. Climate, life form and family jointly control variation of leaf traits. *Plants* **2019**, *8*, 286. [CrossRef] [PubMed]
36. Zhao, M.; Lian, J.Y.; Liu, X.R.; Liu, H.; Ye, Q. Comparison studies on water transport and nutrient acquisition of trees with different mycorrhiza types in subtropical forest. *J. Trop. Subtrop. Bot.* **2021**, *29*, 589–596.
37. Rodriguez-Caton, M.; Andreu-Hayles, L.; Morales, M.S.; Daux, V.; Christie, D.A.; Coopman, R.E.; Alvarez, C.; Rao, M.P.; Aliste, D.; Flores, F.; et al. Different climate sensitivity for radial growth, but uniform for tree-ring stable isotopes along an aridity gradient in *Polylepis tarapacana*, the world's highest elevation tree species. *Tree Physiol.* **2021**, *41*, 1353–1371. [CrossRef]
38. Zadworny, M.; Mucha, J.; Bagniewska-Zadworna, A.; Zytkowiak, R.; Maderek, E.; Danusevicius, D.; Oleksyn, J.; Wyka, T.P.; McCormack, M.l. Higher biomass partitioning to absorptive roots improves needle nutrition but does not alleviate stomatal limitation of northern Scots pine. *Global Chang. Biol.* **2021**, *27*, 3859–3869. [CrossRef]
39. Diefendorf, A.F.; Mueller, K.E.; Wing, S.L.; Koch, P.L.; Freeman, K.H. Global patterns in leaf δ13C discrimination and implications for studies of past and future climate. *Proc. Natl. Acad. Sci. USA* **2010**, *107*, 5738–5743. [CrossRef]
40. Averill, C.; Dietze, M.C.; Bhatnagar, J.M. Continental-scale nitrogen pollution is shifting forest mycorrhizal associations and soil carbon stocks. *Global Chang. Biol.* **2018**, *24*, 4544–4553. [CrossRef]
41. Lu, S.C.; Shi, Z.Y.; Zhang, M.G.; Yang, M.; Wang, X.G.; Xu, X.F. Discrepancies of leaf ash concentration in arbuscular and etco-mycorrhizal different plants associated arbuscular and etco-mycorrhizas and their response to climate change. *Ecol Environ.* **2020**, *29*, 35–40.
42. Rumman, R.; Atkin, O.K.; Bloomfield, K.J.; Eamus, D. Variation in bulk-leaf $\delta^{13}C$ discrimination, leaf traits and water-use efficiency-trait relationships along a continental-scale climate gradient in Australia. *Global Chang. Biol.* **2018**, *24*, 1186–1200. [CrossRef] [PubMed]

43. Zhao, N.; Meng, P.; He, Y.B.; Yu, X.X. Interaction of CO_2 concentrations and water stress in semiarid plants causes diverging response in instantaneous water use efficiency and carbon isotope composition. *Biogeosciences* **2017**, *14*, 3431–3444. [CrossRef]
44. Deveautour, C.; Donn, S.; Power, S.A.; Bennett, A.E.; Powell, J.R. Experimentally altered rainfall regimes and host root traits affect grassland arbuscular mycorrhizal fungal communities. *Mol. Ecol.* **2018**, *27*, 2152–2163. [CrossRef] [PubMed]
45. Yan, G.Y.; Han, S.J.; Zhou, M.X.; Sun, W.J.; Huang, B.B.; Wang, H.L.; Xing, Y.J.; Wang, Q.G. Variations in the natural $\delta^{13}C$ and $\delta^{15}N$ abundance of plants and soils under long-term N addition and precipitation reduction: Interpretation of C and N dynamics. *For. Ecosyst.* **2020**, *7*, 49. [CrossRef]
46. Liese, R.; Leuschner, C.; Meier, I.C. The effect of drought and season on root life span in temperate arbuscular mycorrhizal and ectomycorrhizal tree species. *J. Ecol.* **2019**, *107*, 2226–2239. [CrossRef]
47. Shi, Z.Y.; Zhang, K.; Miao, Y.F.; Wang, F.Y. Responses of net primary productivity to precipitation in forests dominated by different mycorrhizal types. *Bull. Soil Water Conserv.* **2014**, *34*, 14–19.
48. Al-Karaki, G.N.; Williams, M. Mycorrhizal mixtures affect the growth, nutrition, and physiological responses of soybean to water deficit. *Acta Physiol. Plant* **2021**, *43*, 75. [CrossRef]
49. Iqbal, M.T.; Ahmed, I.A.M.; Isik, M.; Sultana, F.; Ortas, I. Role of mycorrhizae inoculations on nutrient uptake in rice grown under aerobic and anaerobic water management. *J. Plant Nutr.* **2021**, *44*, 550–568. [CrossRef]
50. Bloomfield, K.J.; Prentice, I.C.; Cernusak, L.A.; Eamus, D.; Medlyn, B.E.; Rumman, R.; Wright, J.; Boer, M.M.; Cale, P.; Cleverly, J.; et al. The validity of optimal leaf traits modelled on environmental conditions. *New Phytol.* **2019**, *221*, 1409–1423. [CrossRef]
51. Bitterlich, M.; Franken, P.; Graefe, J. Atmospheric drought and low light impede mycorrhizal effects on leaf photosynthesis—A glasshouse study on tomato under naturally fluctuating environmental conditions. *Mycorrhiza* **2018**, *29*, 13–28. [CrossRef]
52. Gavito, M.E.; Jakobsen, I.; Mikkelsen, T.N.; Mora, F. Direct evidence for modulation of photosynthesis by an arbuscular mycorrhiza-induced carbon sink strength. *New Phytol.* **2019**, *223*, 896–907. [CrossRef] [PubMed]
53. Wang, X.Y.; Peng, L.Q.; Jin, Z.X. Effects of AMF inoculation on growth and photosynthetic physiological characteristics of *Sinocalycanthus chinensis* under conditions of simulated warming. *Acta Ecol. Sin.* **2016**, *36*, 5204–5214.

plants

Article

Root-Shoot Nutrient Dynamics of Huanglongbing-Affected Grapefruit Trees

Lukas M. Hallman [1], Davie M. Kadyampakeni [2], John-Paul Fox [1], Alan L. Wright [3] and Lorenzo Rossi [1,*]

1 Indian River Research and Education Center, Horticultural Sciences Department, Institute of Food and Agricultural Sciences, University of Florida, Fort Pierce, FL 34945, USA
2 Citrus Research and Education Center, Soil, Water and Ecosystem Sciences Department, Institute of Food and Agricultural Sciences, University of Florida, Lake Alfred, FL 33850, USA
3 Indian River Research and Education Center, Soil, Water and Ecosystem Sciences Department, Institute of Food and Agricultural Sciences, University of Florida, Fort Pierce, FL 34945, USA
* Correspondence: l.rossi@ufl.edu

Abstract: With huanglongbing (HLB) causing a reduction in fine root mass early in disease progression, HLB-affected trees have lower nutrient uptake capability. Questions regarding the uptake efficiency of certain fertilizer application methods have been raised. Therefore, the goals of this study are to determine if nutrient management methods impact nutrient translocation and identify where in the tree nutrients are translocated. Destructive nutrient and biomass analysis were conducted on field grown HLB-affected grapefruit trees (*Citrus* × *paradisi*) grafted on 'sour orange' (*Citrus* × *aurantium*) rootstock under different fertilizer application methods. Fertilizer was applied in the form of either 100% soluble granular fertilizer, controlled release fertilizer (CRF), or liquid fertilizer. After three years, the entire tree was removed from the grove, dissected into eight different components (feeder roots, lateral roots, structural roots, trunk, primary branches, secondary branches, twigs, and leaves), weighed, and then analyzed for nutrient contents. Overall, application methods showed differences in nutrient allocation in leaf, twig, and feeder root; however, no consistent pattern was observed. Additionally, leaf, twig, and feeder roots had higher amount of nutrients compared to the other tree components. This study showed that fertilization methods do impact nutrient contents in different components of HLB-affected trees. Further research should be conducted on the impact of different fertilizer application methods and rates on HLB-affected trees.

Keywords: citrus greening; *Citrus paradisi*; destructive analysis; nutrient translocation; fertilizer application methods

Citation: Hallman, L.M.; Kadyampakeni, D.M.; Fox, J.-P.; Wright, A.L.; Rossi, L. Root-Shoot Nutrient Dynamics of Huanglongbing-Affected Grapefruit Trees. *Plants* **2022**, *11*, 3226. https://doi.org/10.3390/plants11233226

Academic Editor: Ivana Puglisi

Received: 30 October 2022
Accepted: 24 November 2022
Published: 24 November 2022

1. Introduction

Proper nutrient management is integral to increase both the operational profitability and environmental sustainability of Florida's citrus industry. Citrus trees require 17 elements for optimum growth and production [1]; of which, nitrogen (N), phosphorus (P), potassium (K), magnesium (Mg), calcium (Ca), sulfur (S), iron (Fe), zinc (Zn), manganese (Mn), copper (Cu), and boron (B) are commonly supplied via inorganic fertilizers. If any essential element is deficient, tree growth, development, and yield is reduced [2–4]. Although fertilization is commonplace, supplying the optimum levels of nutrients can be challenging due to several factors such as edaphic conditions, environmental issues, and disease pressure [5,6].

Oftentimes sandy soils in citrus growing regions, such as those found in Florida, are low in natural fertility, resulting in required higher and more frequent fertilizer inputs [7–9]. This management method is not without risk since the over application of fertilizers can result in toxicity effects on the plants and nutrient leaching into groundwater [10,11].

Huanglongbing (HLB; Citrus greening) impacts nutrient uptake and translocation. This disease is caused by the phloem limited bacteria *Candidatus* Liberibacter asiaticus

(CLas) and vectored by the Asian citrus Psyllid *Diaphorina citri;* moreover, it is commonly associated with root and canopy dieback, poor fruit yield and quality, and tree death [12,13]. Research has also shown that HLB alters nutrient contents within affected trees [14,15]. For example, a study by da Silva et al. [16] found that HLB-affected sweet orange (*Citrus* × *aurantium*) trees had lower levels of N, Mg, and S in leaves and sap extracts. Similarly, a study by Shahzad et al. [17] found that HLB-affected sweet orange trees had lower leaf Ca, Mg, and S compared to healthy trees.

The studies described above clearly demonstrate that HLB changes how nutrients are taken up by and translocated within affected trees. Although much research has been focused on how different nutrient rates impact HLB-affected tree health [18–21], there is a lack of understanding regarding where in the trees these nutrients are translocated. For example, a previous study conducted by [22] analyzed nutrient contents on different components of sweet orange trees, but this was done with non HLB-affected trees.

In addition to applying the correct rate of fertilizers, the method in which fertilizers are applied can be equally important. In the Florida citrus industry, both liquid and solid granular fertilizers are readily available and commonly used [23]. Granular fertilizers can be formulated as 100% soluble or controlled release fertilizers (CRF). The nutrients in CRF have polymer coatings which better control nutrient release timing and rate compared to 100% soluble granular fertilizers [24,25]. Liquid fertilization is also popular since it can be integrated into existing irrigation systems. This integration allows for more precise control over fertilizer rate, timing, and location. The use of liquid fertilization has been shown to reduce nutrient leaching, increase tree growth rates, and increase yield [23,26]. Very few studies exist that compare the use of 100% soluble granular fertilizers, CRF, and liquid fertilization on HLB-affected grapefruits (*Citrus* × *paradisi*) grown on flatwood soils; furthermore, none of the existing studies conducted a full tree destructive biomass and nutrient content analysis.

As HLB continues to hinder Florida's citrus industry, it is necessary to create more accurate nutrient application guidelines to improve both HLB-affected tree health and profitability. Although nutrient application methods such as CRF and liquid treatments are likely to lead to higher nutrient uptake compared to 100% soluble granular fertilizers, there is a lack of published literature on nutrient dynamics in HLB-affected grapefruit. Therefore, the goal of this study was to determine if nutrient management methods impact nutrient translocation and to identify where nutrients are translocated within the tree.

2. Results

2.1. Macronutrient Concentration

Differences in leaf macronutrients were observed between application methods (Figure 1A). The CRF treatment led to 12.1% higher leaf N compared to the LW treatment, higher leaf Ca compared to both the L (27.5%) and LW (25.0%) treatment, and higher S compared to the L (18.25%) and LW (26.43%) treatments. Additionally, 23.8% higher leaf K was observed in the L treatment compared to the control.

In the twigs, the only difference observed was in Mg levels, where the L treatment led to 37.5% higher Mg compared to the control and 45.2% higher Mg compared to the CRF treatment (Figure 1B).

No differences were observed in macronutrient allocation between treatments in the secondary branches (Figure 1C); however, in the primary branches, K levels were 33.3% higher in the L treatment compared to the control treatment (Figure 1D). In the trunk, the L treatment had 29.0% higher N compared to the LW treatment.

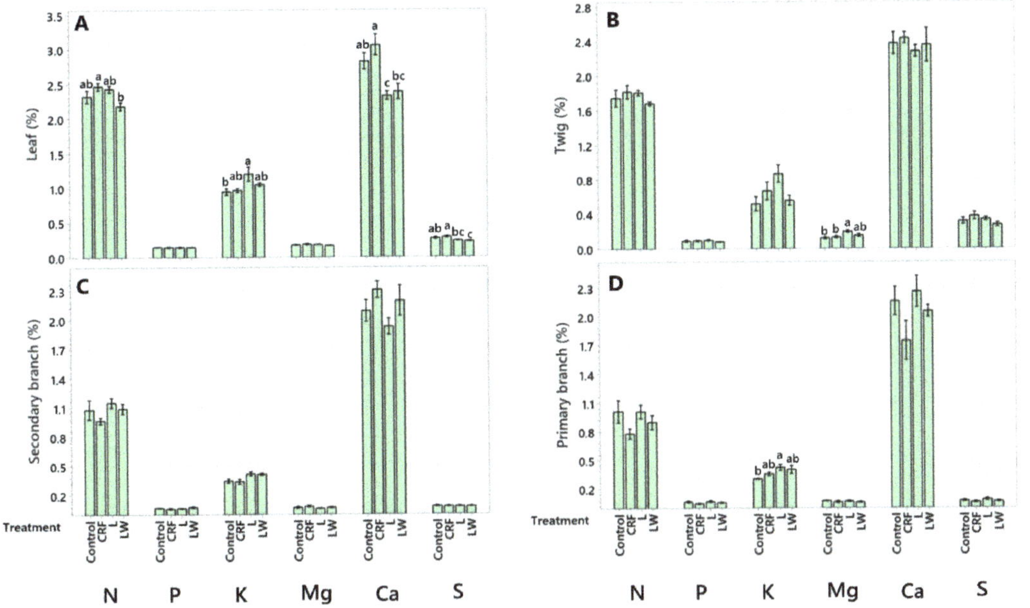

Figure 1. Macronutrient concentration (%) in leaf (**A**), twig (**B**), secondary branch (**C**), and primary branch (**D**). Six-year-old Huanglongbing-affected 'Ruby Red' grapefruit trees grafted on sour orange rootstock were used. Treatments consisted of two liquid (L and LW) and two granular fertilizers (Control and controlled-release fertilizer (CRF)). Treatments were applied three times a year (control and CRF), biweekly (L), or weekly (LW), for three years. A one-way variance (ANOVA) with a Tukey's honestly significant difference (HSD) test was used to determine significant differences between means. Lowercase letters (a, b, c) indicate statistically significant differences ($p \leq 0.05$). Data represents means (n = 4) ± standard error.

In the structural roots, differences in N and Ca levels were observed between treatments (Figure 2B). The L treatment led to 20% more N compared to the LW treatment. However, the control resulted in 23.1% more Ca compared to the L treatment and 18.8% more Ca compared to the LW treatment. No differences in macronutrients were observed between any of the treatments in the lateral roots (Figure 2C). In the feeder roots, the only difference between macronutrient concentrations was observed in P (Figure 2D). The control led to 19.4% more P compared to the CRF and LW treatments and 34.5% more P compared to the L treatment.

2.2. Micronutrient Concentration

In the leaves, B was 50.7% higher in the CRF treatment compared to the L treatment and 42.2% higher compared to the LW treatment (Figure 3A). No differences in micronutrient concentrations were detected between treatments in the twigs (Figure 3B), secondary branches (Figure 3C), primary branches (Figure 3D), trunk (Figure 4A), structural roots (Figure 4B), or the lateral roots (Figure 4C). Like the leaves, B in the feeder roots was 19.8% and 22.2% higher in the CRF treatment compared to the control and LW treatment, respectively (Figure 4D).

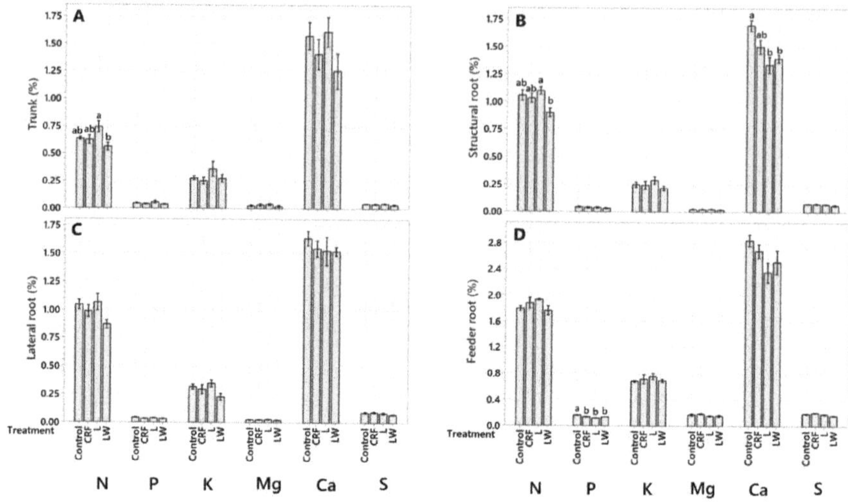

Figure 2. Macronutrient concentration (%) in trunk (**A**), structural root (**B**), lateral root (**C**), and feeder root (**D**). Six-year-old Huanglongbing-affected 'Ruby Red' grapefruit trees grafted on sour orange rootstock were used. Treatments consisted of two liquid (L and LW) and two granular fertilizers (Control and controlled-release fertilizer (CRF)). Treatments were applied three times a year (control and CRF), biweekly (L), or weekly (LW), for three years. A one-way analysis of variance (ANOVA) with a Tukey's honestly significant difference (HSD) test was used to determine significant differences between means. Lowercase letters (a, b) indicate statistically significant differences ($p \leq 0.05$). Data represents means (n = 4) ± standard error.

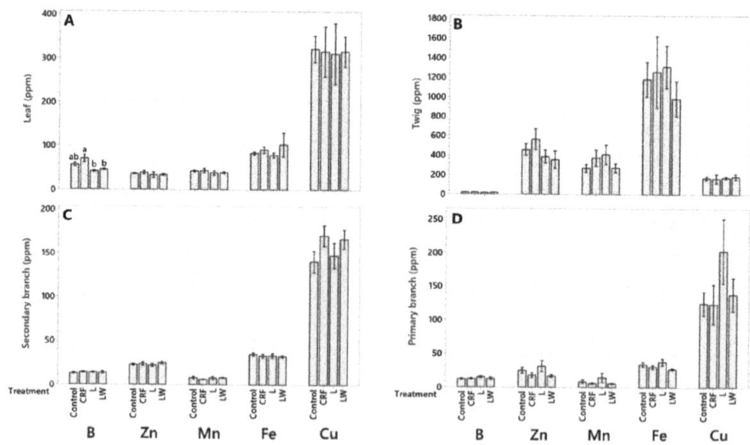

Figure 3. Micronutrient concentration (ppm) in leaf (**A**), twig (**B**), secondary branch (**C**), and primary branch (**D**). Six-year-old Huanglongbing-affected 'Ruby Red' grapefruit trees grafted on sour orange rootstock were used. Treatments consisted of two liquid (L and LW) and two granular fertilizers (Control and controlled-release fertilizer (CRF)). Treatments were applied three times a year (control and CRF), biweekly (L), or weekly (LW), for three years. A one-way variance (ANOVA) with a Tukey's honestly significant difference (HSD) test was used to determine significant differences between means. Lowercase letters (a, b) indicate statistically significant differences ($p \leq 0.05$). Data represents means (n = 4) ± standard error.

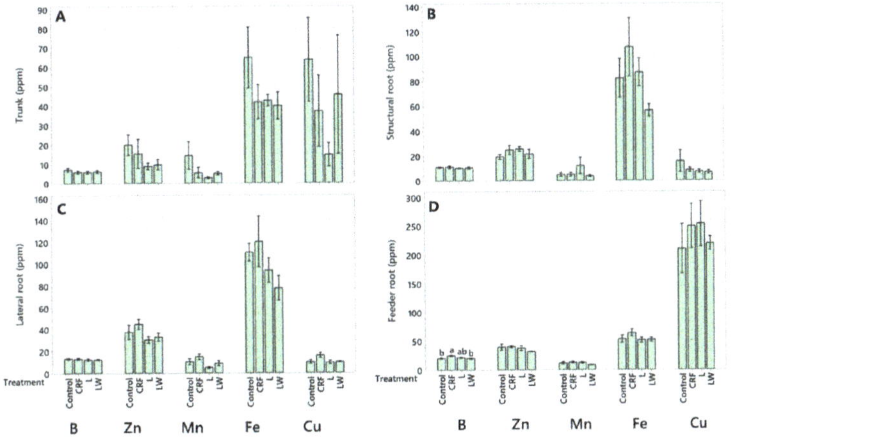

Figure 4. Micronutrient concentration (ppm) in trunk (**A**), structural root (**B**), lateral root (**C**), and feeder root (**D**). Six-year-old Huanglongbing-affected 'Ruby Red' grapefruit trees grafted on sour orange rootstock were used. Treatments consisted of two liquid (L and LW) and two granular fertilizers (Control and controlled-release fertilizer (CRF)). Treatments were applied three times a year (control and CRF), biweekly (L), or weekly (LW), for three years. A one-way variance (ANOVA) with a Tukey's honestly significant difference (HSD) test was used to determine significant differences between means. Lowercase letters (a, b) indicate statistically significant differences ($p \leq 0.05$). Data represents means (n = 4) ± standard error.

2.3. Total Tree Nutrient Content and Biomass

No differences in individual tree component (Figure 5) and total N, P, K, Mg, Ca, and S content were observed between any of the treatments (Tables 1 and 2). Additionally, no differences were observed in total plant N, P, K, Mg, Ca, and S between any of the treatments. Differences in macronutrient levels were observed between different plant components. The greatest amount of total N, P, K, Mg, Ca, and S tended to be found in the leaves, secondary branches, primary branches, and structural roots in all treatments (Tables 1 and 2).

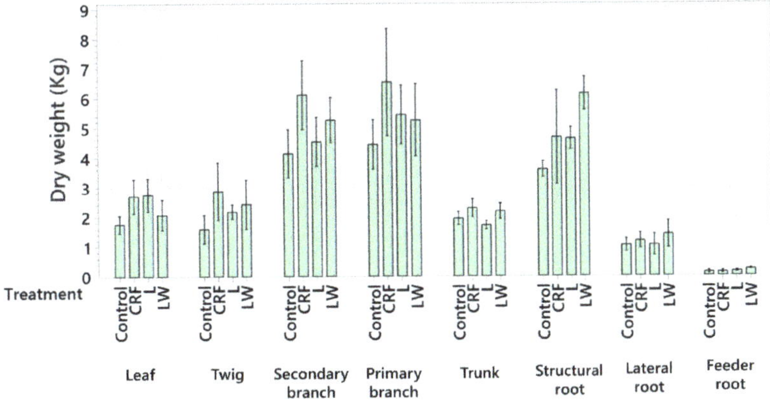

Figure 5. Dry weight (kg) of each tree component. Six-year-old Huanglongbing-affected 'Ruby Red' grapefruit trees grafted on sour orange rootstock were used. Treatments consisted of two liquid (L and LW) and two granular fertilizers (Control and controlled-release fertilizer (CRF)). Treatments were applied three times a year (control and CRF), biweekly (L), or weekly (LW), for three years. Data represents means (n = 4) ± standard error.

Table 1. Nitrogen (N), phosphorus (P), and potassium (K) contents per individual tree components and tree total. Six-year-old Huanglongbing -affected 'Ruby Red' grapefruit trees grafted on sour orange rootstock were used. Treatments consisted of two liquid (L and LW) and two granular fertilizers (Control and controlled-release fertilizer (CRF)). A Kruskal–Wallis test was used in combination with a Tukey's honestly significant difference (HSD) test to identify and test significant differences in total nutrients in individual tree components (differences signified by letters). Data represents means (n = 4) ± standard error.

	N g tree^{-1}			
Tree component	Control	CRF	L	LW
Leaf	40.64 ± 7.14 abc	67.51 ± 14.9 a	66.42 ± 13.31 a	45.35 ± 10.88 abc
Twig	26.81 ± 7.11 abcd	50.75 ± 15.27 abc	39.4 ± 4.89 ab	40.89 ± 13.86 abc
Secondary Branch	46.67 ± 13.03 ab	59.59 ± 12.84 ab	51.57 ± 8.57 a	56.72 ± 7.17 a
Primary Branch	47.43 ± 12.17 a	48.55 ± 9.36 abc	54.07 ± 8.53 a	48.38 ± 14.41 ab
Trunk	12.34 ± 1.51 bcd	14.34 ± 2.16 bc	12.67 ± 1.33 bc	12.40 ± 1.99 bc
Structural Root	38.31 ± 4.22 abc	46.38 ± 15.56 abc	50.68 ± 3.46 a	55.69 ± 6.78 a
Lateral Root	10.68 ± 2.22 cd	11.30 ± 2.22 bc	10.50 ± 2.87 bc	12.40 ± 4.37 bc
Feeder Root	2.15 ± 0.84 d	2.19 ± 0.76 c	2.88 ± 0.65 c	3.9 ± 0.24 c
Total	225.07 ± 1.81	300.64 ± 5.55	288.22 ± 5.17	275.77 ± 4.60
	P g tree^{-1}			
Tree component	Control	CRF	L	LW
Leaf	2.6 ± 0.40 ab	4.00 ± 0.87 a	4.11 ± 0.97 a	3.0 ± 0.76 a
Twig	1.34 ± 0.31 abc	2.78 ± 0.91 ab	2.2 ± 0.41 ab	1.98 ± 0.65 abc
Secondary Branch	2.83 ± 0.68 ab	3.87 1.18 a	2.95 ± 0.58 ab	3.91 ± 0.31 a
Primary Branch	3.14 ± 0.77 a	2.96 ± 0.62 ab	4.00 ± 1.09 a	2.94 ± 0.72 ab
Trunk	0.93 ± 0.16 bc	0.91 ± 0.15 ab	0.98± 0.24 b	0.81± 0.11 bc
Structural Root	1.64 ± 0.28 abc	1.90 ± 0.82 ab	1.88 ± 0.44 ab	2.02 ± 0.40 abc
Lateral Root	0.46 ± 0.09 c	0.42 ± 0.10 b	0.45 ± 0.20 b	0.47 ± 0.13 c
Feeder Root	0.2 ± 0.08 c	0.16 ± 0.05 b	0.18 ± 0.04 b	0.31 ± 0.03 c
Total	13.19 ± 0.23	17.03 ± 1.50	16.78 ± 1.45	15.53 ± 3.38
	K g tree^{-1}			
Tree component	Control	CRF	L	
Leaf	16.33 ± 2.55 a	26.19 ± 5.58 a	33.91 ± 8.82 a	
Twig	7.59 ± 2.23 bcd	19.17 ± 6.31 abc	19.4 ± 4.48 abc	
Secondary Branch	13.88 ± 1.90 ab	21.04 ± 4.92 ab	18.96 ± 3.55 ab	
Primary Branch	13.64 ± 2.67 abc	22.52 ± 5.22 ab	23.63 ± 5.69 abc	
Trunk	5.52 ± 0.95 cd	5.83 ± 1.06 bc	6.18 ± 1.16 bc	
Structural Root	9.05 ± 1.16 abcd	10.04 ± 3.64 abc	13.36 ± 1.63 bc	
Lateral Root	3.29 ± 0.66 d	3.33 ± 0.70 bc	3.59 ± 1.22 bc	
Feeder Root	0.83 ± 0.32 d	0.78 ± 0.25 c	1.17 ± 0.29 c	
Total	70.17 ± 0.58	108.96 ± 3.93	120.25 ± 4.01	

Table 2. Magnesium (Mg), calcium (Ca), and potassium (S) contents per individual components and tree total. Six-year-old Huanglongbing-affected 'Ruby Red' grapefruit trees grafted on sour orange rootstock were used. Treatments consisted of two liquid (L and LW) and two granular fertilizers (Control and controlled-release fertilizer (CRF)). A Kruskal–Wallis test was used in combination with a Tukey's honestly significant difference (HSD) test to identify and test significant differences in total nutrients in individual tree components (differences signified by letters). Data represents means (n = 4) ± standard error.

	Mg g tree^{-1}			
Tree component	Control	CRF	L	LW
Leaf	3.09 ± 0.50 a	5.17 ± 1.22 a	4.98 ± 0.97 a	3.43 ± 0.78 a
Twig	1.89 ± 0.59 abc	3.5 ± 0.91 abc	4.33 ± 0.68 a	3.59 ± 1.03 a
Secondary Branch	2.97 ± 0.71 ab	5.19 ± 1.07 a	2.79 ± 0.44 a	3.93 ± 0.87 a
Primary Branch	3.33 ± 0.60 a	3.66 ± 0.51 ab	3.65 ± 0.47 a	3.02 ± 0.67 ab
Trunk	0.57 ± 0.22 c	0.82 ± 0.30 bc	0.67 ± 0.16 bc	0.49 ± 0.19 b
Primary Root	0.90 ± 0.12 bc	1.10 ± 0.41 bc	1.19 ± 0.22 bc	1.24 ± 0.32 ab
Secondary Root	0.27 ± 0.04 c	0.31 ± 0.06 c	0.31 ± 0.11 c	0.31 ± 0.07 b
Feeder Root	0.22 ± 0.09 c	0.22 ± 0.07 c	0.23 ± 0.05 c	0.35 ± 0.03 b
Total	13.268 ± 0.33	20.01 ± 1.96	18.17 ± 1.90	16.40 ± 1.67
	Ca g tree^{-1}			
Tree component	Control	CRF	L	LW
Leaf	49.71 ± 9.34 abc	83.65 ± 18.45 abc	63.39 ± 12.60 bc	47.77 ± 9.53 bcd
Twig	36.42 ± 9.76 bc	71.09 ± 24.22 abcd	49.9 ± 7.06 bcd	54.23 ± 15.39 abcd
Secondary Branch	85.05 ± 14.59 a	138.67 ± 23.22 a	87.72 ± 17.63 ab	118.75 ± 25.01 a
Primary Branch	97.40 ± 19.79 a	103.83 ± 13.01 ab	122.57 ± 22.45 a	106.77 ± 23.41 ab
Trunk	30.46 ± 3.73 bc	32.32 ± 5.41 bcd	27.63 ± 3.29 cd	27.69 ± 5.44 cd
Structural Root	60.93 ± 4.34 ab	72.91 ± 24.80 abcd	62.56 ± 7.43 bc	85.98 ± 9.17 abc
Lateral Root	16.99 ± 3.68 bc	18.38 ± 4.14 cd	15.64 ± 5.04 cd	21.44 ± 7.05 cd
Feeder Root	3.54 ± 1.46 c	3.15 ± 1.06 d	3.34 ± 0.65 d	5.59 ± 0.61 d
Total	380.53 ± 1.24	524.05 ± 8.08	432.79 ± 7.60	468.24 ± 7.57
	S g tree^{-1}			
Tree component	Control	CRF	L	LW
Leaf	4.78 ± 0.87 a	7.96 ± 1.90 ab	6.59 ± 1.39 ab	4.64 ± 1.01 ab
Twig	4.92 ± 1.52 a	9.69 ± 2.31 a	7.42 ± 1.19 a	6.19 ± 1.72 a
Secondary Branch	3.63 ± 0.89 ab	5.05 ± 1.09 abc	3.67 ± 0.57 bc	4.39 ± 0.76 abc
Primary Branch	3.49 ± 0.87 ab	3.54 ± 0.44 bc	4.73 ± 1.02 ab	3.55 ± 1.05 abc
Trunk	0.93 ± 0.13 b	1.03 ± 0.17 c	0.81 ± 0.09 c	0.82 ± 0.15 bc
Structural Root	2.71 ± 0.27 ab	3.43 ± 1.19 bc	3.37 ± 0.34 bc	3.87 ± 0.63 abc
Lateral Root	0.92 ± 0.16 b	1.07 ± 0.20 c	0.88 ± 0.27 c	1.03 ± 0.30 abc
Feeder Root	0.22 ± 0.07 b	0.24 ± 0.08 c	0.27 ± 0.06 c	0.36 ± 0.04 c
Total	21.63 ± 0.53	32.04 ± 2.50	27.78 ± 2.12	24.89 ± 1.63

Total B, Zn, Mn, and Fe contents per individual tree components were consistent across all treatments except for Mn in the lateral roots (Table 3). The CRF trees had significantly more Mn in the lateral roots compared to the F treatment; however, no differences were observed in total tree micronutrients regardless of treatment (Tables 3 and 4).

Table 3. Boron (B), zinc (Zn), and manganese (Mn) contents per individual components and tree total. Six-year-old Huanglongbing-affected 'Ruby Red' grapefruit trees grafted on sour orange rootstock were used. Treatments consisted of two liquid (L and LW) and two granular fertilizers (Control and controlled-release fertilizer (CRF)). A Kruskal–Wallis test was used in combination with a Tukey's honestly significant difference (HSD)test to identify and test significant differences in total nutrients in individual tree components (differences signified by letters). A one-way analysis of variance (ANOVA) with a Tukey HSD test was used to determine significant differences between the means of between different fertilizer application methods (differences signified by *). Data represents means (n = 4) ± standard error.

		B mg kg^{-1}		
Tree component	Control	CRF	L	LW
Leaf	99.77 ± 20.85 a	199.19 ± 58.39 a	115.65 ± 24.16 a	92.58 ± 20.11 a
Twig	29.16 ± 9.14 bc	52.38 ± 17.25 b	35.68 ± 3.57 bcd	41.02 ± 12.78 abc
Secondary Branch	52.70 ± 8.74 b	85.73 ± 18.14 b	61.16 ± 9.01 bc	74.04 ± 17.64 ab
Primary Branch	55.58 ± 13.56 ab	77.05 ± 14.13 b	81.80 ± 14.40 ab	70.92 ± 26.73 abc
Trunk	14.00 ± 2.14 bc	13.77 ± 2.40 b	10.28 ± 1.43 d	13.69 ± 2.10 bc
Structural Root	40.42 ± 3.32 bc	56.37 ± 20.04 b	48.99 ± 5.16 bcd	66.23 ± 9.86 abc
Lateral Root	13.67 ± 2.58 bc	15.54 ± 3.43 b	12.51 ± 3.65 cd	17.44 ± 5.42 bc
Feeder Root	2.35 ± 0.82 c	3.09 ± 1.15 b	3.22 ± 0.73 d	4.43 ± 0.35 c
Total	307.70 ± 1.77	503.17 ± 9.18	369.32 ± 6.85	380.40 ± 5.79
		Zn mg kg^{-1}		
Tree component	Control	CRF	L	LW
Leaf	62.45 ± 8.47 b	109.01 ± 29.73 b	93.43 ± 23.72 b	71.35 ± 20.47 b
Twig	756.91 ± 254.33 a	1467.75 ± 403.15 a	815.33 ± 128.33 a	870.69 ± 342.09 a
Secondary Branch	92.43 ± 15.89 b	142.7 6± 30 31 b	96.53 ± 13.51 b	127.33 12.56 b
Primary Branch	113.91 ± 27.63 b	102.58 ± 11.14 b	169.62 ± 54.63 b	86.82 ± 24.17 b
Trunk	37.67 ± 9.94 b	36.44 ± 17.89 b	15.70 ± 3.66 b	21.72 ± 6.13 b
Structural Root	70.97 ± 9.83 b	127.51 ± 45.64 b	119.87 ± 14.28 b	131.17 ± 21.74 b
Lateral Root	34.87 ±c4.05 b	49.10 ± 7.81 b	28.78 ± 6.13 b	43.17 ± 10.13 b
Feeder Root	4.46 ± 1.39 b	5.05 ± 1.93 b	5.39 ± 1.13 b	7.08 ± 0.65 b
Total	1173.31 ± 7.36	2040.23 ± 31.68	1344.69 ± 21.45	1359.36 ± 22.08
		Mn mg kg^{-1}		
Tree component	Control	CRF	L	LW
Leaf	73.07 ± 11.10 b	121.10 ± 34.10 b	104.64 ± 22.53 b	80.04 ± 19.03 b
Twig	453.87 ± 157.22 a	931.26 ± 265.62 a	843.84 ± 157.15 a	686.03 ± 246.98 a
Secondary Branch	32.95 ± 11.28 b	35.23 ± 4.32 b	32.16 ± 7.24 b	39.76 ± 3.60 b
Primary Branch	38.26 ± 10.90 b	32.56 ± 1.24 b	79.98 ± 44.31 b	29.90 ± 9.83 b
Trunk	27.62 ± 13.46 b	12.91 ± 6.97 b	4.82 ± 1.13 b	11.43 ± 2.36 b
Structural Root	18.95 ± 5.85 b	23.25 ± 11.78 b	61.36 ± 36.51 b	23.47 ± 4.25 b
Lateral Root *	9.60 ± 2.30 b	16.26 ± 3.50 b	4.67 ± 0.69 b	9.98 ± 1.15 b
Feeder Root	1.26 ± 0.33 b	1.67 ± 0.66 b	1.75 ± 0.32 b	1.89 ± 0.17 b
Total	655.63 ± 6.34	1174.27 ± 29.55	1133.25 ± 26.47	882.53 ± 24.53

Table 4. Iron (Fe) and copper (Cu) contents per individual components and tree total. Six-year-old Huanglongbing-affected 'Ruby Red' grapefruit trees grafted on sour orange rootstock were used. Treatments consisted of two liquid (L and LW) and two granular fertilizers (Control and controlled-release fertilizer (CRF)). A Kruskal–Wallis test was used in combination with a Tukey's honestly significant difference (HSD) test to identify and test significant differences in total nutrients in individual tree components (differences signified by letters). Data represents means (n = 4) ± standard error.

		Fe mg kg^{-1}		
Tree component	Control	CRF	L	LW
Leaf	144.04 ± 21.07 b	251.31 ± 59.80 b	215.68 ± 40.79 b	218.14 ± 73.26 b
Twig	2060 ± 863.63 a	3305.2 ± 1182.37 a	2827.4 ± 507.11 a	2380.39 ± 818.77 a
Secondary Branch	140.61 ± 22.13 b	197.23 ± 33.46 b	149.03 ± 28.58 b	170.09 ± 25.40 b
Primary Branch	151.01 ± 28.07 b	204.10 ± 67.91 b	197.80 ± 35.09 b	137.59 ± 25.97 b
Trunk	122.70 ± 29.65 b	98.34 ± 24.10 b	74.21 ± 10.74 b	87.74 ±17.99 b
Structural Root	304.55 ± 77.35 b	473.65 ± 202.92 b	403.70 ± 57.21 b	342.15 ± 31.18 b
Lateral Root	115.10 ± 24.12 b	123.89 ± 14.00 b	88.32 ± 18.88 b	104.99 ± 29.36 b
Feeder Root	6.22 ± 2.07 b	7.34 ± 2.68 b	7.31 ± 1.53 b	11.49 ± 0.86 b
Total	3044.27 ± 16.42	4661.09 ± 47.00	3963.51 ± 43.75	3452.63 ± 39.62
		Cu mg kg^{-1}		
Tree component	Control	CRF	L	LW
Leaf	551.08 ± 77.48 a	908.36 ± 310.29 a	880.74 ± 249.09 a	660.71 ± 194.10 ab
Twig	260.7 ± 69.64 ab	425.37 ± 149.26 abc	377.82 ± 52.99 ab	377.4 ± 94.32 ab
Secondary Branch	574.28 ± 109.89 a	1000.56 ± 127.06 a	653.89 ± 114.90 ab	859.96 ± 102.49 a
Primary Branch	570.21 ± 131.41 a	663.72 ± 81.37 ab	1113.84 ± 346.38 a	790.24 ± 325.53 a
Trunk	123.32 ± 41.76 b	86.529 ± 44.81 bc	25.97 ± 11.06 b	98.32 ± 62.54 b
Structural Root	58.39 ± 32.95 b	42.88 ± 16.66 bc	35.77 ± 8.15 b	40.88 ± 3.62 b
Lateral Root	9.49 ± 1.34 b	17.98 ± 4.28 c	9.83 ± 3.34 b	13.64 ± 3.90 b
Feeder Root	22.64 ± 6.00 b	33.97 ± 14.97 c	35.89 ± 8.29 b	49.85 ± 7.69 b
Total	2170.15 ± 198.15	3179.4 ± 1649.66	3133.80 ± 1735.76	2891.04 ± 1520.87

Differences were observed in total micronutrient contents between plant components. Twigs consistently contained the largest amounts of Zn (61–72% of total plant Zn), Mn (69–79% of total Mn), and Fe (67–71% of total Fe) compared to all other plant components (Tables 3 and 4). Additionally, the leaves contained between 24–39% of the total tree B in all treatments, making the leaves the greatest pool of B in the entire tree (Table 3). Finally, no differences were observed in total tree dry weight between any treatments (Table 5).

Table 5. Total mean dry weight (DW) per treatment. Six-year-old HLB-affected 'Ruby Red' grapefruit trees grafted on sour orange rootstock were used. Treatments consisted of two liquid (L and LW) and two granular fertilizers (Control and controlled-release fertilizer (CRF)). Data represents means ± standard error.

Treatment	Total (DW)
	kg
Control	18.61 ± 1.81
CRF	26.47 ± 4.88
L	18.45 ± 2.74
LW	20.98 ± 3.99

3. Discussion

Overall, the application methods showed differences in uptake and translocation, particularly in the leaves and feeder roots. Regardless of the treatment, leaves and feeder roots contained higher concentrations of N and K compared to the other tree components sampled. The leaves also contained higher B concentration compared to any other organ. All other nutrient concentrations were consistent across components.

The CRF treatment had higher N, Ca, S, and B in the leaves compared to the other treatments, and these increases in leaf nutrient concentrations are consistent with the current literature. It is well established that CRF's outperform conventional granular fertilizers in nutrient uptake and fruit production [27–29]. In HLB-affected sweet orange trees, [30] found that CRF formulations resulted in exceptionally high yields. Although no differences in biomass were reported between any of the treatments, the higher macronutrient concentrations in the leaves of the CRF treatment supports the increased effectiveness of CRF's over conventional fertilizers. Research by [3] did show that a CRF led to higher plant biomass compared to fertigation and water soluble granular treatments; however, the study was conducted on 32-month-old sweet orange trees. The lack of differences observed in total biomass between treatments could be a result of the three-year time frame of this study. This may not have been long enough to detect changes due to fertilizer treatments.

The lower levels of N, Ca, S, and B found in the leaves of the LW treatment compared to the other treatments is important to note. The LW method was chosen as a treatment due to grower reports that supplying nutrients in smaller doses but at shorter intervals led to higher uptake [31]; however, this was not observed. Our results were consistent with those of [32,33] which found that increased fertilizer frequency led to no differences in growth parameters or leaf nutrient concentration of sweet orange trees. The lower uptake of leaf macronutrients in the LW treatment could be due to increased leaching compared to the other treatments. Liquid treatments can lead to a better response compared to granular fertilizers; however, if the grove experiences high rainfall and/or irrigation management is poor, leaching can occur [34]. Additionally, the trees used in the study were severely HLB-affected and had a depleted root system. Logically, a depleted root system will intercept less nutrients and thus the benefits of more frequent fertilization may not be realized.

Higher nutrient contents in the leaves and feeder roots were expected. This was similar to research conducted by [22], which showed that the leaves of 32-month old sweet orange trees had the highest content of total N compared to other tree components. Plants require large amounts of N due to its roles in amino acids, sugars, and proteins [7,35]. In citrus trees, large amounts of N are found in young tender tissues such as leaves [7]. Levels of K are often higher in the leaves due to its role in regulation of stomatal opening and closing [36]. Additionally, citrus fruits utilize large amount of K [3]. Although this study did not sample any fruits due to fruit drop before analysis, the higher contents of K observed in the leaves compared to other plant components could indicate that leaves are a reservoir for K export into fruit. Lastly, the higher levels of B in the leaves compared to other components are likely due to its role in photosynthesis and carbohydrate metabolism [37].

When considering the biomass of each component in relation to nutrient percentage, the leaves, twigs, and secondary branches contained most of the total macronutrients in the trees. Similar findings were reported by [38,39], which showed higher percentages of N in newer organs of citrus trees. The root system, particularly the feeder and lateral roots, contained the lowest macronutrient content. Although the feeder root nutrient samples tend to have higher levels of all nutrients compared to nutrient samples taken from the woody sections of the trees (trunk, branches, lateral and structural roots), the feeder root biomass is much lower compared to the other components. As a result, feeder roots often constitute the lowest amount of nutrients, particularly macronutrients, to the overall nutrient total of the trees.

No other investigators have conducted a destructive study coupled with nutrient analysis of field-grown HLB-affected grapefruit trees. Furthermore, the trees were sub sectioned into 8 different components: five above ground and three below ground components. This level of detail is challenging to obtain due to the inaccessibility of fruit bearing citrus trees that can be removed from the ground and dissected. This provides a greater amount of detailed data on the nutrient uptake and translocation inside the different components and the tree as a whole. Although the aforementioned peculiarities of our study represents strengths, some limitations also need to be acknowledged: (i) the lack of timepoints to compare different seasons and (ii) the method used to remove the trees from the field. Obtaining seasonal timepoints would have required much more resources and available trees to be excavated and dissected over time. This data would have allowed for better inferences regarding the impact of seasonality on nutrient dynamics in HLB-affected citrus trees. On the other hand, the method by which trees were extracted from the field allowed for a greater number of replications to be analyzed but may have resulted in a loss of fine roots. The utilization of a better and more gentle excavation method would have limited fine root loss at the expense of time and resources.

4. Materials and Methods

4.1. Site Description

A 3-year nutrition trial was conducted at the University of Florida, Institute of Food and Agricultural Sciences (UF/IFAS), Indian River Research and Education Center (IRREC) located in Fort Pierce, Florida, USA. Tree material consisted of 6-year-old field grown 'Ruby Red' grapefruit trees (*Citrus* × *paradisi*) grafted on 'sour orange' (*Citrus* × *aurantium*) rootstock (Figure 6). All trees examined were HLB-affected, confirmed by both Ct values and visual analysis.

Figure 6. Satellite image of the area in which the study was conducted. The experimental grove from which the grapefruit trees were excavated is highlighted in red. The grapefruit grove is located at the University of Florida, Institute of Food and Agricultural Sciences (UF/IFAS), Indian River Research and Education Center (IRREC) located in Fort Pierce, Florida, USA. Image was acquired from Google Maps on 16 November 2022.

The trees were grown on 1-m-high raised beds for drainage purposes and irrigated with 39.7 L per hour microjet sprinklers (Maxijet, Dundee, FL, USA). The grove soils were sandy Alfisols classified as loamy, siliceous, active, hyperthermic Arenic Glossaqualfs with less than 1% organic matter. Soil pH was 5.8 and cation exchange capacity (C.E.C.) was 3.5 cmol kg^{-1} [40].

Treatments consisted of four different fertilizer application methods (Control, controlled release fertilizer (CRF), liquid (L) and liquid weekly (LW). The control treatment was an industry standard 100% soluble dry granular fertilizer applied three times throughout the growing season. The CRF treatment was a granular fertilizer with 50% solubility and contained polymer coated nutrients. Like the control treatment, the CRF treatment was applied three times throughout the growing season. The L and LW were both liquid treatments applied at a rate of 11.36 L per tree using a ~1100-L single axle admire mobile spray tank (Chemical Containers, Inc., Lake Wales, FL, USA). The L treatment was applied bi-weekly throughout the growing season while the LW treatment was applied weekly. The same yearly amount of nutrients were applied to all treatments regardless of application method. These amounts were calculated using the current UF/IFAS recommendations [41].

4.2. Tree Excavation and Dissection

Trees were excavated in September 2021 using a John Deere 7030 tractor and root rake implement. The base of the tree trunk was securely grasped by the rake and slowly lifted from the soil. During the lifting process, the trees were gently shaken to assist in removal of sand from the root mass. Once removed from the soil, the entire tree was moved to a covered location for dissection.

Each tree was divided into eight different components: leaf, twig, secondary branch, primary branch, trunk, structural root, lateral root, and feeder root (Figure 7). All leaves, including both immature and fully expanded leaves, were collected by hand. Twigs were defined as the woody portions connected to the leaves. Secondary branches were defined as those directly connected to the twigs but not directly connected to the main trunk. The primary branches were those that connected directly to the main trunk and supported the secondary branches. The main trunk was defined as the central woody structure connecting the root mass to the branches.

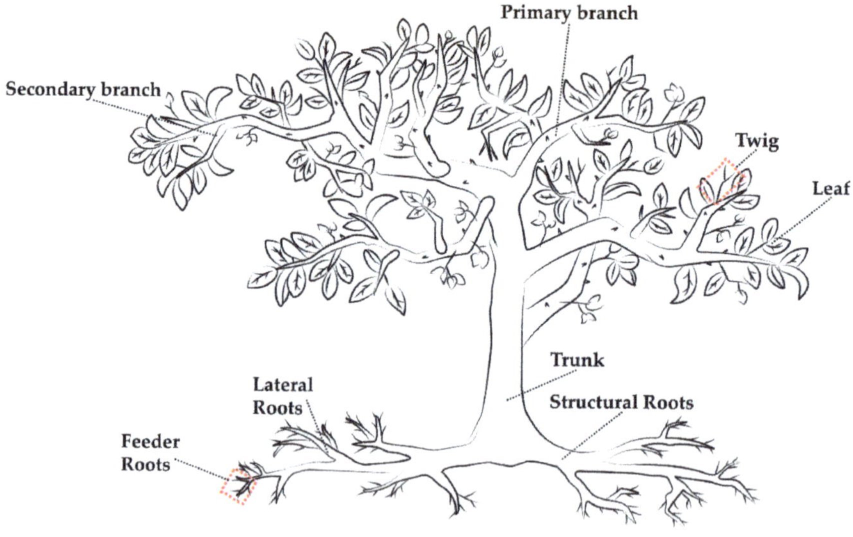

Figure 7. Schematic representation of a six-year-old grapefruit tree divided into eight different components: leaf, twig, secondary branch, primary branch, trunk, structural root, lateral root, and feeder root.

Roots were divided into structural, lateral, and feeder roots. The structural roots were directly connected to the trunk and ranged from 10 mm to 20 mm. Lateral roots connected the primary roots to the feeder roots and ranged from 2 mm to 10 mm. Feeder roots were defined as the non-woody roots less than 2 mm in diameter protruding from the lateral roots.

The different components were removed from one another using Felco 2 pruners (Les Geneveys-sur-Coffrane, Switzerland), Fiskars 28″ loppers (Helsinki, Finland), and a Ryobi 40V HP Brushless 14in. battery operated chainsaw (Fuchu, Hiroshima, Japan). Once the components were removed from the tree and separated from each other, the total weight of each component, as well as nutrient analysis, was conducted.

4.3. Tree Biomass and Mineral Analysis

The entire fresh weight of each component was collected and weighed using a digital field scale (Ohaus Corporation, Parsippany, NJ, USA). Subsamples were then collected from each component and weighed. The subsamples were then dried at 60 °C for 3 to 15 days depending on the component size. Once dried, the subsample dry weight was collected using an analytical scale (Sartorius AG, Göttingen, Germany). Dried subsamples were then ground to pass through 1.0 mm mesh screen and 5 mL of nitric acid (HNO_3) was added. Samples were then heated to 95 °C for 90 min and 4 mL of 30% Hydrogen Peroxide (H_2O_2) was added. After 20 min of cooling, 50 mL of deionized water was added to each sample. Analysis of N, P, K, Mg, Ca, S, B, Zn, Mn, and Fe concentration was conducted using inductively coupled argon plasma emission (ICP-MS) spectrophotometer (Spectro Ciros CCD, Fitzburg, MA, USA) [42].

4.4. Experimental Design and Statistical Analysis

The experiment was organized into a completely randomized design with split plot arrangement. Each treatment was replicated 4 times and each replicant consisted of 10 trees. One tree from each replication was randomly selected to be excavated and dissected. A one-way Analysis of Variance (ANOVA) was used to determine significant differences between fertilizer treatments. When differences were detected ($p < 0.05$), a Tukey's honestly significant difference (HSD) test was conducted. A Kruskal–Wallis test was used in combination with a Tukey's HSD test to identify and test significant differences in total nutrients in individual tree components. All statistical analyses were conducted using the software R with 'agricolae' package [43]. Figures were generated using the software Minitab 17 (Minitab, LCC, State College, PA, USA).

5. Conclusions

Although nutrient use has been extensively studied in citrus, the uptake and distribution of nutrients in field grown HLB-affected grapefruit trees is less understood. This was the first study to conduct destructive nutrient analysis on field grown HLB-affected grapefruit trees. The research above shows where in the trees nutrients are stored and how much of each given nutrient are present in the entire tree. Additionally, this study showed that fertilization methods do impact nutrient contents in different components of HLB-affected trees. With these results in mind, further research should be conducted on how foliar nutrient application methods impact nutrient allocation within HLB-affected trees. Future studies should include internal movements of nutrients and tree destructive analysis at multiple time points throughout the year to account for seasonal variations. These continued evaluations of nutrient uptake and allocation could improve the efficiency of nutrient management programs in the age of HLB.

Author Contributions: Conceptualization, L.R. and D.M.K.; methodology, D.M.K., L.R. and A.L.W.; statistical analysis, L.M.H.; resources, L.R. and D.M.K.; writing—original draft preparation, L.M.H.; writing—review and editing, L.M.H., D.M.K., J.-P.F., L.R. and A.L.W.; data visualization, L.R. and L.M.H.; supervision, L.R.; funding acquisition, L.R. and D.M.K. All authors have read and agreed to the published version of the manuscript.

Funding: This research was funded by the Citrus Research and Development Foundation project #18-042C "Development of Root Nutrient and Fertilization Guidelines for Huanglongbing (HLB)-Affected Orange and Grapefruit". This work was also supported by the U.S. Department of Agriculture National Institute of Food and Agriculture, Hatch project #FLA-IRC-005743.

Institutional Review Board Statement: Not applicable.

Informed Consent Statement: Not applicable.

Data Availability Statement: Raw data will be available by requesting them via email to l.rossi@ufl.edu.

Acknowledgments: The authors are grateful to Jacob Lang, Felix Palencia and Laura Muschweck for assistance in treatment application and data collection. Lastly, the authors would like to thank H. Thomas James, III and Randy G. Burton for their management of the UF/IFAS Indian River Research and Education Center experimental citrus grove.

Conflicts of Interest: The authors declare no conflict of interest. The funders had no role in the design of the study; in the collection, analyses, or interpretation of data; in the writing of the manuscript; or in the decision to publish the results.

References

1. Obreza, T.A.; Tucker, D.P.H. Nutrient Management. In *Florida Citrus: A Comprehensive Guide*; Bird, C.J., Ed.; Cooperative Extension Service, University of Florida, University of Florida Institute of Food and Agricultural Sciences: Gainesville, FL, USA, 2006; pp. 167–182.
2. Srivastava, A.K.; Singh, S. Diagnosis of Nutrient Constraints in Citrus Orchards of Humid Tropical India. *J. Plant Nutr.* **2006**, *29*, 1061–1076. [CrossRef]
3. Alva, A.K.; Mattos, D.; Paramasivam, S.; Patil, B.; Dou, H.; Sajwan, K.S. Potassium Management for Optimizing Citrus Production and Quality. *Int. J. Fruit Sci.* **2006**, *6*, 3–43. [CrossRef]
4. Mattos-Jr, D.; Kadyampakeni, D.M.; da Silva, J.R.; Vashisth, T.; Boaretto, R.M. Reciprocal effects of huanglongbing infection and nutritional status of citrus trees: A review. *Trop. Plant Pathol.* **2020**, *45*, 586–596. [CrossRef]
5. Bungau, S.; Behl, T.; Aleya, L.; Bourgeade, P.; Aloui-Sossé, B.; Purza, A.L.; Abid, A.; Samuel, A.D. Expatiating the impact of anthropogenic aspects and climatic factors on long-term soil monitoring and management. *Environ. Sci. Pollut. Res.* **2021**, *28*, 30528–30550. [CrossRef]
6. Gitea, M.A.; Gitea, D.; Tit, D.M.; Purza, L.; Samuel, A.D.; Bungau, S.; Badea, G.E.; Aleya, L. Orchard management under the effects of climate change: Implications for apple, plum, and almond growing. *Environ. Sci. Pollut. Res.* **2019**, *26*, 9908–9915. [CrossRef]
7. Kadyampakeni, D.M.; Chinyukwi, T. Are macronutrients and micronutrients therapeutic for restoring performance of trees affected by citrus greening? A discussion of current practices and future research opportunities. *J. Plant Nutr.* **2021**, *44*, 2949–2969. [CrossRef]
8. Hendricks, G.S.; Shukla, S. Water and Nitrogen Management Effects on Water and Nitrogen Fluxes in Florida Flatwoods. *J. Environ. Qual.* **2011**, *40*, 1844–1856. [CrossRef]
9. Obreza, T.A.; Collins, M.E. Common Soils Used for Citrus Production in Florida. 2002. Available online: https://ufdcimages.uflib.ufl.edu/IR/00/00/31/34/00001/SS40300.pdf (accessed on 18 October 2022).
10. Obreza, T.A.; Schumann, A. Keeping Water and Nutrients in the Florida Citrus Tree Root Zone. *HortTechnology* **2010**, *20*, 67–73. [CrossRef]
11. Huang, J.-H.; Cai, Z.-J.; Wen, S.-X.; Guo, P.; Ye, X.; Lin, G.-Z.; Chen, L.-S. Effects of boron toxicity on root and leaf anatomy in two Citrus species differing in boron tolerance. *Trees* **2014**, *28*, 1653–1666. [CrossRef]
12. Etxeberria, E.; Gonzalez, P.; Achor, D.; Albrigo, G. Anatomical distribution of abnormally high levels of starch in HLB-affected Valencia orange trees. *Physiol. Mol. Plant Pathol.* **2009**, *74*, 76–83. [CrossRef]
13. Gottwald, T.R. Current Epidemiological Understanding of Citrus Huanglongbing. *Annu. Rev. Phytopathol.* **2010**, *48*, 119–139. [CrossRef] [PubMed]
14. Johnson, E.G.; Wu, J.; Bright, D.B.; Graham, J.H. Association of '*Candidatus* Liberibacter asiaticus' root infection, but not phloem plugging with root loss on huanglongbing-affected trees prior to appearance of foliar symptoms. *Plant Pathol.* **2013**, *63*, 290–298. [CrossRef]
15. Graham, J.; Gottwald, T.; Irey, M. Balancing Resources for Management of Root Health in HLB Affected Groves. 2012. Available online: https://citrusagents.ifas.ufl.edu/media/crecifasufledu/citrus-agents/growers-institutes/2012/2012-Graham.pdf (accessed on 18 October 2022).
16. da Silva, J.R.; De Alvarenga, F.V.; Boaretto, R.M.; Lopes, J.R.S.; Quaggio, J.A.; Coletta-Filho, H.; Mattos, D. Following the effects of micronutrient supply in HLB-infected trees: Plant responses and '*Candidatus* Liberibacter asiaticus' acquisition by the Asian citrus psyllid. *Trop. Plant Pathol.* **2020**, *45*, 597–610. [CrossRef]
17. Shahzad, F.; Chun, C.; Schumann, A.; Vashisth, T. Nutrient Uptake in Huanglongbing-affected Sweet Orange: Transcriptomic and Physiological Analysis. *J. Am. Soc. Hortic. Sci.* **2020**, *145*, 349–362. [CrossRef]
18. Hallman, L.M.; Kadyampakeni, D.M.; Ferrarezi, R.S.; Wright, A.L.; Ritenour, M.A.; Johnson, E.G.; Rossi, L. Impact of Ground Applied Micronutrients on Root Growth and Fruit Yield of Severely Huanglongbing-Affected Grapefruit Trees. *Horticulturae* **2022**, *8*, 763. [CrossRef]
19. Esteves, E.; Kadyampakeni, D.M.; Zambon, F.; Ferrarezi, R.S.; Maltais-Landry, G. Magnesium fertilization has a greater impact on soil and leaf nutrient concentrations than nitrogen or calcium fertilization in Florida orange production. *Nutr. Cycl. Agroecosyst.* **2022**, *122*, 73–87. [CrossRef]

20. Ferrarezi, R.S.; Jani, A.D.; James, H.T.; Gil, C.; Ritenour, M.A.; Wright, A.L. Sweet Orange Orchard Architecture Design, Fertilizer, and Irrigation Management Strategies under Huanglongbing-endemic Conditions in the Indian River Citrus District. *HortScience* **2020**, *55*, 2028–2036. [CrossRef]
21. Phuyal, D.; Nogueira, T.A.R.; Jani, A.D.; Kadyampakeni, D.M.; Morgan, K.T.; Ferrarezi, R.S. 'Ray Ruby' Grapefruit Affected by Huanglongbing II. Planting Density, Soil, and Foliar Nutrient Management. *HortScience* **2020**, *55*, 1420–1432. [CrossRef]
22. Alva, A.K.; Fares, A.; Dou, H. Managing Citrus Trees to Optimize Dry Mass and Nutrient Partitioning. *J. Plant Nutr.* **2003**, *26*, 1541–1559. [CrossRef]
23. Zekri, M.; Schumann, A.W.; Vashisth, T.; Kadyampakeni, D.M.; Morgan, K.T.; Boman, B.; Obreza, T. *2019–2020 Florida Citrus Production Guide: Fertilizer Application Methods*; University of Florida George A Smathers Libraries: Gainesville, FL, USA, 2019.
24. Sempeho, S.I.; Kim, H.T.; Mubofu, E.; Hilonga, A. Meticulous Overview on the Controlled Release Fertilizers. *Adv. Chem.* **2014**, *2014*, 363071. [CrossRef]
25. Dou, H.; Alva, A.K. Nitrogen uptake and growth of two citrus rootstock seedlings in a sandy soil receiving different controlled-release fertilizer sources. *Biol. Fertil. Soils* **1998**, *26*, 169–172. [CrossRef]
26. Schumann, A. Optimizing Citrus Fertigation. 2014. Available online: https://citrusagents.ifas.ufl.edu/media/crecifasufledu/citrus-agents/growers-institutes/2014/Schumann--Citrus-Institute-2014V2.pdf (accessed on 18 October 2022).
27. El Bey, N.; Aounallah, M.K.; Chammam, M.R.; Bettaieb, T. Effect of controlled-release fertilizers (CRF) on vegetative growth, nutritional status, fruit yield and quality of 'Maltaise Ballerin' orange trees. *Plant Physiol. Rep.* **2021**, *26*, 699–708. [CrossRef]
28. Gao, Y.; Song, X.; Liu, K.; Li, T.; Zheng, W.; Wang, Y.; Liu, Z.; Zhang, M.; Chen, Q.; Li, Z.; et al. Mixture of controlled-release and conventional urea fertilizer application changed soil aggregate stability, humic acid molecular composition, and maize nitrogen uptake. *Sci. Total Environ.* **2021**, *789*, 147778. [CrossRef] [PubMed]
29. Bettaga, N.; Ben Mimoun, M. Effects of controlled-release fertilizer on fruit yield and quality of clementine citrus trees. *Acta Hortic.* **2010**, *868*, 429–432. [CrossRef]
30. Vashisth, T.; Grosser, J. Comparison of Controlled Release Fertilizer (CRF) for Newly Planted Sweet Orange Trees under Huanglongbing Prevalent Conditions. *J. Hortic.* **2018**, *5*, 244. [CrossRef]
31. Singerman, A. *Cost of Producing Processed Oranges in Southwest Florida in 2017/18*; Citrus Research and Education Center, University of Florida, IFAS: Lake Alfred, FL, USA, 2019.
32. Willis, L.E.; Davies, F.S.; Graetz, D.A. Fertigation and Growth of Young 'Hamlin' Orange Trees in Florida. *HortScience* **1991**, *26*, 106–109. [CrossRef]
33. Syvertsen, J.; Jifon, J. Frequent fertigation does not affect citrus tree growth, fruit yield, nitrogen uptake, and leaching losses. *Annu. Meet. Fla. State Hortic. Soc.* **2001**, *114*, 88–93.
34. Obreza, T.A.; Sartain, J.B. Improving Nitrogen and Phosphorus Fertilizer Use Efficiency for Florida's Horticultural Crops. *HortTechnology* **2010**, *20*, 23–33. [CrossRef]
35. de Bang, T.C.; Husted, S.; Laursen, K.H.; Persson, D.P.; Schjoerring, J.K. The molecular–physiological functions of mineral macronutrients and their consequences for deficiency symptoms in plants. *New Phytol.* **2020**, *229*, 2446–2469. [CrossRef] [PubMed]
36. Humble, G.D.; Hsiao, T.C. Light-dependent Influx and Efflux of Potassium of Guard Cells during Stomatal Opening and Closing. *Plant Physiol.* **1970**, *46*, 483–487. [CrossRef]
37. Chen, M.; Mishra, S.; Heckathorn, S.A.; Frantz, J.M.; Krause, C. Proteomic analysis of Arabidopsis thaliana leaves in response to acute boron deficiency and toxicity reveals effects on photosynthesis, carbohydrate metabolism, and protein synthesis. *J. Plant Physiol.* **2014**, *171*, 235–242. [CrossRef] [PubMed]
38. Martínez-Alcántara, B.; Quiñones, A.; Primo-Millo, E.; Legaz, F. Nitrogen remobilization response to current supply in young citrus trees. *Plant Soil* **2011**, *342*, 433–443. [CrossRef]
39. Feigenbaum, S.; Bielorai, H.; Erner, Y.; Dasberg, S. The fate of ^{15}N labeled nitrogen applied to mature citrus trees. *Plant Soil* **1987**, *97*, 179–187. [CrossRef]
40. USDA. *Supplement to the Soil Survey of Osceola County Area, Florida*; United States Department of Agriculture: Washington, DC, USA, 2011; p. 35.
41. Morgan, K.T.; Kadyampakeni, D.M.; Zekri, M.; Schumann, A.W.; Vashisth, T.; Obreza, T.A. 2020–2021 Florida Citrus Production Guide: Nutrition Management for Citrus Trees. 2020. Available online: https://www.researchgate.net/publication/344061579_2020-2021_Florida_Citrus_Production_Guide_Nutrition_Management_for_Citrus_Trees (accessed on 18 October 2022).
42. Isaac, R.A.; Johnson, W.C. Elemental Analysis of Plant Tissue by Plasma Emission Spectroscopy: Collaborative Study. *J. AOAC Int.* **1985**, *68*, 499–505. [CrossRef]
43. R Development Core Team. *R: A Language and Environment for Statistical Computing*; R Development Core Team: Vienna, Austria, 2022.

plants

MDPI

Article

Plant Growth Stimulators Improve Two Wheat Cultivars Salt-Tolerance: Insights into Their Physiological and Nutritional Responses

Neveen B. Talaat * and Alaa M. A. Hanafy

Department of Plant Physiology, Faculty of Agriculture, Cairo University, Giza 12613, Egypt
* Correspondence: neveen.talaat@agr.cu.edu.eg

Abstract: Spermine (SPM) and salicylic acid (SA), plant growth stimulators, are involved in various biological processes and responses to environmental cues in plants. However, the function of their combined treatment on wheat salt tolerance is unclear. In this study, wheat (*Triticum aestivum* L. cvs. Shandawel 1 and Sids 14) plants were grown under non-saline and saline (6.0 and 12.0 dS m^{-1}) conditions and were foliar sprayed with 100 mgL^{-1} SA and/or 30 mgL^{-1} SPM. Exogenously applied SA and/or SPM relieved the adverse effects caused by salt stress and significantly improved wheat growth and production by inducing higher photosynthetic pigment (chlorophyll *a*, chlorophyll *b*, carotenoids) content, nutrient (N, P, K^+, Ca^{2+}, Mg^{2+}, Fe, Zn, Cu) acquisition, ionic (K^+/Na^+, Ca^{2+}/Na^+, Mg^{2+}/Na^+) homeostatics, osmolyte (soluble sugars, free amino acids, proline, glycinebetaine) accumulation, protein content, along with significantly lower Na^+ accumulation and chlorophyll *a*/*b* ratio. The best response was registered with SA and SPM combined treatment, especially in Shandawel 1. This study highlighted the recovery impact of SA and SPM combined treatment on salinity-damaged wheat plants. The newly discovered data demonstrate that this treatment significantly improved the photosynthetic pigment content, mineral homeostasis, and osmoprotector solutes buildup in salinity-damaged wheat plants. Therefore, it can be a better strategy for ameliorating salt toxicity in sustainable agricultural systems.

Keywords: photosynthetic pigments; ionic balance; osmotic adjustment; salicylic acid; salt tolerance; spermine; wheat (*Triticum aestivum* L.)

Citation: Talaat, N.B.; Hanafy, A.M.A. Plant Growth Stimulators Improve Two Wheat Cultivars Salt-Tolerance: Insights into Their Physiological and Nutritional Responses. *Plants* **2022**, *11*, 3198. https://doi.org/10.3390/plants11233198

Academic Editor: Lorenzo Rossi

Received: 20 September 2022
Accepted: 20 November 2022
Published: 22 November 2022

1. Introduction

Soil salinization is one of the most important environmental hazards that inhibits plant growth and development, causing significant yield losses. Approximately 50% of irrigated lands are suffered from the deleterious impact of salt stress [1]. Salinity affects plant growth through osmotic effect, nutritional imbalance, ionic toxicity, and oxidative stress [2]. The absorption of high concentrations of Na^+ in cells limits K^+ uptake, resulting in Na^+ toxicity and nutrient imbalance during salt stress [1]. In order to establish the defense, plants restrict the accumulation of salt ions in the cytosol and enhance the synthesize of osmolytes [3–5]. Osmolyte biosynthesis is critical for maintaining osmotic potential, metabolic activity, and water uptake under saline conditions [6]. Additionally, enzymatic and non-enzymatic antioxidants can scavenge effectively the excessive reactive oxygen species (ROS) generated during salt stress [7–11]. Ionic status inside the plant cell is also very important for plant salt tolerance because the excess of salt ions in the cytoplasm disrupts ion homeostasis, inhibits plant growth, and affects water transport [7,12]. Excessive soil Na^+ transported to the above ground organs disturbs intracellular ionic homeostasis in plants, damages photosynthetic membrane structure, promotes chlorophyll degradation, reduces photosynthetic efficiency, affects cytosolic enzyme activity, and inhibits cellular metabolism, all of which result in restrained plant growth and development [7,9,11]. The ability of plants to maintain a high cytosolic K^+/Na^+ ratio under ion toxicity is another salt tolerance mechanism [3,7].

Many plant growth regulators have been used to improve the salt tolerance of crops such as salicylic acid (SA). SA is a natural phenolic compound regulating plant growth and development under both normal and stressful conditions [13,14]. It has been reported that pretreatment with exogenous SA under saline environments prevents the chlorophyll degradation and increases the photosynthetic efficiency [13,15], reduces the ion toxicity, maintains the osmotic potential [7,13,16], stimulates the antioxidant enzymes, and scavenges the ROS in plants [10,15], as well as regulates the cellular signaling, senescence, and overall cellular redox homeostasis [15,16]. Moreover, an exogenous application of SA prevents lowering of indole acetic acid and cytokinin levels in salt-stressed wheat plants, thereby maintaining cell division and elongation processes in apical meristem of the roots resulting in an increase in growth and productivity of plants [17]. Previous report provides evidence that SA could alleviate salt stress through accelerating plant growth and improving leaf physiological processes [18]. In addition, exogenous SA balanced the osmotic potential and lowered the osmotic damage to the plasma membrane by mediating the accumulation profile of ions, such as Na^+, K^+, and Ca^{2+}, as well as compatible metabolites, such as proline and soluble sugar [19]. Exogenous application of SA to wheat plants exposed to salt stress alleviated the deleterious effects of salinity by improving the photosynthetic pigment content and inducing the accumulation of certain osmolytes, such as soluble sugars, proline, and glycinebetaine [13]. A similar trend has been found in baby corn under salt stress in response to SA seed priming, which has been attributed to an increased ion content and an enhanced accumulation of certain osmolytes, such as proline [20]. Evidence also shows a promising role of SA foliar application against soil salinization in wheat plants through its effects on the content of polyamines [14]. Although there have been a lot of investigations study the effect of SA under salt stress, the physiological and biochemical mechanisms of SA in regulating wheat response to saline environments are still not fully understood.

Another class of biomolecules involved in plant stress response is composed of polyamines (PAs), which are low molecular weight aliphatic cations that are ubiquitous cellular components [21]. In plants, the major PAs—putrescine, spermidine, and spermine—have been shown to be involved in many aspects of plant growth and development, such as organogenesis, embryogenesis, flower initiation and development, leaf senescence, fruit development and ripening, as well as abiotic and biotic plant stress responses [21–23]. They can also act as anti-senescence and anti-stress agents due to their acid neutralizing and antioxidant properties, as well as for their membrane stabilizing abilities [24]. Among PAs, emphasis has been placed on the unique role of spermine (SPM, tetraamine) in the regulation of various defensive processes in plants. It has been described as a 'potent plant defense activator with broad-spectrum protective effects' when used exogenously [25]. The protective role of SPM in plants under salt stress has been reported. For example, the exogenously applied SPM alleviated the adverse effect of saline environments on plant development via maintaining the osmotic adjustment, protecting the structure and function of photosynthetic apparatus, maintaining the cationic-anionic stability, reducing the ethylene production, enhancing the protein content, modulating the endogenous phytohormone level, and inducing the organic solutes accumulation [23,26,27]. Additionally, previous study provides evidence that the SPM application could maintain significantly higher root vitality, leaf relative water content, photosynthesis, water use efficiency, osmolytes accumulation, K^+/Na^+ ratio, and antioxidant enzyme activity, as well as lower osmotic potential, Na^+ accumulation, and oxidative damage in salt-stressed creeping bentgrass plants [28]. Recently, it has been shown that exogenous SPM treatment alleviated the inhibition of maize plant growth and productivity under stressful conditions by improving the antioxidant enzyme activity, antioxidant molecules content, and cell membrane integrity [29]. Although several researchers have studied the effects of exogenous SPM on enhancing plant abiotic stress tolerance, not much is known about the mechanism of SPM associated with wheat salt tolerance.

Among various food crops, wheat (*Triticum aestivum* L.) is a staple food in 30% of the world. However, it faces severe losses in its productivity due to salt stress [30,31]. Development of salt-tolerant wheat cultivars adapted to saline conditions is the most effective and economic strategy to combat this detrimental phenomenon. Exogenous treatment with plant defense elicitors, such as SA and SPM, is the best way to enhance plant salt tolerance. Some studies have looked into the effect of a single application of SA or SPM on plants exposed to salt stress. However, no one has looked into the impact of their combined treatment on wheat salt tolerance. To fill this gap, as a first investigation, we carried out this study to investigate the ameliorative impact of SA and SPM dual application to two wheat cultivars (*Triticum aestivum* L. cvs. Shandawel 1 and Sids 14) grown in salty soils through measuring the growth, productivity, photosynthetic pigment (chlorophyll *a*, chlorophyll *b*, carotenoids, total pigments) content, chlorophyll *a*/*b* ratio, nutrient (N, P, K^+, Na^+, Ca^{2+}, Mg^{2+}, Fe, Zn, Cu) acquisition, ionic (K^+/Na^+, Ca^{2+}/Na^+, Mg^{2+}/Na^+) homeostatic, osmoprotector solutes (total soluble sugars, total free amino acids, proline, glycinebetaine) accumulation, as well as grain carbohydrate and protein content. Accordingly, it has been hypothesized that the foliar application of SA and SPM combined treatment can help wheat plants to overcome the deleterious effects of salt stress on plant growth and production by enhancing photosynthetic pigment content, maintaining ionic homeostasis, adjusting osmotic balance, and increasing grain carbohydrate and protein content. This research provides novel insights into the mechanisms of SA- and SPM-mediated amelioration of salt stress on wheat.

2. Results

2.1. Foliar Applications of SA and/or SPM Ameliorate the Growth and Yield Reduction Caused by Salinity

Salinity is one of the major constraints in wheat growth, development, and production. To investigate the ameliorative effect of exogenous SA and/or SPM on the growth and productivity of two wheat cultivars (Shandawel 1 and Sids 14) grown under saline circumstances, we measured the changes of leaf area, shoot dry weight, root dry weight, grain number, and grain yield in the SA- and/or SPM-treated plants subjected to non-saline and saline conditions.

Salt stress (6.0 and 12.0 dS m^{-1}) has a major impact on wheat development. It suppressed wheat growth, resulting in a sharp reduction in the leaf area, shoot dry weight, and root dry weight, in both cultivars. These deleterious effects of saline conditions were prominent for cv. Sids 14 compared to cv. Shandawel 1. On the contrary, foliar applications of SA and/or SPM significantly mitigated this reduction compared with the salt-stressed plants that had not received any supplementations (Figure 1a–c). The combined treatment of SA and SPM yielded the best results, especially in Shandawel 1. It significantly ($p < 0.05$) elevated the leaf area (30.7%, 42.2%, and 55.0%), shoot dry weight (27.4%, 36.4%, and 49.8%), and root dry weight (27.7%, 38.0%, and 50.0%), in Shandawel 1 wheat plants subjected to 0.1, 6.0, and 12.0 dS m^{-1} salinity levels, respectively, compared with untreated plants.

In addition, as shown in Figure 2a,b, saline environments significantly reduced the productivity of cultivars Shandawel 1 and Sids 14 in terms of grain number and grain yield. These detrimental impacts of salt stress were prominent for cv. Sids 14 compared to cv. Shandawel 1. On the contrary, foliage applications of SA and/or SPM significantly alleviated the salt toxicity and attenuated the inhibitory impact of salt on these parameters. The best response was registered with SA and SPM combined treatment, especially in Shandawel 1. When compared to untreated plants, co-application of SA and SPM significantly ($p < 0.05$) increased the grain number by 44.8%, 49.6%, and 60.0% and grain yield by 40.0%, 47.5%, and 59.5% in Shandawel 1 wheat plants at 0.1, 6.0, and 12.0 dS m^{-1} salinity levels, respectively.

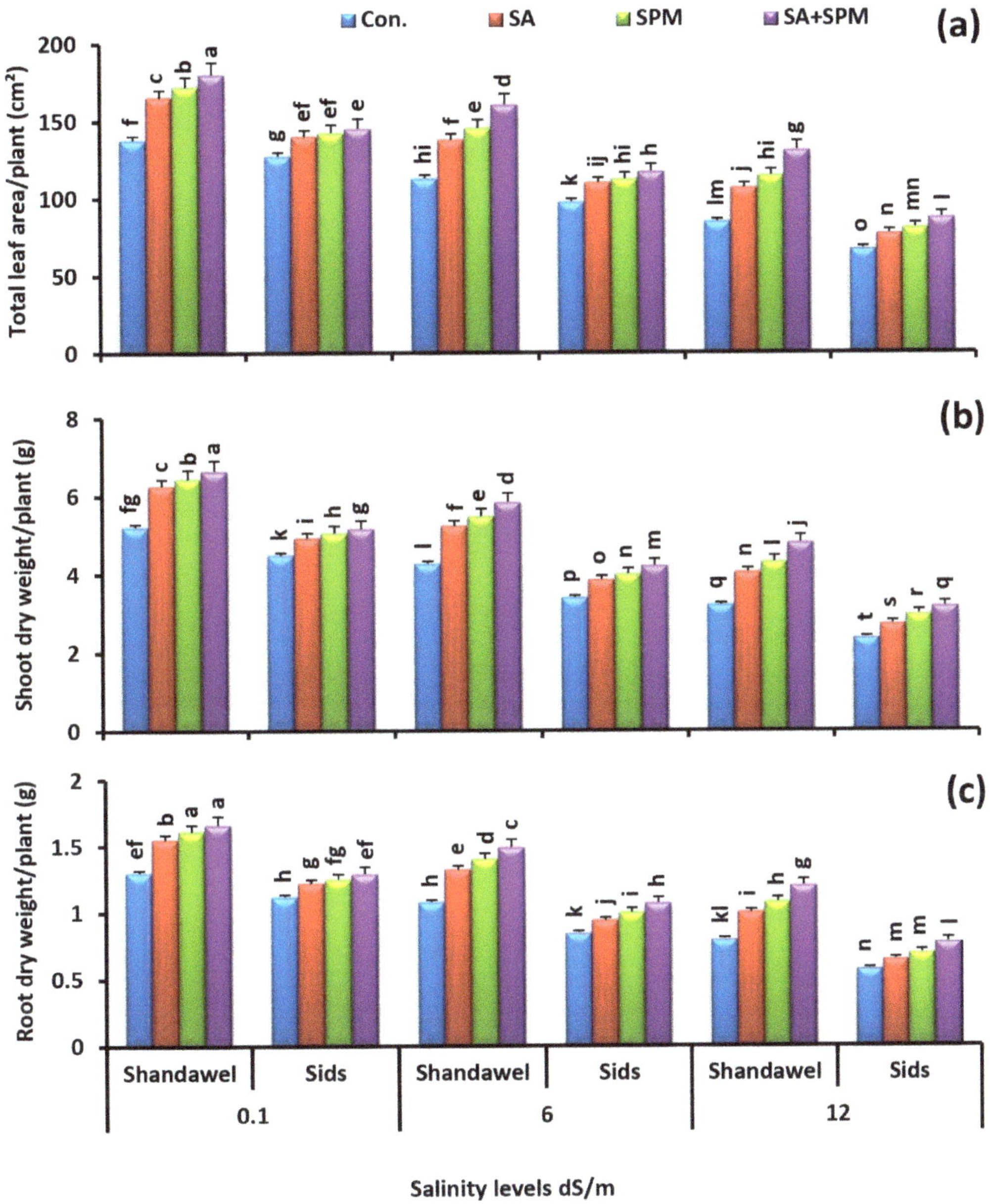

Figure 1. Influence of salicylic acid (SA; 100 mgL^{-1}) and/or spermine (SPM; 30 mgL^{-1}) on the (**a**) total leaf area, (**b**) shoot dry weight, and (**c**) root dry weight of two wheat cultivars (Shandawel 1 and Sids 14) grown under 0.1, 6, and 12 dS m^{-1} salinity levels. Data are mean of four replicates (n = 4). Different letters indicate significant differences at ($p < 0.05$) level, according to Duncan's test.

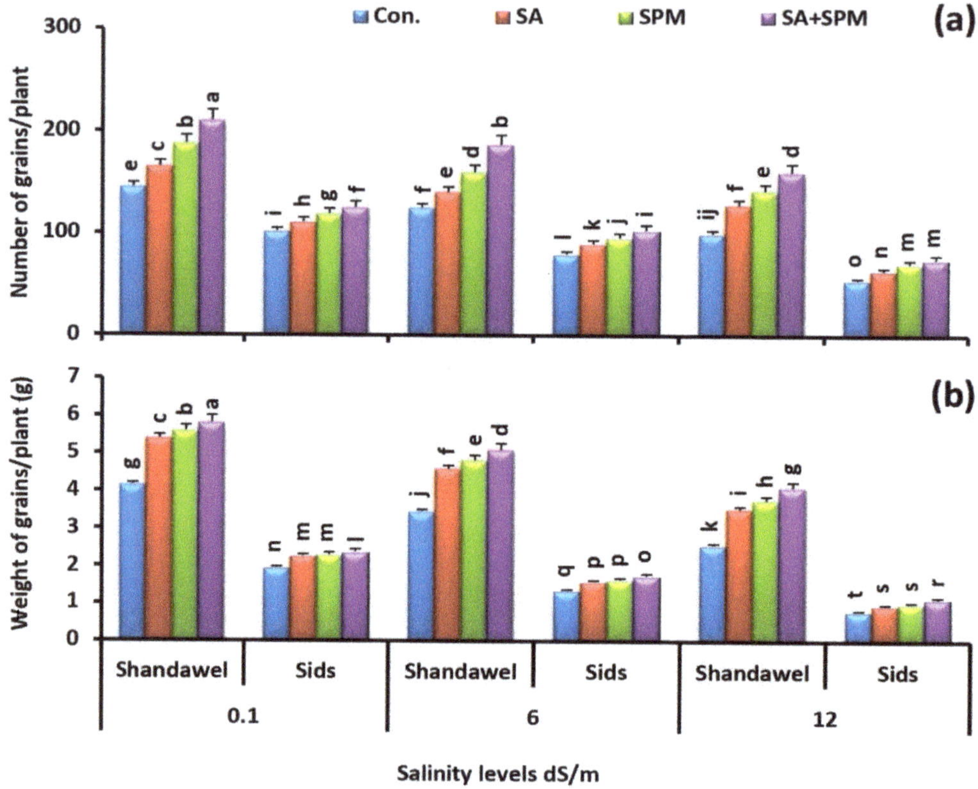

Figure 2. Influence of salicylic acid (SA; 100 mgL^{-1}) and/or spermine (SPM; 30 mgL^{-1}) on the (**a**) number of grains $plant^{-1}$ and (**b**) weight of grains $plant^{-1}$ of two wheat cultivars (Shandawel 1 and Sids 14) grown under 0.1, 6, and 12 dS m^{-1} salinity levels. Data are mean of four replicates (n = 4). Different letters indicate significant differences at ($p < 0.05$) level according to Duncan's test.

2.2. *SA and/or SPM Treatments Enhance Photosynthetic Pigment Concentration in Salt-Stressed Wheat Plants*

One of the most serious consequences of saline conditions is the decrease in the amount of photosynthetic pigments. To better understand whether SA and/or SPM foliage applications alleviate the damage of salt stress to the photosynthetic pigments, we measured the concentration of chlorophyll *a*, chlorophyll *b*, carotenoids, and total pigments along with the ratio of chlorophyll *a*/*b* in leaves of treated plants grown under non-saline and saline conditions.

Soil salinization sharply decreased the concentration of chlorophyll *a*, chlorophyll *b*, carotenoids, and total pigments in the leaves of the two wheat cultivars (Shandawel 1 and Sids 14). The deleterious effect of salt stress was prominent for cv. Sids 14 compared to cv. Shandawel 1. However, exogenously-applied SA and/or SPM minimized the detrimental effect of saline environments and increased the photosynthetic pigment level in the absence and presence of salt stress (Figure 3a–d). Co-application of SA and SPM had the greatest ameliorative effect, especially in Shandawel 1. When compared to untreated plants, SA and SPM combined treatment significantly ($p < 0.05$) increased the concentration of chlorophyll *a* by 16.7%, 26.0%, and 62.8%, chlorophyll *b* by 30.0%, 48.9%, and 114.9%, carotenoids by 36.0%, 56.1%, and 93.8%, and total pigments by 24.7%, 38.9%, and 83.6% in Shandawel 1 wheat plants at 0.1, 6.0, and 12.0 dS m^{-1} salinity levels, respectively.

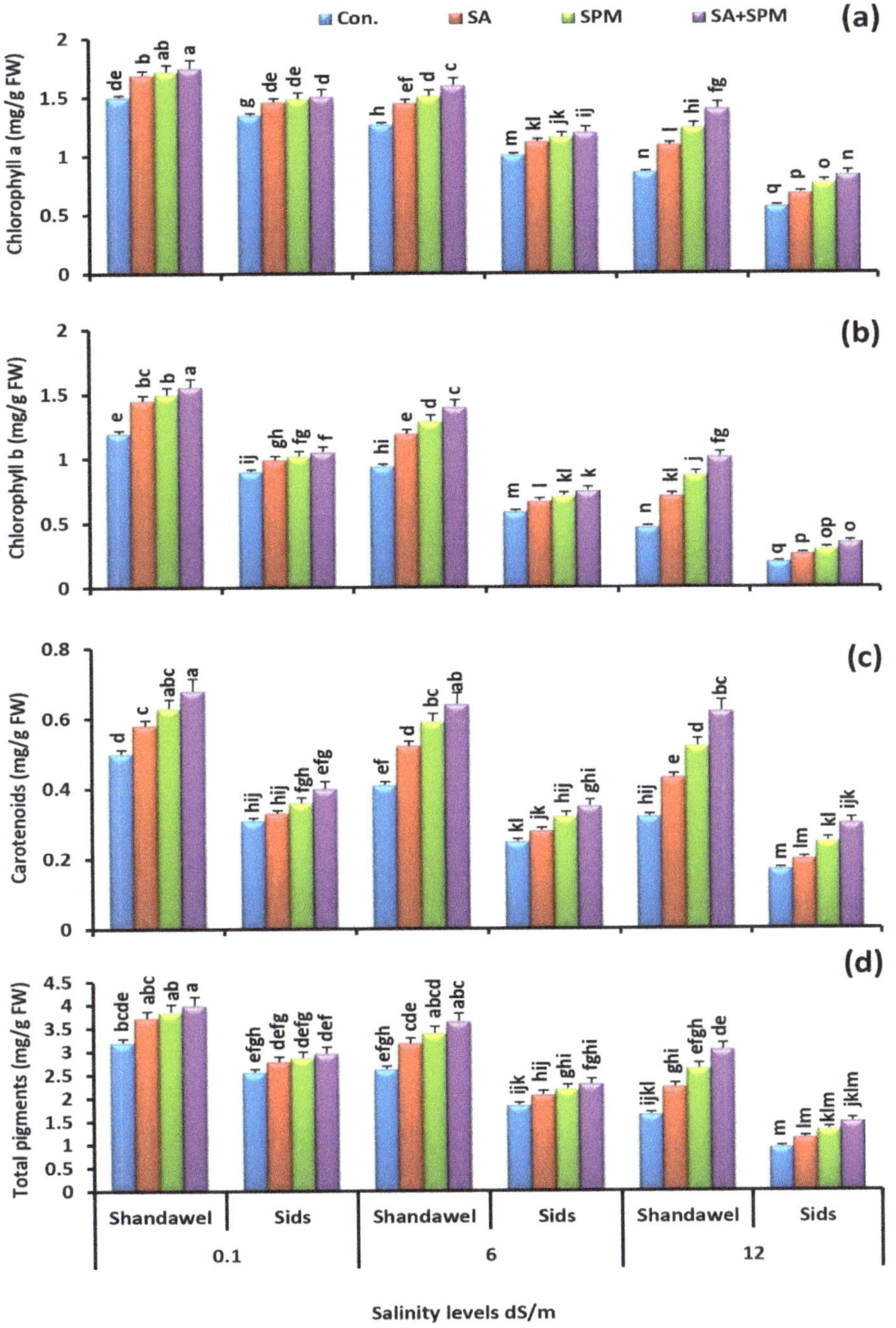

Figure 3. Influence of salicylic acid (SA; 100 mgL^{-1}) and/or spermine (SPM; 30 mgL^{-1}) on the concentration of (**a**) chlorophyll *a*, (**b**) chlorophyll *b*, (**c**) carotenoids, and (**d**) total pigments in leaves of two wheat cultivars (Shandawel 1 and Sids 14) grown under 0.1, 6, and 12 dS m^{-1} salinity levels. Data are mean of four replicates (n = 4). Different letters indicate significant differences at ($p < 0.05$) level according to Duncan's test.

Furthermore, salt stress treatments (6.0 and 12.0 dS m^{-1}) increased the chlorophyll *a*/*b* ratio in the leaves of both cultivars. On the contrary, comparing to the untreated stressed plants, the foliar treatments with SA and/or SPM significantly reduced the chlorophyll *a*/*b* ratio in the leaves of salt-stressed plants (Figure 4). The combined treatment of SA and SPM yielded the best results, especially in Shandawel 1. It significantly ($p < 0.05$) decreased the chlorophyll *a*/*b* ratio by 10.4%, 15.6%, and 24.0% in Shandawel 1 wheat plants subjected to 0.1, 6.0, and 12.0 dS m^{-1} salinity levels, respectively, compared with untreated plants.

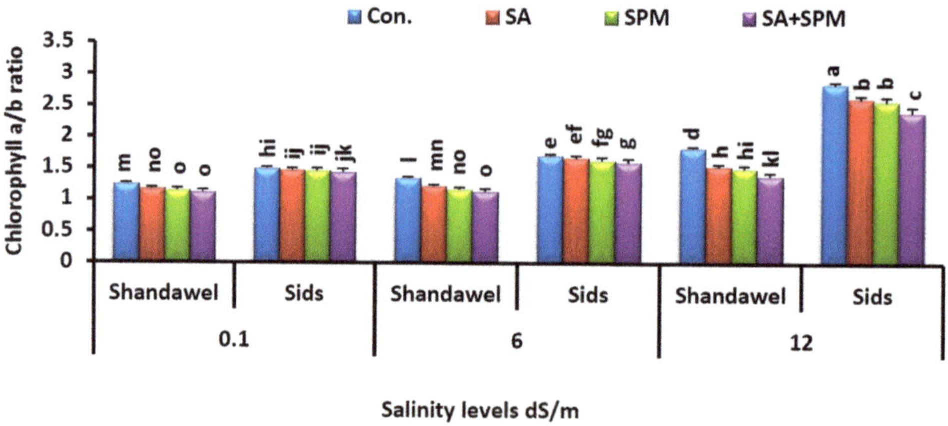

Figure 4. Influence of salicylic acid (SA; 100 mgL^{-1}) and/or spermine (SPM; 30 mgL^{-1}) on the chlorophyll *a*/*b* ratio in leaves of two wheat cultivars (Shandawel 1 and Sids 14) grown under 0.1, 6, and 12 dS m^{-1} salinity levels. Data are mean of four replicates (n = 4). Different letters indicate significant differences at ($p < 0.05$) level, according to Duncan's test.

2.3. *Spraying of SA and/or SPM Maintain Ionic Balance in Salt-Stressed Wheat Plants*

Ion toxicity is an important factor triggering severe damage to the growing plants under saline conditions. For exploring the influence of exogenous SA and/or SPM treatments on ions accumulation, we measured the accumulation profile of N, P, K^+, Na^+, Ca^{2+}, Mg^{2+}, Fe, Zn, and Cu in the grains of two wheat cultivars (Shandawel 1 and Sids 14) grown under non-saline and saline circumstances and sprayed with SA and/or SPM.

Under saline conditions, the concentrations of N, P, and K^+ were negatively affected. Moreover, this effect was more pronounced at high salinity level in both cultivars. Conversely, exogenous SA and/or SPM applications alleviated the deleterious injures of salt stress on mineral acquisition and increased the N, P, and K^+ concentrations, especially in Shandawel 1 (Figure 5a–c). Co-application of SA and SPM yielded the best response and significantly ($p < 0.05$) increased the N concentration (23.5%, 34.9%, and 53.9%), P concentration (36.4%, 42.1%, and 53.3%), and K^+ concentration (38.1%, 50.0%, and 60.0%) in grains of Shandawel 1 plants subjected to 0.1, 6.0, and 12.0 dS m^{-1} salinity levels, respectively, compared with untreated plants.

Moreover, the Na^+, Ca^{2+}, and Mg^{2+} concentrations were increased by increasing the salt doses in the soil. However, the treatments with SA and/or SPM decreased the accumulation of Na^+ while increasing the accumulation of Ca^{2+} and Mg^{2+} in the wheat grains (Figure 6a–c). The best response was registered with SA and SPM combined treatment, especially in Shandawel 1. At 6.0 and 12.0 dS m^{-1} salinity levels, the combination treatment significantly ($p < 0.05$) reduced Na^+ values in grains of treated Shandawel 1 plants by 22.0% and 31.3%, respectively, compared to untreated plants.

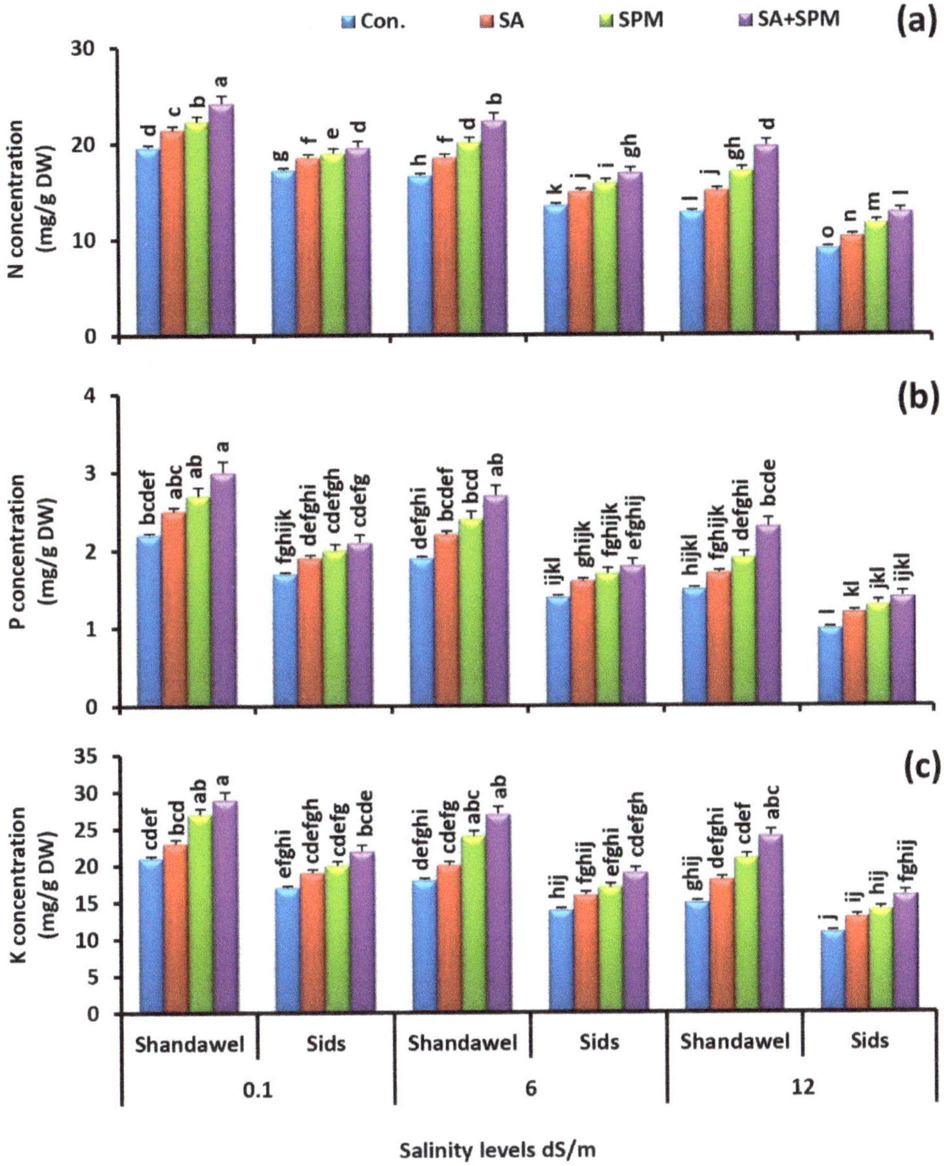

Figure 5. Influence of salicylic acid (SA; 100 mgL^{-1}) and/or spermine (SPM; 30 mgL^{-1}) on the concentration of (**a**) nitrogen, (**b**) phosphorus, and (**c**) potassium in grains of two wheat cultivars (Shandawel 1 and Sids 14) grown under 0.1, 6, and 12 dS m^{-1} salinity levels. Data are mean of four replicates (n = 4). Different letters indicate significant differences at ($p < 0.05$) level according to Duncan's test.

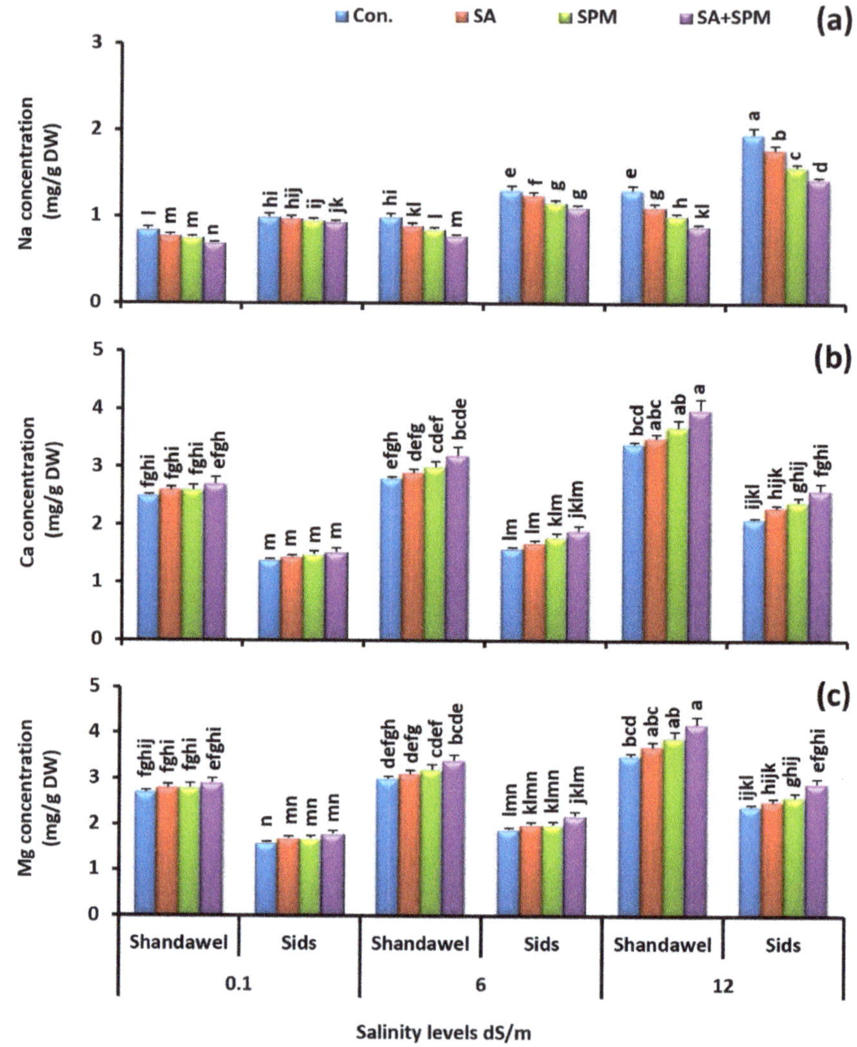

Figure 6. Influence of salicylic acid (SA; 100 mgL^{-1}) and/or spermine (SPM; 30 mgL^{-1}) on the concentration of (**a**) sodium, (**b**) calcium, and (**c**) magnesium in grains of two wheat cultivars (Shandawel 1 and Sids 14) grown under 0.1, 6, and 12 dS m^{-1} salinity levels. Data are mean of four replicates (n = 4). Different letters indicate significant differences at ($p < 0.05$) level, according to Duncan's test.

Additionally, under soil salinization, the concentrations of Fe, Zn, and Cu were sharply decreased with increasing the salinity level. Conversely, foliar SA and/or SPM treatments ameliorated the deleterious injures of saline conditions on mineral acquisition and increased the Fe, Zn, and Cu concentrations, especially in Shandawel 1 (Figure 7a–c). Co-application of SA and SPM yielded the best response and significantly ($p < 0.05$) increased the Fe concentration by 31.1%, 35.8%, and 47.6%; Zn concentration by 23.5%, 34.5%, and 47.8%; and Cu concentration by 23.2%, 28.0%, and 37.5%, in grains of Shandawel 1 plants grown under 0.1, 6.0, and 12.0 dS m^{-1} salinity levels, respectively, compared with untreated plants.

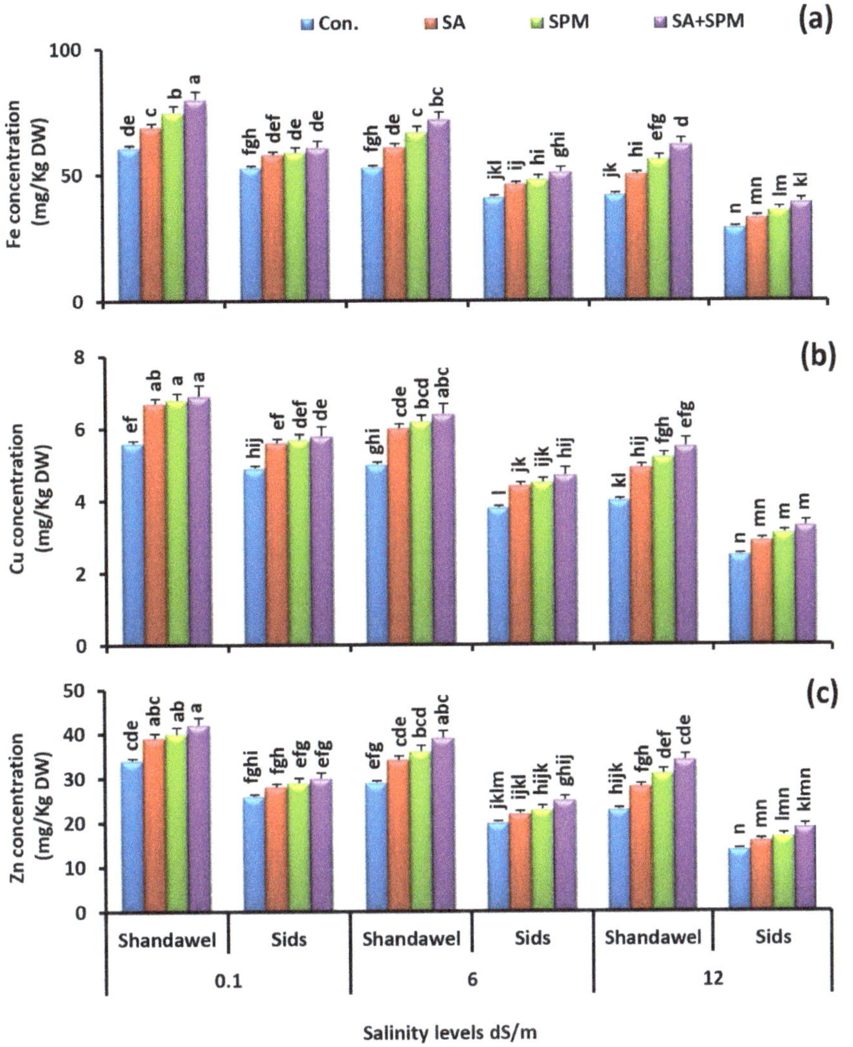

Figure 7. Influence of salicylic acid (SA; 100 mgL^{-1}) and/or spermine (SPM; 30 mgL^{-1}) on the concentration of (**a**) iron, (**b**) copper, and (**c**) zinc in grains of two wheat cultivars (Shandawel 1 and Sids 14) grown under 0.1, 6, and 12 dS m^{-1} salinity levels. Data are mean of four replicates (n = 4). Different letters indicate significant differences at ($p < 0.05$) level, according to Duncan's test.

2.4. Exogenously Applied SA and/or SPM Preserve K^+/Na^+, Ca^{2+}/Na^+, and Mg^{2+}/Na^+ Ratios under Saline Conditions

Salinity stress altered the ionic composition in wheat grains. To investigate whether the salt tolerance conferred by exogenous SA and/or SPM treatments, we quantified the K^+/Na^+, Ca^{2+}/Na^+, and Mg^{2+}/Na^+ ratios of two wheat cultivars (Shandawel 1 and Sids 14). Our findings showed that the K^+/Na^+, Ca^{2+}/Na^+, and Mg^{2+}/Na^+ ratios in the grains of salt-stressed plants were considerably lower than those in the grains of unstressed ones. On the contrary, SA and/or SPM applications sharply increased the ratios of K^+/Na^+, Ca^{2+}/Na^+, and Mg^{2+}/Na^+ in grains of stressed treated plants comparing to the stressed untreated ones (Figure 8a–c). Co-application of SA and SPM had the greatest ameliorative effect, especially in Shandawel 1. In comparison to values of untreated plants at salinity levels of 0.1, 6.0, and 12.0 dS m^{-1}, coupling SA with SPM significantly ($p < 0.05$) increased the K^+/Na^+ ratio in grains of treated Shandawel 1 plants by 67.6%, 92.2%, and 132.2%; the Ca^{2+}/Na^+ ratio by 34.5%, 46.4%, and 69.2%; and the Mg^{2+}/Na^+ ratio by 28.1%, 46.7%, and 74.1%, respectively.

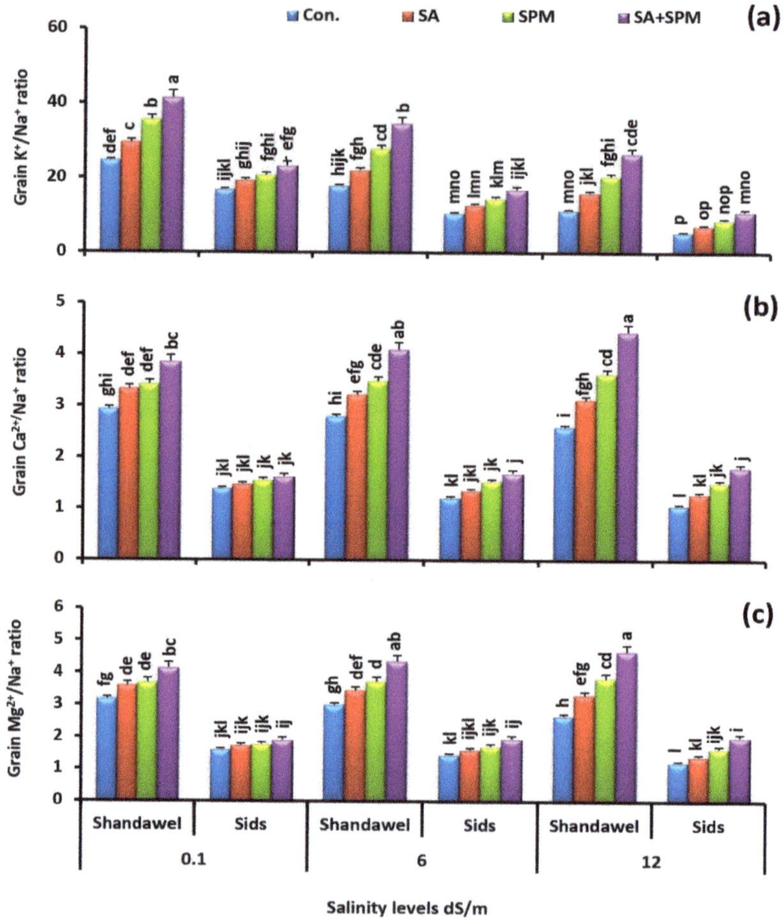

Figure 8. Influence of salicylic acid (SA; 100 mgL^{-1}) and/or spermine (SPM; 30 mgL^{-1}) on the ratio of (**a**) K^+/Na^+, (**b**) Ca^{2+}/Na^+, and (**c**) Mg^{2+}/Na^+ in grains of two wheat cultivars (Shandawel 1 and Sids 14) grown under 0.1, 6, and 12 dS m^{-1} salinity levels. Data are mean of four replicates (n = 4). Different letters indicate significant differences at ($p < 0.05$) level, according to Duncan's test.

2.5. *SA and/or SPM Treatments Improve Organic Solutes Accumulation under Salt Stress*

Exposure to salinity induces the synthesis of organic osmolytes, which play a crucial role in counterbalancing Na^+-induced reduction of water potential. Accumulation of osmoprotector solutes can protect plants from osmotic stress caused by salinity. To observe the change of compatible metabolites in the two wheat cultivars (Shandawel 1 and Sids 14) grown under different saline conditions and foliar sprayed with SA and/or SPM, we measured the concentration of total soluble sugars, total free amino acids, proline, and glycinebetaine in the leaves of treated plants grown under non-saline and saline environments.

The accumulation of total soluble sugars, total free amino acids, proline, and glycinebetaine was increased dramatically in response to salinity treatments, as well as exogenous SA and/or SPM treatments in both cultivars (Figure 9a–d). The highest values were recorded in salt-stressed plants supplemented with a combination of SA and SPM, especially in

Shandawel 1. Application of combined treatment under 6.0 and 12.0 dS m^{-1} salinity levels led to an increase in Shandawel 1 plants total soluble sugars concentration by 14.9% and 17.3%, total free amino acids concentration by 12.5% and 20.3%, proline concentration by 21.2% and 32.2%, and glycinebetaine concentration by 32.0% and 44.0%, respectively, relative to untreated plants.

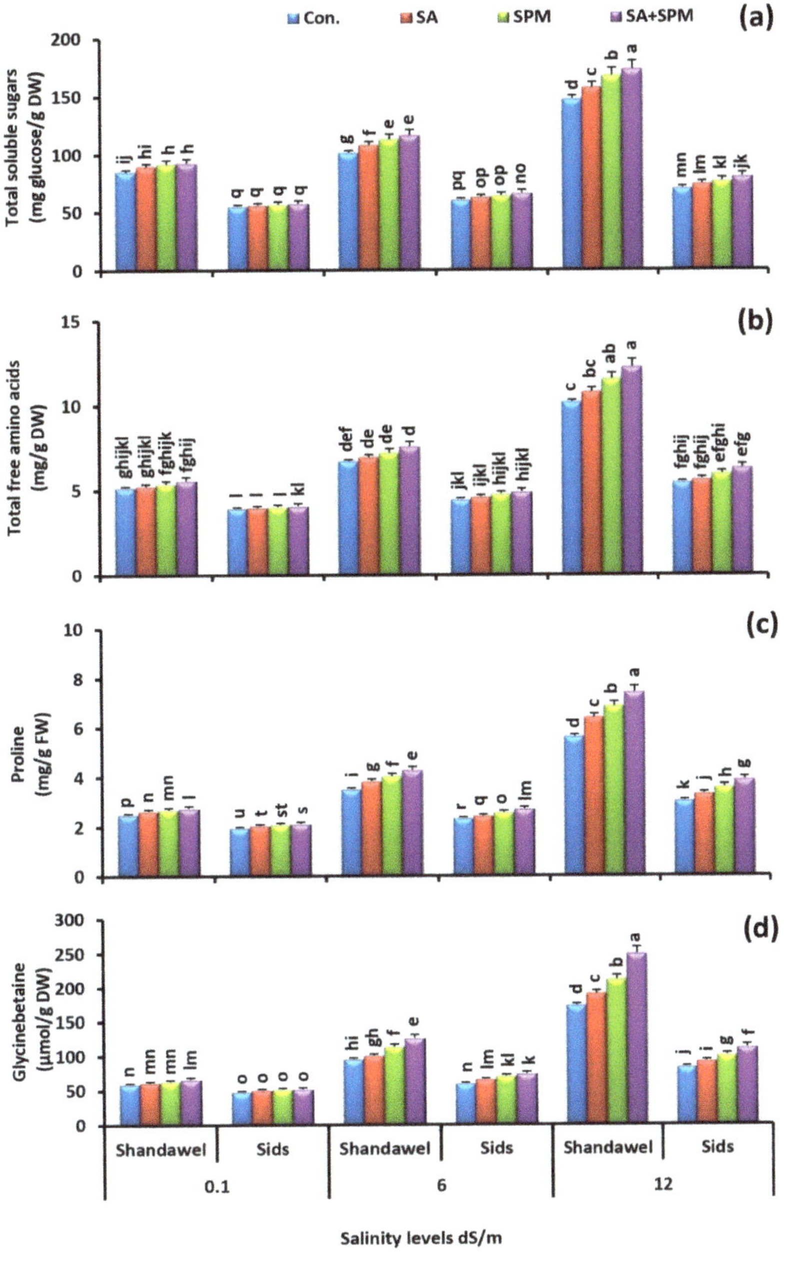

Figure 9. Influence of salicylic acid (SA; 100 mgL^{-1}) and/or spermine (SPM; 30 mgL^{-1}) on the concentration of (**a**) total soluble sugars, (**b**) total free amino acids, (**c**) proline, and (**d**) glycinebetaine

in leaves of two wheat cultivars (Shandawel 1 and Sids 14) grown under 0.1, 6, and 12 dS m^{-1} salinity levels. Data are mean of four replicates (n = 4). Different letters indicate significant differences at ($p < 0.05$) level, according to Duncan's test.

2.6. *Spraying of SA and/or SPM Enhance Grain Carbohydrate and Protein Content under Saline Conditions*

Grains carbohydrate and protein content is important indicator of wheat grain quality, which is influenced by salinity stress, especially when wheat plants are exposed to saline environments at the grain filling stage. To investigate the ameliorative effect of exogenous SA and/or SPM on the grain quality of two wheat cultivars (Shandawel 1 and Sids 14) grown under saline circumstances, we measured the content of protein and carbohydrate in wheat grains of treated plants subjected to non-saline and saline conditions.

In view of the effect of salt treatments on grain carbohydrate and protein content, it was postulated that saline environments considerably reduced their values in both cultivars. The grains of salt-stressed plants had much lower amounts of carbohydrate and protein than those of unstressed ones. This deleterious effect of saline conditions was prominent for cv. Sids 14 compared to cv. Shandawel 1. On the contrary, under both saline and non-saline conditions, application of SA and/or SPM significantly ($p < 0.05$) enhanced their contents, especially in Shandawel 1 (Figure 10a,b). The combined treatment of SA and SPM had the greatest impact. When compared to untreated plants, co-application of SA and SPM significantly ($p < 0.05$) boosted carbohydrate content by 20.8%, 31.4%, and 48.2% and the protein content by 23.5%, 34.9%, and 53.9% in Shandawel 1 wheat plants at 0.1, 6.0, and 12.0 dS m^{-1} salinity levels, respectively.

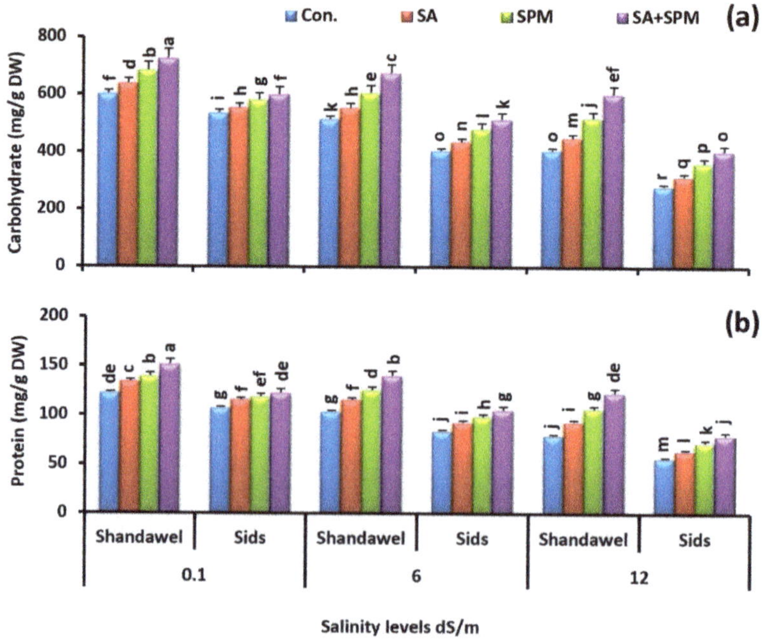

Figure 10. Influence of salicylic acid (SA; 100 mgL^{-1}) and/or spermine (SPM; 30 mgL^{-1}) on the content of (**a**) carbohydrate and (**b**) protein in grains of two wheat cultivars (Shandawel 1 and Sids 14) grown under 0.1, 6, and 12 dS m^{-1} salinity levels. Data are mean of four replicates (n = 4). Different letters indicate significant differences at ($p < 0.05$) level, according to Duncan's test.

3. Discussion

Among all abiotic stresses, soil salinity is one of the major environmental constraints challenging crop production worldwide [1,4]. The excess amount of salts present in the agricultural land is considered as an immense threat to global food security [2,3]. Wheat constitutes pivotal position for ensuring food and nutritional security; however, rapidly rising soil and water salinity pose a serious threat to its production globally. Salinity stress negatively affects the growth and development of wheat resulting in restrained grain yield and quality [30,31]. Hence, improving wheat salt tolerance and exploiting arable areas of saline soils are becoming the most important scientific research and agricultural practices. Plant growth stimulators such as SA and SPM can play a key role in diverse plant growth and the development of physiological processes under different environmental stresses [13,14,23,26]. Some studies have looked into the effect of a single application of SA or SPM on plants exposed to salt stress. However, no one has looked into the impact of their combined treatment on wheat salt tolerance. To fill this gap, as a first investigation, we examined the effect of exogenously applied SA and/or SPM on photosynthetic pigment content, nutritional homeostasis, osmoprotectant synthesis, as well as grain carbohydrate and protein content in two wheat cultivars (Shandawel 1 and Sids 14) grown under non-saline and saline (6.0 and 12.0 dS m^{-1}) conditions. Our findings clearly show that using SA and SPM together can reduce the negative effects of salt stress on wheat growth and production by enhancing photosynthetic pigment content, improving mineral acquisition, and promoting osmolytes accumulation (Figure 11). This research provides novel insights into the mechanisms of SA- and SPM-mediated amelioration of salt stress on wheat.

Growth and yield reduction can be used to determine the extent of salt-induced damage to the plants [4,32]. In the current study, data showed that salt stress resulted in a significant decline in the growth and yield of wheat plants in terms of total leaf area, dry weights of shoot and root, grain number, and grain weight, and this reduction increased with the increase in salinity levels, especially in cv. Sids 14 compared to cv. Shandawel 1 (Figures 1a–c and 2a,b). This deleterious effect of saline environments on plant growth and development could be the result of (a) reducing the photosynthetic pigment concentration that can suppresses the photosynthetic efficiency, (b) inducing the nutritional imbalance that was closely associated with the reduction of N, P, K^+, Fe, Zn, and Cu acquisition, along with the improvement of Na^+ accumulation, (c) inducing the specific ion toxicity, as indicated by the disturbance in K^+/Na^+, Ca^{2+}/Na^+, and Mg^{2+}/Na^+ ratios, as well as (d) decreasing the synthesis and translocation of photosynthates (metabolites) from source to sink organs as indicated by the reduction in grain carbohydrate and protein content. Previous studies have shown that salinity affects plant growth through nutritional imbalances and ionic toxicity and causes membrane dysfunction and attenuation of metabolic activity [32,33]. By contrast, in agreement with previous reports [7,10,15,16,23,26,27], we observed that foliage applications of SA and/or SPM significantly ameliorated the negative impacts of soil salinization on wheat growth and production via upregulating of the photosynthetic pigment content, maintaining the optimal mineral nutrition, motivating the plant osmotic adjustment, and enhancing the grain carbohydrate and protein content. These findings clearly support the effectiveness of SA and/or SPM foliar treatments in attenuating the inhibitory effect of salt stress on plant development. In line with our findings, previous research by Miao [18] suggested that SA not only enhanced absorption range and capacity of water and nutrition by accelerating root growth, but also increased carbohydrate accumulation by promoting leaf photosynthetic ability, thus contributing to dry matter accumulation in salt-stressed plants. Moreover, the SPM application reduced the salt-induced growth inhibition by improving the chlorophyll content, compatible solutes accumulation, K^+ content, and K^+/Na^+ balance [28].

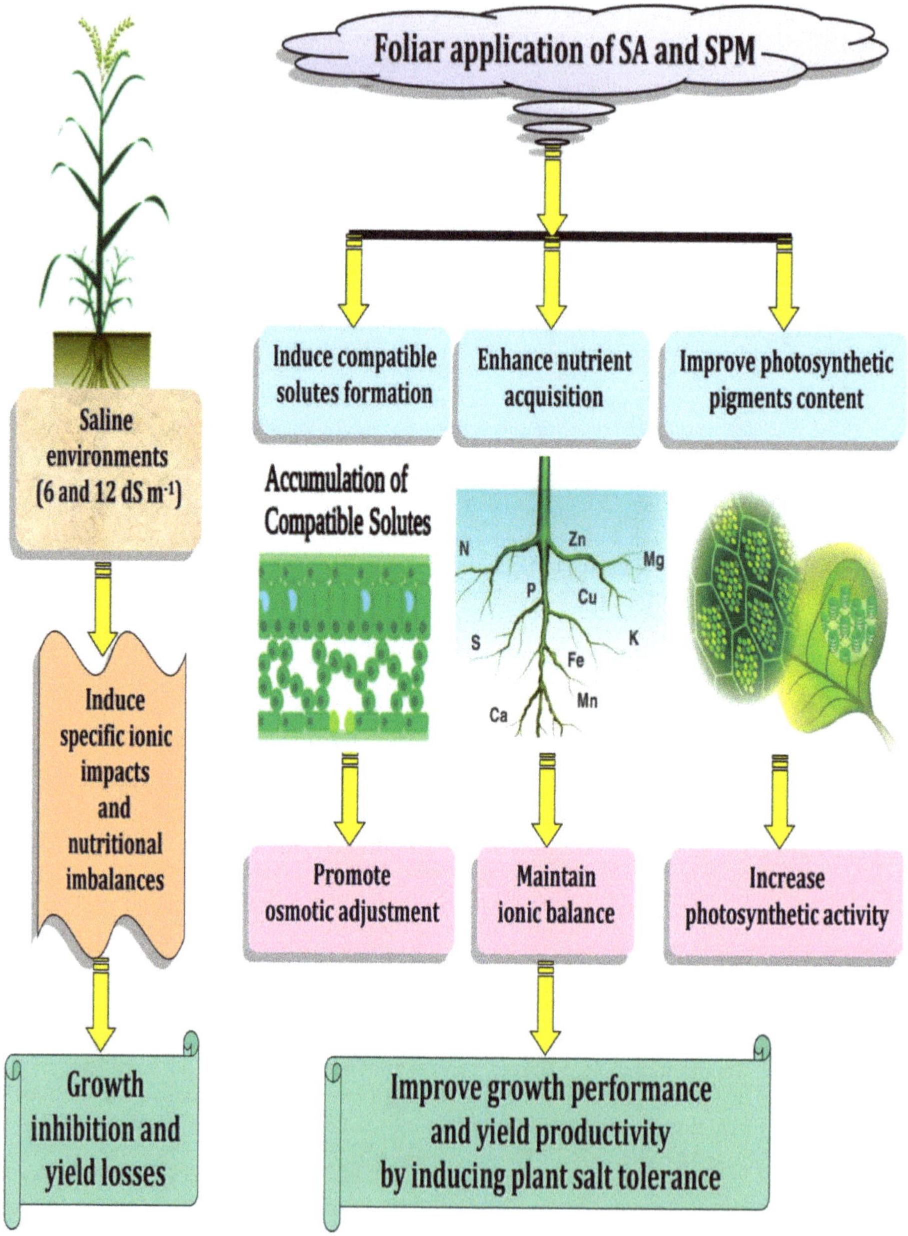

Figure 11. A model showing saline environments inhibit growth and productivity of wheat plants by inducing specific ionic effects and nutritional imbalances. Meanwhile, salicylic acid (SA) and spermine (SPM) reduce salt stress damage to the plant by improving photosynthetic pigment content, enhancing nutrient acquisition, and inducing osmolytes accumulation, thereby helping plants maintain ionic and osmotic balances.

Photosynthetic pigment concentration is an indicator of the photosynthetic machinery integrity, with positive correlations to photosynthetic activity [22,34,35]. In the present investigation, we showed that the concentration of photosynthetic pigment was sharply decreased as the salt concentration in soil increased (Figure 3a–d), implying that salt stress may cause pigment oxidation and degradation, and hence reduce pigment concentration [36]. Noteworthily, exogenously applied SA and/or SPM proved favorable by reducing the negative impacts of saline conditions and enhanced chlorophyll content in the absence, as well as the presence of salt stress. Previously, several scientists reported that foliage applications of SA and/or SPM improved chlorophyll levels in both unstressed and stressed plants by protecting thylakoid membranes and regulating chlorophyll biosynthesis and degradation pathways [13,26]. It has also been reported that SA and SPM may boost the activity of enzymes involved in the manufacture of chlorophyll or reduce the malfunction of the photosynthetic system, hence reducing chlorophyll degradation [15,27]. The results obtained in this trial also revealed that SA and/or SPM applied to salt-stressed plants prevented further salt damage and preserved the carotenoid concentrations. Carotenoids are important non-enzymatic lipid soluble antioxidants that play multiple functions in plant metabolism, including oxidative stress tolerance, as they protect the chloroplast form the harmful ROS [37]. Indeed, SA and SPM's beneficial effects on maintaining carotenoid amount may be directly related to their capacity to control cellular signaling, activate redox-sensitive regulatory pathways, and regulate processes in carotenoid production [13,22]. It is important to note that the growth improvement caused by SA and/or SPM treatments may be the result of an increase in the amount of photosynthetic pigments. Indeed, these applications could function as regulators to avert degradation of chlorophyll and carotenoid pigments and protect the photosynthetic apparatus, thus enhancing photosynthetic efficiency.

Another effective strategy to resist salt stress employed by plants is to keep ion homeostasis and relieve ionic toxicity. Excessive soil Na^+ absorbed and transported to the aboveground organs disturbs the intracellular ionic homeostasis in plants, damages the photosynthetic membrane structure, promotes the chlorophyll degradation, and affects the cytosolic enzyme activity, resulting in the restraint of plant growth and productivity [7,28]. In the current study, we found that saline conditions induced alteration of ion homeostasis, as shown by higher Na^+ accumulation (Figure 6a) and lower N, P, K^+, Fe, Zn, and Cu acquisition (Figures 5a–c and 7a–c). Moreover, this effect was more pronounced at high salinity level. Excessive accumulation of Na^+ interferes with various physiological processes in plants, thus changing the ion balance in cells and triggering ion damage to the plant [38]. By contrast, in agreement with the previous reports [23,26,28,39,40], we observed that foliage applications of SA and/or SPM relieved the adverse effects caused by salt stress and significantly improved N, P, K^+, Fe, Zn, and Cu acquisition in wheat grains, indicating their regulatory role in enhancing mineral nutrition uptake, accumulation, and translocation. Furthermore, the results obtained in this trial also demonstrated that exogenous SA and/or SPM applications alleviated the deleterious injures of salt stress and reduced the Na^+ accumulation, implying that the ability of plants to reduce ions influx into the cytoplasm is of great importance to salinity tolerance. Strong evidence has demonstrated that SPM can directly block non-selective cation channels from the cytosolic side, restricting Na^+ penetration into cell [41]. SPM was also reported to improve the intracellular ion homeostasis by enhancing the salt overly sensitive (SOS) pathway (*AsSOS1*, *AsSOS2*, *AsSOS3*) and upregulating the transcript levels of the high affinity K^+ transporters (*HKTs*: *AsHKT1*, *AsHKT2*, *AsHKT4*, *AsHKT6*, and *AsHKT7*) [28]. Plant high affinity K^+ transporters can unload Na^+ from the xylem, thus improving plant salt tolerance [42]. A pervious study has also shown that SPM can act as an endogenous regulator of cell K^+ transport [43]. Furthermore, SA's beneficial effect on maintaining ionic homeostasis could be directly linked to SA ability to induce H^+-ATPase activity [44]. SA was also reported to prevent salt-induced K^+ leakage through depolarization-activated-outward-rectifying K channels [45]. Generally, retaining high levels of K^+ and lowering accumulation of Na^+ in the cytosol are essential for increasing the salinity tolerance in plants [46]. Bearing in mind that, maintaining ion homeostasis by SA and/or SPM treatments could be contributed to better salt tolerance. Generally, our study's results reveal that SA and/or SPM increase the nutrient acquisition in wheat grains, highlighting a substantial protective role of SA and/or SPM in maintaining the grain nutritional value under salt stress.

Salt stress induces specific ion toxicity, which was measured by the buildup of Na^+, K^+, Ca^{2+}, and Mg^{2+}. Na^+ is the foremost toxic ion that typically accumulates during salt stress conditions interfering with K^+, Ca^{2+}, and Mg^{2+} uptake and transport, which causes disturbance to K^+/Na^+, Ca^{2+}/Na^+, and Mg^{2+}/Na^+ ratios [7,20]. A plant's capacity to tolerate salt is directly correlated with its ability to keep adequate K^+/Na^+, Ca^{2+}/Na^+, and Mg^{2+}/Na^+ ratios [7,9]. In the present study, we found that, with increasing salt doses, these ratios were negatively affected (Figure 8a–c), resulting in poor plant growth and productivity. Under conditions of salt stress, Na^+ enters the cytosol and depolarizes the plasma membrane, which results in continuous outflow of K^+ and raises the Na^+/K^+ ratio in the cytosol [1]. Conversely, our results revealed significant increases in K^+/Na^+, Ca^{2+}/Na^+, and Mg^{2+}/Na^+ ratios in plants subjected to salty soils and sprayed with SA and/or SPM. This is consistent with a previous report, which demonstrated that the treatment with SA reduced Na^+ accumulation, increased K^+/Na^+ ratio, and mitigated the negative effects of salt stress on plant growth and development [20]. SA's beneficial impact on maintaining ionic homeostasis may be directly related to its ability to increase root H^+-pump activity, which stimulates the input of K^+ and improves K^+/Na^+ balance [7]. Previous investigation also provides evidence that SPM improves K^+/Na^+ homeostasis in barely by blocking Na^+ influx into root epidermal and cortical cells and restricting K^+ loss from shoots [47]. Most probably, the ability of stress treated plants to mitigate salt damage and survive under saline environments is achieved by maintaining a high selectivity for K^+, Ca^{2+}, and Mg^{2+} ions despite an excess of Na^+ ions. In sum, our findings suggest that preserving ionic balance may be the key action of SA and/or SAM in preventing salt toxicity. Furthermore, the growth–promoter effects of SA and SPM might be the result of their role in ion homeostasis.

To overcome the osmotic stress induced by salinity, plants synthesize and accumulate compatible solutes, such as total soluble sugars, total free amino acids, proline, and glycinebetaine [13,39]. These organic solutes can induce osmotic balance and act as antioxidants [48,49]. Proline possesses antioxidant properties that act as chaperones to protect the structure of macromolecules from destruction when the cell is dehydrated [49]. In fact, proline can act as a free radical scavenger, a cell redox balancer, an enzyme protectant, and a cytosolic pH buffer stabilizer for subcellular structures [50,51]. Furthermore, glycinebetaine has a potential role in cell osmotic adjustment, membrane stabilization, and the detoxification of toxic ions in plants exposed to salt stress [49]. Similarly, a higher accumulation of soluble sugar is also beneficial to the osmotic regulation [52]. In this study, a high level of salt stress induced a sharp increase in the total soluble sugars, total free amino acids, proline, and glycinebetaine accumulation, and this increase was more pronounced in the leaves of Shandawel 1 plants than that in the leaves of Sids 14. Furthermore, the treatments with SA and/or SPM further increased their accumulation in wheat leaves, especially in Shandawel 1 (Figure 9a–d). In fact, the osmoprotector solute accumulation under salt stress has been linked to the stress tolerance in many plant species. Moreover, their concentrations have been reported to be greater in salt-tolerant plants than in salt-sensitive ones [20]. Much evidence showed that exogenous application of SA or SPM increases the synthesis of compatible osmolytes in the plant cells [13,22,26,39,51]. A previous report provides evidence that SA induces glycinebetaine accumulation under salinity stress and increases Na^+ flux from cytoplasm to vacuole [53]. In addition, an exogenous application of SA induces proline accumulation to alleviate the deleterious effects of salinity [20]. According to earlier studies, SA causes the building of glycinebetaine by inhibiting the generation of ethylene [53] and causes the production of proline by boosting the activity of its biosynthesis enzymes [54]. SPM has been also proved to be an important regulator of tolerance to salt stress by inducing the buildup of the compatible osmotic substances [28], which is fairly in agreement with our findings. Based on these determined parameters, it is worth underlining that the application of SA and SPM to wheat plants' exposure to salt stress improves their ability to synthesize osmolytes that helps them to survive under this harsh condition.

Excessive soil Na^+ absorbed and transported to the aboveground organs disturbs intracellular ionic homeostasis in plants and reduces photosynthetic rate, biomass accumulation, and source-sink activity, which hastens the reproductive organs' senescence [55]. It has also been observed that the unavailability of sufficient photo-assimilates during the reproductive stage is the leading cause for losing yield potential and affecting the grain quality of wheat [56]. Protein and carbohydrate content is the most important indicator of wheat grain quality and, hence, it governs and determines the endue quality. In the present investigation, we showed that the contents of proteins and carbohydrates in wheat grains were significantly decreased as a result of soil salinization (Figure 10a,b). Previous investigation provides evidence that the decrease in protein content of grains was due to the accumulation of salts in the root zone, which deteriorated the grain quality [55]. Similarly, higher concentration of Na^+ in the external environment interferes with the absorption of N, which leads to lower protein content in wheat grains [56]. Furthermore, salt stress interfered with photosynthesis and reduced P concentration in plant shoots, which interfered with grain development and affected the cereals' grain carbohydrate content [57]. Perhaps this reaffirms our hypothesis that the nutritional imbalance could be the major factor behind the deteriorated grain quality of wheat under salt stress. Interestingly, the results obtained in this trial revealed that, under both non-saline and saline circumstances, the SA and/or SPM treatments significantly increased the grain carbohydrate and protein content. It could possibly be due to their positive impact on (i) the photosynthetic pigment concentration that can improve the photosynthetic efficiency and (ii) the nutritional balance that was closely associated with the improvement of N, P, K^+, Fe, Zn, and Cu acquisition, along with the reduction of Na^+ accumulation. It is praiseworthy that increasing grain carbohydrate and protein content with SA and/or SPM applications sustains yield of wheat plants and maintains grain nutritional quality under salt stress.

Different wheat cultivars have different defense responses to salt stress, so it is important to investigate the beneficial impact of using the alleviating stressor agents, such as SA and SPM on different wheat cultivars. It is worth noting that wheat cultivars' responses to SA and/or SPM applications in saline environments may vary. In our study, the magnitude of salt stress mitigation by SA and/or SPM appears to be stronger for Shandawel 1 than Sids 14, as evidenced by a higher increase in photosynthetic pigment (chlorophyll *a*, chlorophyll *b*, and carotenoids) content, nutrient (N, P, K^+, Fe, Zn, and Cu) acquisition, osmolytes (total soluble sugars, total free amino acids, proline, and glycinebetaine) accumulation, grain carbohydrate and protein content, and a decrease in Na^+ accumulation. As a result, we can suggest that, compared to Sids 14, Shandawel 1 was more responsive to exogenous applications of SA and/or SPM under salinity, resulting in greater improvement in photosynthetic pigment amount, ionic balance, osmolytes adjustment, and grain quality. It is widely accepted that positive ionic and osmotic responses to salt stress are widely accepted as signs of tolerance.

4. Materials and Methods

4.1. Plant Material and Growth Conditions

Wheat (*Triticum aestivum* L. cvs. Shandawel 1 and Sids 14) grains were obtained from the Wheat Research Department, Agriculture Research Center, Ministry of Agriculture, Egypt. Shandawel 1 and Sids 14 cultivars were selected based on their high yield productivity, and we tried to increase their salt tolerance by using SA and SPM foliar applications. Grains were sown in the plastic pots with 30 cm diameter and 35 cm height. Pots were filled with 15 kg of clay loamy soil (sand 37%, silt 28%, and clay 35%). For each pot, twelve grains thinned to six after germination were planted. Ammonium nitrate (33.5% N), calcium superphosphate (15.5% P_2O_5), and potassium sulfate (48% K_2O) were applied at rates of 2.0, 2.0, and 0.5 g pot^{-1}, respectively. In addition, 2.0 g pot^{-1} ammonium nitrate was added 30 days after planting. The pots were placed in the greenhouse of the Department of

Plant Physiology, Faculty of Agriculture, Cairo University, Egypt, under natural light and temperature conditions, with average day/night temperature conditions of 22/16 ±2 °C and average humidity of 65%. The experiment was performed twice, on 10 September of 2019 and 2020, with consistent results. Before sowing, pots were divided into three groups. The first one was assigned as a control (non-saline; 0.1 dS m^{-1}), and the other two groups were assigned as two levels of salinity treatment (6.0 and 12.0 dS m^{-1} salinity level; which obtained by adding to the soil a mixture of NaCl, $CaCl_2$, and $MgSO_4$ at the molar ratio of 2:2:1, respectively). Soil chemical analysis was carried out following the procedures of [58] and presented in Table 1.

Table 1. Chemical properties of the soil under different salinity levels.

Salinity Levels EC (dS m^{-1})	pH	HCO_3^- + CO_3^{2-} (mg kg^{-1})	Cl^- (mg kg^{-1})	SO_4^{2-} (mg kg^{-1})	Ca^{2+} (mg kg^{-1})	Mg^{2+} (mg kg^{-1})	Na^+ (mg kg^{-1})	K^+ (mg kg^{-1})
0.1	7.2	213.5	324.0	430.7	92.2	41.4	3.7	31.4
6.0	7.5	263.6	1173.4	996.9	398.5	173.9	306.7	39.7
12.0	7.8	275.4	1987.8	1686.1	886.5	314.5	808.6	52.6

The wheat plants at 50 days old (vegetative stage) and 100 days old (grain filling stage) from each salinity level were foliar sprayed with 0.00 (distilled water; DW), 100 mgL^{-1} SA, 30 mgL^{-1} SPM, and 100 mgL^{-1} SA + 30 mgL^{-1} SPM. SA and SPM concentrations were chosen according to the results of a preliminary experiment. Tween-20 (0.05%) was added as a surfactant at the time of treatment.

The experimental layout was a completely randomized design with three factors: two wheat cultivars (Shandawel 1 and Sids 14), three levels of salinity [0.1 dS m^{-1} (control), 6.0 and 12.0 dS m^{-1}], and four spraying treatments [0.00 (distilled water; DW), 100 mgL^{-1} SA, 30 mgL^{-1} SPM, and 100 mgL^{-1} SA + 30 mgL^{-1} SPM]. Each treatment had four replicates.

4.2. Plant Growth and Productivity Analysis

The 75-days old plants were sampled (after 25 days of SA and/or SPM first applications) to measure total leaf area, shoot dry weight, and root dry weight. Total leaf area was estimated using a portable leaf area meter (LI-COR 3000, Lambda Instruments Corporation, Lincoln, NE, USA). For dry weight determination, plants were dried at 70°C for 48 h until a constant weight was obtained. At maturity, the number of grains and grain yield were recorded. Each treatment included four replicates, and each replication consists of six plants gathered from the same pot.

4.3. Determination of Photosynthetic Pigments Concentration

The concentration of photosynthetic pigments was determined in the upper leaves of 75-day-old wheat plants. Data were collected from four replicates, each of which contained six plants gathered from the same pot. Photosynthetic pigments from fresh leaves were extracted in 80% (*v*/*v*) acetone, and the concentrations of chlorophyll *a*, chlorophyll *b*, and carotenoids were determined spectrophotometrically according to the method described by [59] using a UV-1750 spectrophotometer (Shimadzu, Kyoto, Japan).

4.4. Determination of N, P, K, Na, Ca, Mg, Fe, Zn, and Cu Concentrations

The concentration of mineral ions was determined in wheat grains. Data were collected from four replicates, and each replication consists of six plants gathered from the same pot. Dried grains (0.5 g) were ground and digested in a mixture of boiling perchloric acid and hydrogen peroxide for 8 h until a transparent solution is obtained. Nitrogen concentration was obtained with the modified micro-Kjeldahl method following [60]. Phosphorus concentration was performed by the vanadomolybdophosphoric method [61]. Potassium and sodium concentrations were analyzed by a flame photometer (ELE UK). Elemental analyses of calcium, magnesium, iron, zinc, and copper were determined with an atomic–absorption spectrophotometer (Unicam 989-AA Spectrometer-UK).

4.5. Determination of Compatible Solutes Accumulation

The concentration of organic solutes was determined in upper leaves of 75-day-old wheat plants. Data were collected from four replicates, each of which contained six plants gathered from the same pot. Total soluble sugars, total free amino acids, and glycinebetaine concentrations were determined in dried ground leaves by the anthrone reagent method [62], ninhydrin reagent method [63], and the method of [64], respectively. Proline was determined in fresh leaf samples according to [65].

4.6. Estimation of Grain Carbohydrate and Protein Content

Carbohydrate and protein contents were determined in wheat grains. Data were collected from four replicates, and each replication consists of six plants gathered from the same pot. Determination of the content of carbohydrates and proteins in dried ground wheat grains were carried out according to [66,67], respectively.

4.7. Statistical Data Analysis

A completely randomized design was used with four replicates per treatment. Combined analysis was made for the two growing seasons since the results of the two seasons followed a similar trend. Three-way analysis of variance (ANOVA) was conducted using SAS software, version 9.4 (SAS Institute Inc., Cary, NC, USA). Differences between the treatments were tested by the Duncan test at a level of significance ($p < 0.05$).

5. Conclusions

Current findings reveal the important role of SA and SPM in improving the salt tolerance of wheat plants via upregulating the photosynthetic pigment level, motivating the plant osmotic adjustment, promoting the plant nutrient acquisition, and enhancing the grain quality. The SA and/or SPM foliar applications ameliorated the deleterious effect of salt stress on wheat growth and development by reinforcing the photosynthetic pigment (chlorophyll *a*, chlorophyll *b*, and carotenoids) content, improving the nutrient (N, P, K^+, Ca^{2+}, Mg^{2+}, Fe, Zn, and Cu) acquisition, maintaining the ionic (K^+/Na^+, Ca^{2+}/Na^+, and Mg^{2+}/Na^+) homeostatic, inducing the osmolytes (total soluble sugars, total free amino acids, proline, and glycinebetaine) accumulation, enhancing the grain carbohydrate and protein content, and preventing the Na^+ accumulation. The combined treatment of SA and SPM yielded the best results. Hence, this study explained the mechanism involved in salt stress alleviation by SA and/or SPM, which would help the researchers understand the importance of SA and SPM in mitigating salt stress in agronomic crops. Finally, we exposed a new environmentally friendly approach for farmers to stimulate the growth and productivity of their agronomic and horticultural crops under harsh environmental conditions.

Author Contributions: N.B.T. conceptualized and coordinated the research, conceived the idea, designed and carried out the experiments, generated and analyzed the data, and wrote the manuscript. A.M.A.H. carried out the pot experiments. All authors have read and agreed to the published version of the manuscript.

Funding: This research received no external funding.

Institutional Review Board Statement: Not applicable.

Informed Consent Statement: Not applicable.

Data Availability Statement: The data presented in this study are available in the article.

Conflicts of Interest: The authors declare that they have no conflict of interest.

References

1. Zörb, C.; Geilffus, C.M.; Dietz, K.J. Salinity and crop yield. *Plant Biol.* **2019**, *21*, 31–38. [CrossRef]
2. Arzani, A.; Ashraf, M. Smart engineering of genetic resources for enhanced salinity tolerance in crop plants. *Crit. Rev. Plant Sci.* **2016**, *35*, 146–189. [CrossRef]

3. Arzani, A. Improving salinity tolerance in crop plants: A biotechnological view. *In Vitro Cell. Dev. Biol.-Plant* **2008**, *44*, 373–383. [CrossRef]
4. Talaat, N.B.; Shawky, B.T. Influence of arbuscular mycorrhizae on root colonization, growth and productivity of two wheat cultivars under salt stress. *Arch. Agron. Soil. Sci.* **2012**, *58*, 85–100. [CrossRef]
5. Singh, P.; Choudhary, K.K.; Chaudhary, N.; Gupta, S.; Sahu, M.; Tejaswini, B.; Sarkar, S. Salt stress resilience in plants mediated through osmolyte accumulation and its crosstalk mechanism with phytohormones. *Front. Plant Sci.* **2022**, *13*, 1006617. [CrossRef]
6. Munns, R.; Passioura, J.B.; Colmer, T.D.; Byrt, C.S. Osmotic adjustment and energy limitations to plant growth in saline soil. *New Phytol.* **2020**, *225*, 1091–1096. [CrossRef]
7. Talaat, N.B.; Shawky, B.T. Synergistic effects of salicylic acid and melatonin on modulating ion homeostasis in salt-stressed wheat (*Triticum aestivum* L.) plants by enhancing root H^+-pump activity. *Plants* **2022**, *11*, 416. [CrossRef]
8. Kiani, R.; Arzani, A.; Mirmohammady Maibody, S.A.M. Polyphenols, flavonoids, and antioxidant activity involved in salt tolerance in wheat, aegilops cylindrica and their amphidiploids. *Front. Plant Sci.* **2021**, *12*, 646221. [CrossRef]
9. Hinai, M.S.A.; Ullah, A.; Al-Rajhi, R.S.; Farooq, M. Proline accumulation, ion homeostasis and antioxidant defence system alleviate salt stress and protect carbon assimilation in bread wheat genotypes of omani origin. *Environ. Exp. Bot.* **2022**, *193*, 104687. [CrossRef]
10. Talaat, N.B.; Todorova, D. Antioxidant machinery and glyoxalase system regulation confers salt stress tolerance to wheat (*Triticum aestivum* L.) plants treated with melatonin and salicylic Acid. *J. Soil Sci. Plant Nutr.* **2022**, *22*, 3527–3540. [CrossRef]
11. Talaat, N.B.; Mostafa, A.A.; El-Rahman, S.N.A. A novel plant growth-promoting agent mitigates salt toxicity in barley (*Hordeum vulgare* L.) by activating photosynthetic, antioxidant defense, and methylglyoxal detoxification machineries. *J. Soil Sci. Plant Nutr.* **2022**. [CrossRef]
12. Chakraborty, K.; Mondal, S.; Ray, S.; Samal, P.; Pradhan, B.; Chattopadhyay, K.; Kar, M.K.; Swain, P.; Sarkar, R.K. Tissue tolerance coupled with ionic discrimination can potentially minimize the energy cost of salinity tolerance in rice. *Front. Plant Sci.* **2020**, *11*, 265. [CrossRef] [PubMed]
13. Talaat, N.B. Co-Application of melatonin and salicylic acid counteracts salt stress-induced damage in wheat (*Triticum aestivum* L.) photosynthetic machinery. *J. Soil Sci. Plant Nutr.* **2021**, *21*, 2893–2906. [CrossRef]
14. Talaat, N.B. Polyamine and nitrogen metabolism regulation by melatonin and salicylic acid combined treatment as a repressor for salt toxicity in wheat (*Triticum aestivum* L.) plants. *Plant Growth Regul.* **2021**, *95*, 315–329. [CrossRef]
15. Bukhat, S.; Manzoor, H.; Athar, H.; Zafar, Z.U.; Azeem, F.; Rasul, S. Salicylic acid induced photosynthetic adaptability of *Raphanus sativus* to salt stress is associated with antioxidant capacity. *J. Plant Growth Regul.* **2020**, *39*, 809–822. [CrossRef]
16. Hoang, H.L.; de Guzman, C.C.; Cadiz, N.M.; Hoang, T.T.H.; Tran, D.H.; Rehman, H. Salicylic acid and calcium signaling induce physiological and phytochemical changes to improve salinity tolerance in red amaranth (*Amaranthus tricolor* L.). *J. Soil Sci. Plant Nutr.* **2020**, *20*, 1759–1769. [CrossRef]
17. Shakirova, F.M.; Sakhabutdinova, A.R.; Bezrukova, M.V.; Fatkhutdinova, R.A.; Fatkhutdinova, D.R. Changes in the hormonal status of wheat seedlings induced by salicylic acid and salinity. *Plant Sci.* **2003**, *164*, 317–322. [CrossRef]
18. Miao, Y.; Luo, X.; Gao, X.; Wang, W.; Li, B.; Hou, L. Exogenous salicylic acid alleviates salt stress by improving leaf photosynthesis and root system architecture in cucumber seedlings. *Sci. Hortic.* **2020**, *272*, 109577. [CrossRef]
19. Hongna, C.; Leyuan, T.; Junmei, S.; Xiaori, H.; Xianguo, C. Exogenous salicylic acid signal reveals an osmotic regulatory role in priming the seed germination of Leymus chinensis under salt-alkali stress. *Environ. Exp. Bot.* **2021**, *188*, 104498. [CrossRef]
20. Islam, A.T.M.T.; Ullah, H.; Himanshu, S.K.; Tisarum, R.; Cha-um, S.; Datta, A. Effect of salicylic acid seed priming on morpho-physiological responses and yield of baby corn under salt stress. *Sci. Hortic.* **2022**, *304*, 111304. [CrossRef]
21. Todorova, D.; Talaat, N.B.; Katerova, Z.; Alexieva, V.; Shawky, B.T. Polyamines and brassinosteroids in drought stress responses and tolerance in plants. In *Water Stress and Crop Plants: A Sustainable Approach*; Ahmad, P., Ed.; John Wiley & Sons, Ltd.: Chichester, UK, 2016; Volume 2, pp. 608–627.
22. Talaat, N.B. 24-Epibrassinolide and spermine combined treatment sustains maize (*Zea mays* L.) drought tolerance by improving photosynthetic efficiency and altering phytohormones Profile. *J. Soil Sci. Plant Nutr.* **2020**, *20*, 516–529. [CrossRef]
23. Xu, J.; Yang, J.; Xu, Z.; Zhao, D.; Hu, X. Exogenous spermine-induced expression of *SlSPMS* gene improves salinity–alkalinity stress tolerance by regulating the antioxidant enzyme system and ion homeostasis in tomato. *Plant Physiol. Biochem.* **2020**, *157*, 79–92. [CrossRef]
24. Gupta, K.; Dey, A.; Gupta, B. Plant polyamines in abiotic stress responses. *Acta Physiol. Plant.* **2013**, *35*, 2015–2036. [CrossRef]
25. Seifi, H.S.; Shelp, B.J. Spermine differentially refines plant defense responses against biotic and abiotic stresses. *Front. Plant Sci.* **2019**, *10*, 117. [CrossRef] [PubMed]
26. Ahangera, M.A.; Qin, C.; Maodong, Q.; Dong, X.X.; Ahmad, P.; Abd Allah, E.F.; Zhang, L. Spermine application alleviates salinity induced growth and photosynthetic inhibition in *Solanum lycopersicum* by modulating osmolyte and secondary metabolite accumulation and differentially regulating antioxidant metabolism. *Plant Physiol. Biochem.* **2019**, *144*, 1–13. [CrossRef]
27. Islam, M.A.; Jin-huan, P.; Fan-wei, M.; Ya-wen, L.; Ning, X.; Chao, Y.; Jun, L. Putrescine, spermidine, and spermine play distinct roles in rice salt tolerance. *J. Integr. Agric.* **2020**, *19*, 643–655. [CrossRef]
28. Geng, W.; Qiu, Y.; Peng, Y.; Zhang, Y.; Li, Z. Water and oxidative homeostasis, Na^+/K^+ transport, and stress-defensive proteins associated with spermine-induced salt tolerance in creeping bentgrass. *Environ. Exp. Bot.* **2021**, *192*, 104659. [CrossRef]

29. Talaat, N.B.; Ibrahim, A.S.; Shawky, B.T. Enhancement of the expression of *ZmBZR1* and *ZmBES1* regulatory genes and antioxidant defense genes triggers water stress mitigation in maize (*Zea mays* L.) plants treated with 24-epibrassinolide in combination with spermine. *Agronomy* **2022**, *12*, 2517. [CrossRef]
30. Amirbakhtiar, N.; Ismaili, A.; Ghaffari, M.-R.; Mirdar Mansuri, R.; Sanjari, S.; Shobbar, Z.-S. Transcriptome analysis of bread wheat leaves in response to salt stress. *PLoS ONE* **2021**, *16*, e0254189. [CrossRef]
31. Kiani, R.; Arzani, A.; Mirmohammady Maibody, S.A.M.; Rahimmalek, M.; Razavi, K. Morpho-physiological and gene expression responses of wheat by *Aegilops cylindrica* amphidiploids to salt stress. *Plant Cell Tissue Organ Cult.* **2021**, *144*, 619–639. [CrossRef]
32. Acosta-Motos, J.R.; Ortuno, M.F.; Bernal-Vicente, A.; Diaz-Vivancos, P.; Sanchez-Blanco, M.J.; Hernandez, J.A. Plant responses to salt stress: Adaptive mechanisms. *Agronomy* **2017**, *7*, 18. [CrossRef]
33. Farooq, M.; Gogoi, N.; Hussain, M.; Barthakur, S.; Paul, S.; Bharadwaj, N.; Migdadi, H.M.; Alghamdi, S.S.; Siddique, K.H.M. Effects, tolerance mechanisms and management of salt stress in grain legumes. *Plant Physiol. Biochem.* **2017**, *118*, 199–217. [CrossRef]
34. Saha, P.; Sade, N.; Arzani, A.; Wilhelmi, M.R.; Coe, K.M.; Li, B.; Blumwald, E. Effects of abiotic stress on physiological plasticity and water use of *Setaria viridis* (L.). *Plant Sci.* **2016**, *251*, 128–138. [CrossRef]
35. Talaat, N.B. RNAi based simultaneous silencing of all forms of light-dependent NADPH:protochlorophyllide oxidoreductase genes result in the accumulation of protochlorophyllide in tobacco (*Nicotiana tabacum*). *Plant Physiol. Biochem.* **2013**, *71*, 31–36. [CrossRef]
36. Akrami, M.; Arzani, A. Physiological alterations due to field salinity stress in melon (*Cucumis melo* L.). *Acta Physiol Plant.* **2018**, *40*, 91. [CrossRef]
37. Collins, A.R. Carotenoids and genomic stability. *Mutat. Res./Fundam. Mol. Mech. Mutagen.* **2001**, *475*, 21–28. [CrossRef]
38. Ismail, A.; Takeda, S.; Nick, P. Life and death under salt stress: Same players, different timing. *J. Exp. Bot.* **2014**, *65*, 2963–2979. [CrossRef] [PubMed]
39. Kaya, C.; Ashraf, M.; Alyemeni, M.N.; Ahmad, P. The role of endogenous nitric oxide in salicylic acid-induced up-regulation of ascorbate-glutathione cycle involved in salinity tolerance of pepper (*Capsicum annuum* L.) plants. *Plant Physiol. Biochem.* **2020**, *147*, 10–20. [CrossRef] [PubMed]
40. Es-sbihi, F.Z.; Hazzoumi, Z.; Aasfar, A.; Joutei, K.A. Improving salinity tolerance in *Salvia officinalis* L. by foliar application of salicylic acid. *Chem. Biol. Technol. Agric.* **2021**, *8*, 25. [CrossRef]
41. Shabala, S.; Cuin, T.A.; Pottosin, I.I. Polyamines prevent NaCl induced K^+ efflux from pea mesophyll by blocking non-selective cation channels. *FEBS Lett.* **2007**, *581*, 1993–1999. [CrossRef]
42. Hamamoto, S.; Horie, T.; Hauser, F.; Deinlein, U.; Schroeder, J.I.; Uozumi, N. HKT transporters mediate salt stress resistance in plants:from structure and function to the field. *Curr. Opin. Biotechnol.* **2015**, *32*, 113–120. [CrossRef]
43. Suma, A.; Granata, D.; Thomson, A.S.; Carnevale, V.; Rothberg, B.S. Polyamine blockade and binding energetics in the MthK potassium channel. *J. Gen. Physiol.* **2020**, *152*, 1085–1098. [CrossRef]
44. Inada, M.; Ueda, A.; Shi, W.; Takabe, T. A Stress-inducible plasma membrane protein 3 (AcPMP3) in a monocotyledonous halophyte, *Aneurolepidium chinense*, regulates cellular Na^+ and K^+ accumulation under salt stress. *Planta* **2005**, *220*, 395–402. [CrossRef]
45. Gharbi, E.; Lutts, S.; Dailly, H.; Quinet, M. Comparison between the impacts of two different modes of salicylic acid application on tomato (*Solanumlycopersicum*) responses to salinity. *Plant Signal Behav.* **2018**, *13*, e1469361. [CrossRef]
46. Anschütz, U.; Becker, D.; Shabala, S. Going beyond nutrition: Regulation of potassium homoeostasis as a common denominator of plant adaptive responses to environment. *J. Plant Physiol.* **2014**, *171*, 670–687. [CrossRef]
47. Zhao, F.; Song, C.; He, J.; Zhu, H. Polyamines improve K^+/Na^+ homeostasis in barley seedlings by regulating root ion channel activities. *Plant Physiol.* **2007**, *145*, 1061–1072. [CrossRef]
48. Farooq, M.; Hussain, M.; Wakeel, A.; Siddique, K.H.M. Salt stress in maize effects resistance mechanisms and management: A review. *Agron. Sustain. Dev.* **2015**, *35*, 461–481. [CrossRef]
49. Ashraf, M.; Foolad, M.R. Roles of glycine betaine and proline in improving plant abiotic stress resistance. *Environ. Exp. Bot.* **2007**, *59*, 206–216. [CrossRef]
50. Verbruggen, N.; Hermans, C. Proline accumulation in plants: A review. *Amino Acids* **2008**, *35*, 753–759. [CrossRef]
51. Misra, N.; Saxena, P. Effect of salicylic acid on proline metabolism in lentil grown under salinity stress. *Plant Sci.* **2009**, *177*, 181–189. [CrossRef]
52. Rosa, M.; Prado, C.; Podazza, G.; Interdonato, R.; Gonzlez, J.A.; Hilal, M.; Prado, F.E. Soluble sugars: Metabolism, sensing and abiotic stress: A complex network in the life of plants. *Plant Signal. Behav.* **2009**, *4*, 388–393. [CrossRef]
53. Khan, M.I.R.; Asgher, M.; Khan, N.A. Alleviation of salt induced photosynthesis and growth inhibition by salicylic acid involves glycine betaine and ethylene in mung bean (*Vigna radiata* L.). *Plant Physiol. Biochem.* **2014**, *80*, 67–74. [CrossRef]
54. Khan, M.I.R.; Fatma, M.; Per, T.S.; Anjum, N.A.; Khan, N.A. Salicylic acid-induced abiotic stress tolerance and underlying mechanisms in plants. *Front. Plant Sci.* **2015**, *6*, 462–494. [CrossRef]
55. Khataar, M.; Mohammadi, M.H.; Shabani, F. Soil salinity and matric potential interaction on water use, water use efficiency and yield response factor of bean and wheat. *Sci. Rep.* **2018**, *8*, 2679.

56. EL Sabagh, A.; Islam, M.S.; Skalicky, M.; Ali Raza, M.; Singh, K.; Anwar Hossain, M.; Hossain, A.; Mahboob, W.; Iqbal, M.A.; Ratnasekera, D.; et al. Salinity stress in wheat (*Triticum aestivum* L.) in the changing climate: Adaptation and management strategies. *Front. Agron.* **2021**, *3*, 661932. [CrossRef]
57. Fernandez-Figares, I.; Marinetto, J.; Royo, C.; Ramos, J.M.; Del Moral, L.G. Amino-acid composition and protein and carbohydrate accumulation in the grain of triticale grown under terminal water stress simulated by a senescing agent. *J. Cereal Sci.* **2000**, *32*, 249–258. [CrossRef]
58. Cottenie, A.; Verloo, M.; Kiekens, L.; Velghe, G.; Camerlynck, R. Chemical analysis of plants and soils. In *Laboratory of Analytical and Agrochemistry*; State University: Ghent, Belgium, 1982; pp. 14–24.
59. Lichtenthaler, H.K.; Buschmann, C. Chlorophylls and carotenoids: Measurement and characterization by UV-VIS spectroscopy. *Curr. Protoc. Food Anal. Chem.* **2001**, *1*, F4.3.1–F4.3.8. [CrossRef]
60. Pregl, F. *Quantitative Organic Micro Analysis*, 4th ed.; A. Churchill Ltd.: London, UK, 1945.
61. Kacar, B.; Inal, A. Plant analysis. Nobel publication No: 1241. *Appl. Sci.* **2008**, *63*, 879.
62. Irigoyen, J.J.; Emerich, D.W.; Sanchez-Dıaz, M. Water stress induced changes in concentrations of proline and total soluble sugar in nodulated alfalfa (*Medicago sativa*) plants. *Physiol. Plant.* **1992**, *84*, 55–60. [CrossRef]
63. Moore, S.; Stein, W.H. A modified ninhydrin reagent for The photometric determination of amino acids and related compounds. *J. Biol. Chem.* **1954**, *211*, 907–913. [CrossRef]
64. Grieve, C.M.; Grattan, S.R. Rapid assay for determination of water soluble quaternary ammonium compounds. *Plant Soil* **1983**, *70*, 303–307. [CrossRef]
65. Bates, L.S.; Aldren, R.P.; Teare, L.D. Rapid determination of free proline for water stress studies. *Plant Soil* **1973**, *39*, 205–207. [CrossRef]
66. Yih, R.Y.; Clark, H.E. Carbohydrate and protein content of boron deficient tomato root tips in relation to anatomy and growth. *Plant Physiol.* **1965**, *40*, 312–315. [CrossRef]
67. Lowry, O.H.; Rosenbrough, N.J.; Aarr, A.L.; Randaal, R.J. Protein measurement with folin phenol reagent. *J. Biol. Chem.* **1951**, *193*, 265–275. [CrossRef] [PubMed]

plants

MDPI

Article

Effects of Organic Acid Root Exudates of *Malus hupehensis* Rehd. Derived from Soil and Root Leaching Liquor from Orchards with Apple Replant Disease

Nan Sun [1], Chen Yang [1], Xin Qin [1], Yangbo Liu [1], Mengyi Sui [1], Yawen Zhang [1], Xueli Cui [1], Yijun Yin [1], Rong Wang [2], Yanli Hu [1], Xuesen Chen [1], Zhiquan Mao [1], Yunfei Mao [1,*] and Xiang Shen [1,*]

1 State Key Laboratory of Crop Biology, College of Horticulture Science and Engineering, Shandong Agricultural University, Tai'an 271002, China
2 College of Horticulture and Landscape, Tianjin Agricultural Univercity, Tianjin 301799, China
* Correspondence: myfmyf@sdau.edu.cn (Y.M.); shenx@sdau.edu.cn (X.S.)

Abstract: Organic acids secreted by plants, such as p-hydroxybenzoic acid, ferulic acid, cinnamic acid, and benzoic acid, can inhibit seed germination and root growth. The effects of root and soil leaching liquor from orchards on the growth of *M. hupehensis* Rehd. seedlings under sand culture are studied; the seedlings are sampled at 15, 30, 45, and 60 d. Changes in the amount of root exudates are determined using HPLC. Low concentrations of root leaching liquor (A1) and soil leaching liquor (B1) significantly promoted plant growth and chlorophyll synthesis; high concentrations of root leaching liquor (A6) and soil leaching liquor (B4–6) inhibited growth. Low concentrations of soil leaching liquor had no significant effect on the POD, SOD, and CAT activities. A5–6 and B5–6 significantly decreased *Fv/Fm* and qP values, respectively, and increased NPQ values. All root and soil leaching liquor treatments inhibited the secretion of gallic acid, hydroxybenzoic acid, benzoic acid, and phloridzin, and promoted the secretion of caffeic acid. The root leaching liquor treatments inhibited the secretion of catechin and promoted the secretion of phloretin. The soil leaching liquor treatments promoted the secretion of cinnamic acid. The secretion of other phenolic acids is likely associated with the different concentrations of leaching liquor.

Keywords: apple replant disease orchard; leaching liquor; root exudation; phenolic acids; *M. hupehensis* Rehd.

Citation: Sun, N.; Yang, C.; Qin, X.; Liu, Y.; Sui, M.; Zhang, Y.; Cui, X.; Yin, Y.; Wang, R.; Hu, Y.; et al. Effects of Organic Acid Root Exudates of *Malus hupehensis* Rehd. Derived from Soil and Root Leaching Liquor from Orchards with Apple Replant Disease. *Plants* **2022**, *11*, 2968. https://doi.org/10.3390/plants11212968

Academic Editor: Lorenzo Rossi

Received: 16 September 2022
Accepted: 1 November 2022
Published: 3 November 2022

1. Introduction

Replant disease refers to the phenomenon in which the same crop or closely related crops are repeatedly planted in the same soil and, even under normal cultivation and management conditions, their growth potential is weakened, their vulnerability to pests and diseases is aggravated, and yield and quality are reduced [1–4]. Apple (*Malus domestica*) replant disease occurs when old trees are removed from an orchard and the replanted young apple trees are short, tree vigor is weak, and their vulnerability to diseases and insect pests is aggravated, which seriously affects the yield and quality of fruit and impedes the sustainable development of the apple industry in China [5]. Its causes are manifold and include imbalances in the rhizosphere micro-ecosystem [6], the accumulation of phenolic acids [7], and the deterioration of soil physical and chemical properties [8]. Overcoming the challenges of apple replant disease is critically important for ensuring the sustainable development of the apple industry. China is the largest apple producer and consumer in the world [9]. The research on the effect of replant diseases on the environment and crops is important for minimizing losses caused by replant diseases and has major economic implications for China and all countries worldwide.

The importance of plant root secretions has been long recognized [10]. Macromolecular mucilaginous substances in root exudates include polysaccharides, phenolic compounds, and polygalacturonic acid [11]. Plant root exudates contain bioactive substances that inhibit or promote the growth of other adjacent plants or themselves, and these are referred to as allelochemicals. Apple roots secrete a large number of metabolites into the soil during growth and development, and the accumulation and transformation of these metabolites over long periods can cause disservices to trees. Toxins produced by apple roots include rhizomatites, rhizotin, light cinnamic acid, and light benzoic acid, which are associated with apple reimplantation disorders [12]. The effects of phenolic acids on plant growth mainly include the inhibition of seed germination, seedling root growth, seedling ion absorption, the activity of protective enzymes, hormone metabolism, the destruction of the structure of root cell membranes, and interference with DNA replication and RNA transcription, and other physiological and biochemical processes [13,14].

Previous studies have found that acids enter the plant through the cell membrane and alter the activity and function of some enzymes. Chlorogenic acid, caffeic acid, and catechol inhibit the activity of phosphorylase [15]. Its derivatives inhibit the hydrolytic activity of ATPase; tannic acid inhibits the activity of peroxidase (POD), catalase (CAT), and cellulase. Phenolic acids also affect the activity of phenylalanine ammonia-lyase. Callaway et al. (2000) reported that 50 μmol/L cinnamic acid and 4-hydroxy-3-methoxycinnamic acid reduce protein synthesis in lettuce seedlings [16]; 10–30 μmol/L caffeic acid, coumaric acid, ferulic acid, cinnamic acid, and vanillic acid inhibit the photosynthesis and chlorophyll content of soybean, which inhibits growth [17].

Various methods are used to collect root exudates, including the solution culture collection method and water extraction culture medium (vermiculite culture, sand culture, agar culture, and soil culture) collection method according to the culture system of the plant; root exudates can also be collected from the root system or soil [18]. Here, we conduct an experiment in which 20-year-old apple orchard continuous cropping soil and root extracts from replanted orchards are applied at different concentrations to *M. hupehensis* Rehd. seedlings under sand culture and irrigated. The growth of the seedlings is measured at 15, 30, 45, and 60 d. Changes in the content of root exudates, including the main organic phenolic acids, are analyzed. We examine the relationships of root and soil extract concentrations with various physiological indicators of seedlings at different stages, and the effects of different concentrations of soil and root extracts on the content of organic phenolic acids secreted by roots to provide insights that could aid the ability of farmers to overcome the challenges of replanting fruit trees.

2. Results

2.1. Standard Sample Chromatogram

See Figure S3 for a chromatogram of the standard PAH samples and Figure S4 for a chromatogram of the phenolic acid standard samples.

2.2. Effect of Soil and Root Extracts on the Growth of Seedlings of M. hupehensis Rehd.

2.2.1. Effect of Root Extract on Seedling Growth

Figure 1a–c shows the growth status of the aboveground parts of seedlings. Lower concentrations (A1) of root extracts promoted the growth of seedlings (RI > 0), and higher concentrations resulted in stronger inhibition (RI < 0). The allelopathy of root extracts was the strongest at 30 d of seedling growth, especially for treatment A6, and the absolute value of RI reached 0.70. Any concentration of root extracts can significantly inhibit the growth of the ground diameter of seedlings The degree of inhibition increased as the concentration applied increased, and the RI value peaked on the 30th day and then decreased with time. A1 and A2 stimulated chlorophyll synthesis in the leaves, and lower concentrations were associated with stronger stimulation of chlorophyll synthesis. The inhibition of chlorophyll synthesis was highest in A6, and the average RI absolute value reached 0.46 within 60 d.

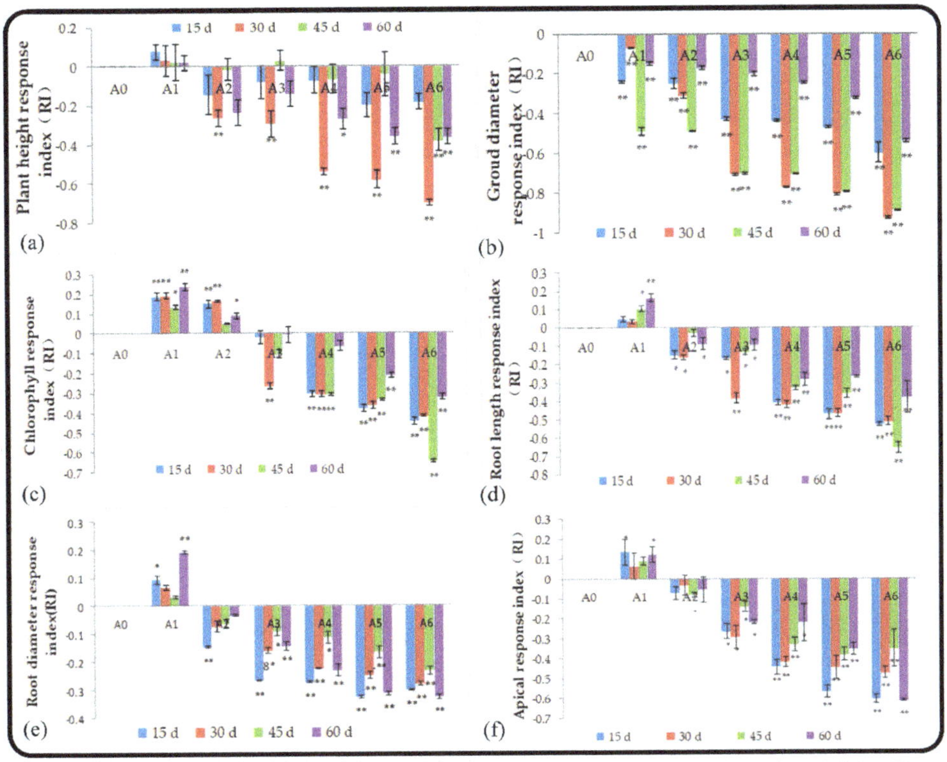

Figure 1. Effect of root extract on the height (**a**), diameter (**b**), and chlorophyll content (**c**), and the effect of root leaching liquor on the root length (**d**), root diameter (**e**), and number of root tips (**f**). Note: The displayed data are presented as mean ± SE (standard error) (n = 3). Error bars represent the standard errors of the means (n = 3), * correlation is significant at the 0.05 level, ** correlation is significant at the 0.01 level by Duncan's new multiple-range test.

The effects of root and soil extracts on seedling roots significantly varied among treatments (Figure 1d–f). The low-concentration root extract treatment (A1) can promote root elongation and thickening and increase the number of root tips; the average RI within 60 d was 0.087, 0.095, and 0.103. When the treatment concentration was higher than that in A1, the root extract had allelopathic effects and significantly inhibited the growth of seedling roots. At 15 d, the allelopathic effects of root extracts on the root growth of seedlings were strongest, and this occurred earlier than when the strength of the allelopathic effects on the shoots was highest.

2.2.2. Effect of Soil Extract on Seedling Growth

The growth status of the aboveground parts is shown in Figure 2a–c. Seedlings treated with concentrations lower than that in B3 showed the same growth pattern as the control during the test period. Low concentrations of soil extract promoted the growth of seedlings, and growth was increased at lower concentrations. The average RI of B1 at 60 d was 0.234. After 2 months, B1 significantly promoted the growth of seedlings, and B6 significantly inhibited the growth of seedlings. The growth rate in the ground diameter of all treatments gradually increased, but was lower in all treatments compared with the control. After 45 d, all treatments had a significant inhibitory effect on the growth in ground diameter, and the absolute value of B6RI reached 0.438. When the treatment concentration was that in B1, the content of synthesized chlorophyll in the leaves was higher than that in the control in the first 30 d; it then remained the same, decreased significantly, and was lower than that of the control at 60 d. At 30 d, the chlorophyll content decreased in the other treatments with the exception of B1, and the decrease was significant in B4, B5, and B6. The low concentration of soil extract had no effect on chlorophyll synthesis over short periods, but the strength of inhibition increased as the concentration increased and the treatment time extended.

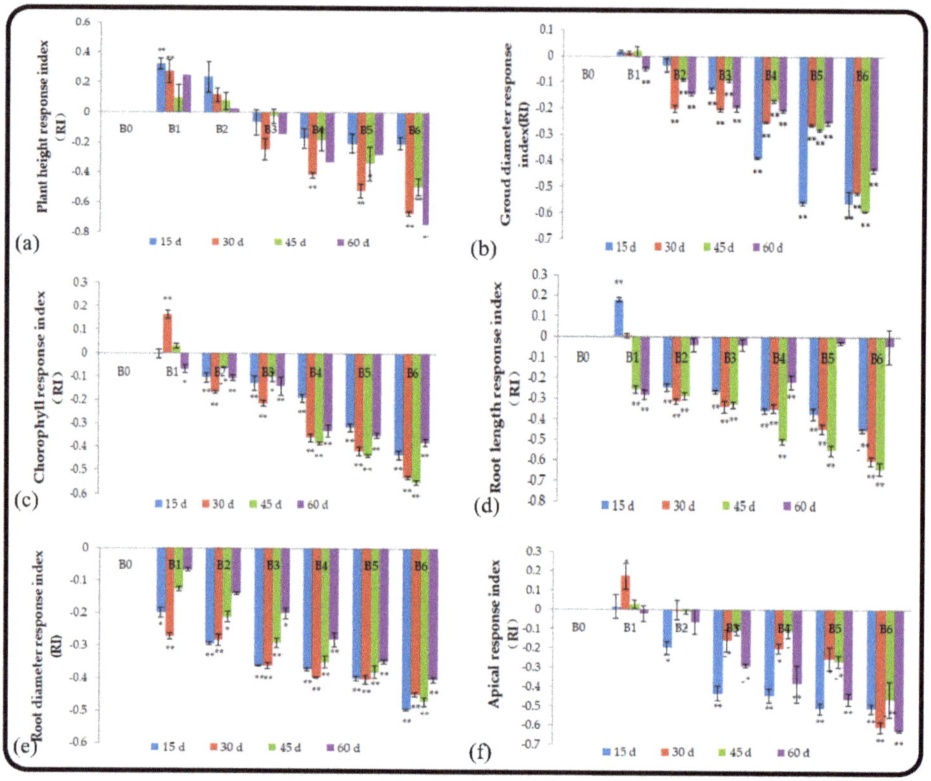

Figure 2. Effect of soil leaching liquor on plant height (**a**), diameter (**b**), and the chlorophyll content (**c**), and on root length (**d**), root diameter (**e**), and number of root tips (**f**). Note: The displayed data are presented as mean ± SE (standard error) (n = 3). Error bars represent the standard errors of the means (n = 3), * correlation is significant at the 0.05 level, ** correlation is significant at the 0.01 level by Duncan's new multiple-range test.

The effects of different soil extract treatments on seedling roots significantly varied (Figure 2d–f). In the first 30 d, the root elongation of *M. hupehensis* Rehd. seedlings was significantly inhibited by all treatments except treatment B1. After 2 months, treatments B1, B2, and B3 significantly inhibited root elongation, and treatments B4, B5, and B6 significantly inhibited root elongation. All treatments inhibited increases in root diameter during the experimental period, and the degree of inhibition increased with the treatment time and the concentration of leaching liquor applied. The inhibition rates of treatment B6 were 49.5%, 44.8%, 47.1%, and 47.7% at 15, 30, 45, and 60 d, respectively, and the inhibition was extremely significant. B1 could stimulate the formation of root tips in the first 45 d, which was beneficial to the absorption of nutrients by roots, and its stimulatory effect was strongest at 30 d when the number of root tips was 20.8% higher compared with that in the control; the stimulatory effect weakened, and no significant effect of leaching liquor on the formation of root tips was observed after 2 months. All the other treatments inhibited the formation of root tips, and the degree of inhibition increased as the concentration of leaching liquor applied increased. The maximum level of inhibition was observed in treatment B6 at 30 d, when the inhibition rate was 60.9% and the RI value was 0.627, indicating that allelopathy was strong.

2.3. *Effect of Soil and Root Extracts on the Activity of Root Protective Enzymes*

2.3.1. Effect on POD

POD mainly catalyzes the decomposition of hydrogen peroxide and some phenolic substances. The high activity of POD indicates the oxidation of membrane lipids and increases in membrane permeability, which negatively affects plant growth. Within 60 d, the root POD activity of *M. hupehensis* Rehd. seedlings treated with root extracts first increased, decreased, and then increased (Figure 3a). The activity of POD in the roots of seedlings treated with A1 and A2 was higher than that of the control, indicating that the low-concentration root extract treatment could stimulate the synthesis of POD enzymes and protect root growth. As the amount of root extract applied increased, the degree of inhibition of seedling root growth increased. The activity of POD in treatment A6 at 15 d, 30, 45, and 60 d was 52.6%, 61.9%, 52.2%, and 61.1% for the control, respectively, and the degree of inhibition was significant. The activity of POD in the roots of seedlings treated with soil and root extracts exhibited the same pattern over the 2-month period. Treatment B1 promoted the activity of POD in the first 45 d, but the effect was not significant; treatment B5 significantly inhibited the activity of POD, and treatment B6 significantly inhibited the activity of POD, indicating that the soil extract had a greater impact on the activity of POD in the roots of seedlings than the root extract.

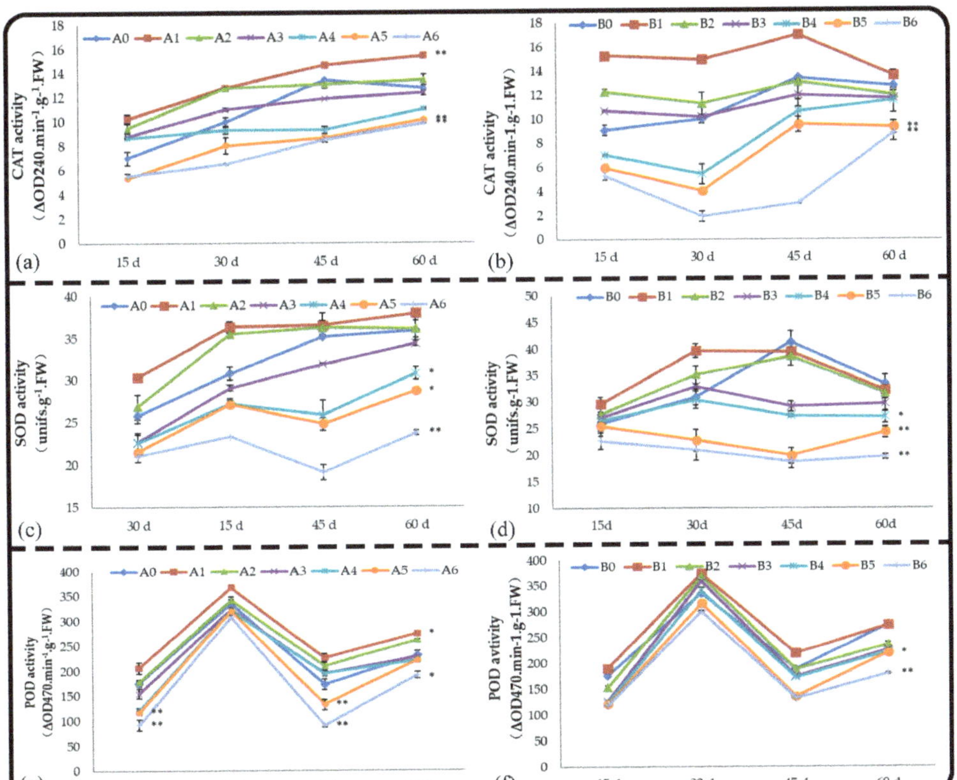

Figure 3. Effect of root and soil leaching liquor on POD activity (**a**,**b**), SOD activity (**c**,**d**), and CAT activity (**e**,**f**). Note: The displayed data are presented as mean ± SE (standard error) ($n = 3$). * Correlation

is significant at the 0.05 level, ** correlation is significant at the 0.01 level by Duncan's new multiple-range test.

2.3.2. Effect on SOD

Many studies have shown that SOD activity increases under moderate stress levels to mediate adaptation to stress. The SOD activity of the root system of *M. hupehensis* Rehd. in the low-concentration root extract treatment and the control treatment increased with treatment time, and the SOD activity of the root system of *M. hupehensis* Rehd. in the high-concentration root extract treatment decreased at 45 d (Figure 3c). Treatments A1 and A2 increased the activity of SOD in the roots in the first 45 d, and treatments A4, A5, and A6 significantly decreased the activity of SOD in the roots by 26.5%, 29.4%, and 45.6%, respectively, at 45 d. B1, B2, and B3 promoted increases in SOD activity in the first 30 d, but as the stress treatment time extended, SOD activity decreased. Treatments B5 and B6 significantly inhibited the SOD activity of roots from the beginning of the experiment, and the degree of inhibition increased with the treatment time (Figure 3d).

2.3.3. Effect on CAT

In the process of scavenging superoxide anion free radicals, SOD forms H_2O_2, which is harmful to cells. CAT has the function of scavenging H_2O_2. CAT, SOD, and POD protect the membrane system from the harm of free radicals. The CAT activity of seedling roots treated with different root extracts increased with treatment time, and this same pattern was observed in the control (Figure 3e). At 15 d, treatments with concentrations lower than A4 promoted the growth of seedlings. At 30 d, treatments with concentrations lower than A3 also promoted seedling growth. At 45 d, the CAT activity of treatment A1 was higher than that of the control. At 60 d, treatment A1 significantly promoted the activity of CAT, whereas treatments A5 and A6 significantly inhibited CAT activity, which affected seedling growth. Treatment B1 significantly promoted CAT activity in the first 45 d, but this stimulatory effect was not observed after 45 d. Treatments B5 and B6 significantly inhibited the activity of CAT in the roots of *M. hupehensis* Rehd. seedlings. The low-concentration soil extract treatments promoted seedling growth and enzyme activities; however, as the treatment time and concentration of leaching liquor applied extended, the degree of stress increased, the strength of allelopathy increased, and the seedlings finally lost their resistance, which led to decreases in root protective enzyme activities.

2.4. *Effect of Soil and Root Extracts on the Chlorophyll Fluorescence of* M. hupehensis *Rehd. Seedlings*

Pn only reflects the apparent photosynthetic capacity, and the specific internal mechanism affecting photosynthesis can be revealed by chlorophyll fluorescence parameters. The chlorophyll fluorescence parameters of *M. hupehensis* Rehd. seedlings in different treatments were measured, and the maximum photochemical efficiency of PSII was measured using the following formula: $F_v/F_m = (F_m - {}_m/F_m)$. The photochemical quenching coefficient was measured using the following formula: $qP = (F_m' - F_s)/(F_m' - {}_m')$. The non-photochemical quenching coefficient was measured using the following formula: $NPQ = (F_m - F_m')/F_m'$.

2.4.1. Effect on F_v/F_m

F_v/F_m represents the primary light-energy-conversion efficiency and the potential quantum efficiency of PSII, which is also known as the maximum photochemical efficiency of PSII and is proportional to the photosynthetic electron-transport activity. The effects of different concentrations of root and soil extracts on F_v/F_m within 60 d were consistent in the treatments and the control (Figure 4). The F_v/F_m values of the seedlings treated with low concentrations of leaching liquor were higher than those of the control (Figure 4a), but no significant difference was observed between treatments and the control. Treatments A5, A6, and A7 had significant inhibitory effects on F_v/F_m throughout the experiment. The

F_v/F_m of seedlings treated with high concentrations of soil extract significantly decreased at the beginning of the treatment compared with the control, and the difference was most significant at 30 d; although this difference subsequently decreased, it remained significant.

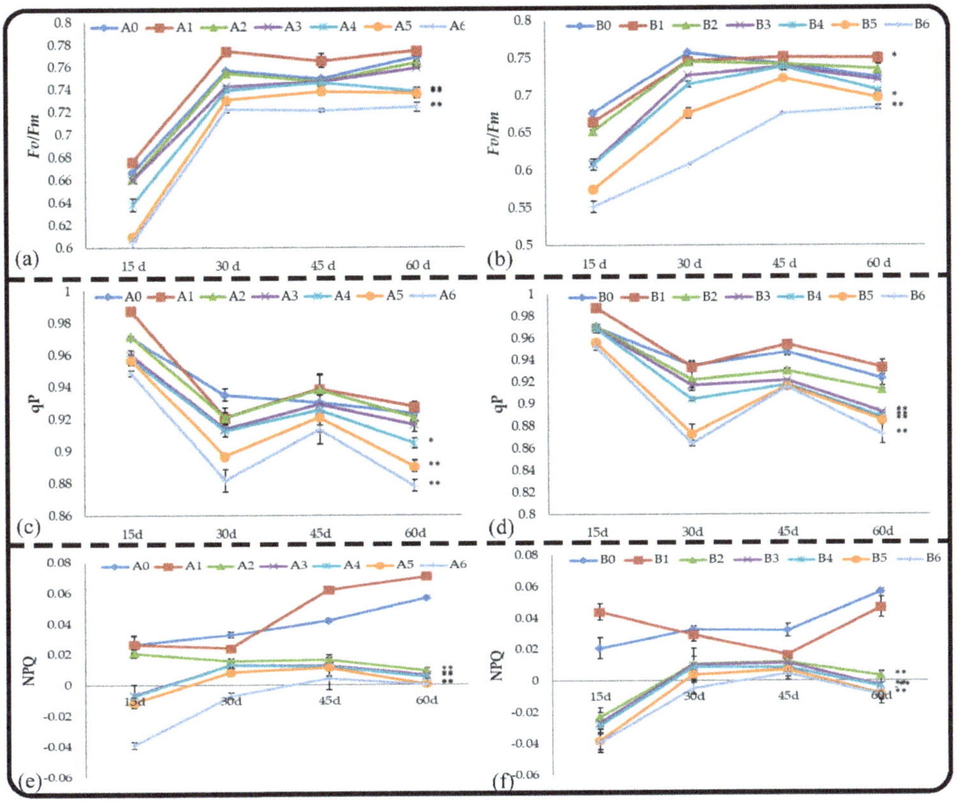

Figure 4. Effect of root and soil leaching liquor on *Fv*/*Fm* (**a**,**b**), qP (**c**,**d**), and NPQ (**e**,**f**). Note: The displayed data are presented as mean ± SE (standard error) (n = 3). * Correlation is significant at the 0.05 level, ** correlation is significant at the 0.01 level by Duncan's new multiple-range test.

2.4.2. Effect on qP

The photochemical quenching coefficient (qP) is the proportion of light energy absorbed by PSII antenna pigments for photochemical reaction electron transfer, that is, the proportion of excitation energy entering the photosynthetic electron-transport system through QA oxidation for photochemical assimilation. It is also a measure of the oxidation state of the primary electron acceptor QA, which reflects the relative number of PSII open centers. Larger qP values indicate a greater amount of QA- is reoxidized to QA and a higher probability of PSII electron transfer. Within 2 months, the effects of root and soil extract treatments on the leaf qP of *M. hupehensis* Rehd. seedlings were similar (Figure 4c,d); leaf qP first decreased, increased, and then decreased. Treatments A5, A6, B3, B4, B5, and B6 significantly inhibited leaf qP.

2.4.3. Effect on NPQ

The non-photochemical quenching coefficient NPQ reflects the light energy absorbed by PSII antenna pigments that cannot be used for photosynthetic electron transport and dissipates in the form of heat. Patterns of change in NPQ were opposite that of qP. The effects of root and soil extracts on NPQ were consistent. Low = concentration leaching liquor treatments had no effect on the NPQ value; the high-concentration treatments A5, A6, B5, and B6 significantly increased the NPQ value (Figure 4e,f).

2.5. Effect of Soil and Root Extracts on the Phenolic Acids Secreted from the Roots of M. hupehensis Rehd. Seedlings

2.5.1. Catechin

Catechins and their oxidation products are ternary compounds of carbon, hydrogen, and oxygen that are synthesized from sugars through a series of enzymes that mediate the formation of benzene-ring compounds through the shikimic acid pathway. Catechins are phenolic substances that can precipitate heavy metals and proteins [19]. Under normal growth conditions, catechin is the main substance secreted by the root system of *M. hupehensis* Rehd. seedlings in the late growth stage, and treatments B3, B4, B5, and B6 had no effect on catechin secretion; however, the other treatments promoted catechin secretion of the root system, as catechin was the main component secreted in the early growth stage of the root system (Figure 5a,b). Treatments B5 and B6 significantly inhibited the secretion of catechins in the roots of *M. hupehensis* Rehd. seedlings, and higher concentrations of leaching liquor resulted in a stronger inhibition of catechin secretion. There were significant differences in the effects of root and soil extracts applied at the same concentration on the catechin secretion of seedlings, and the amount of catechin secreted by the roots of seedlings treated with soil extracts was higher than that secreted by the roots of seedlings treated with root extracts.

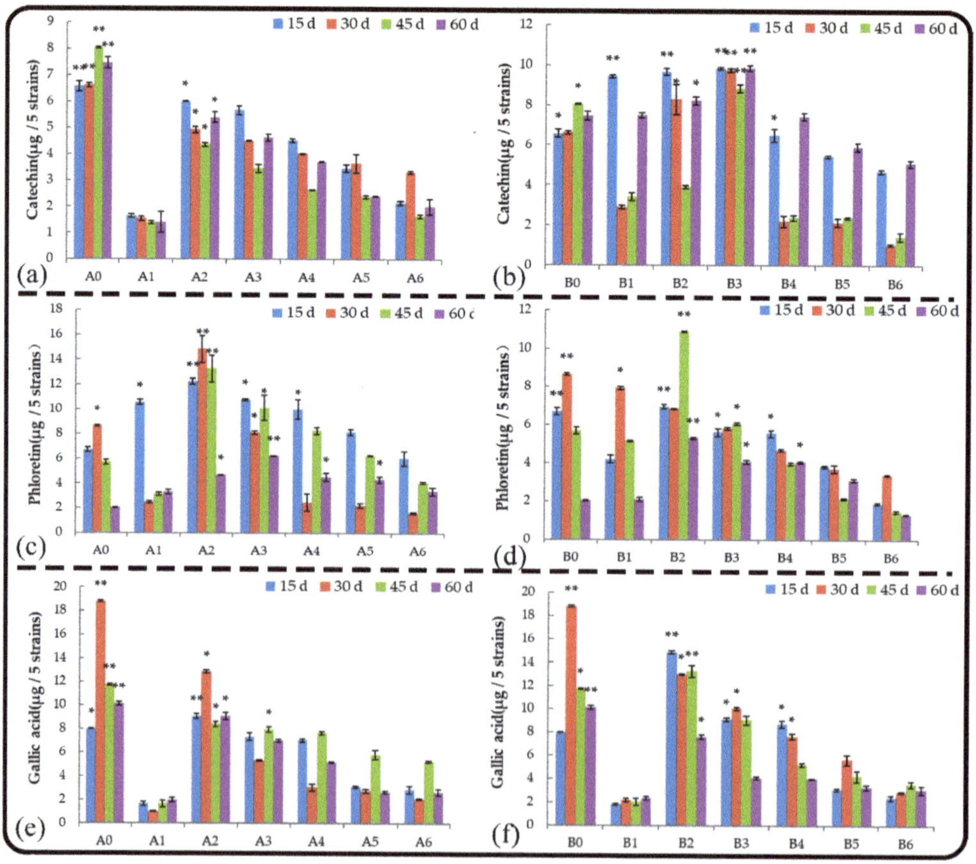

Figure 5. Effect of root and soil leaching liquor on the secretion of catechin (**a**,**b**), phloretin (**c**,**d**), and

gallic acid (**e**,**f**). Note: The displayed data are presented as mean ± SE (standard error) (n = 3). * Correlation is significant at the 0.05 level, ** correlation is significant at the 0.01 level by Duncan's new multiple-range test.

2.5.2. Phloretin

Phloretin mainly exists in the peel and root bark of juicy fruits, such as apples and pears. Under normal growth conditions, phloretin was the main substance secreted by the roots of *M. hupehensis* Rehd. seedlings during the early stage of growth (Figure 5c,d). Treatments B2 and B3 delayed the secretion of phloretin, and the other treatments had no effect on phloretin secretion. Root extracts played a major role in promoting the secretion of phloretin in the roots of *M. hupehensis* Rehd. seedlings. At the end of the treatment, the stimulatory effect first increased and then decreased as the concentration applied increased, and the stimulatory effect was extremely significant. At the end of the treatment period, B1 had no effect on the secretion of phloretin, B6 significantly inhibited the secretion of phloretin, and the other treatments had significant stimulatory effects; the strength of the stimulatory effect decreased as the concentration of leaching liquor applied increased. The effects of root extract on phloretin secretion were stronger than the effects of soil extract when the same concentration of leaching liquor was applied.

2.5.3. Gallic Acid

Gallic acid is an important secondary metabolite of plants, and its secretion is affected by both genetic factors and the environment. Under normal growth conditions, gallic acid is the main substance secreted by the roots of *M. hupehensis* Rehd. seedlings during the early stage of growth (Figure 5e,f). Root extract treatment delayed the secretion of gallic acid by the roots, whereas soil extract treatment had no effect on gallic acid secretion. Root and soil extract treatments mainly inhibited the secretion of gallic acid by the roots. At 15 d, treatments A2, B2, B3, and B4 promoted the secretion of gallic acid from *M. hupehensis* Rehd. roots for a short period, and the secretion of gallic acid was inhibited as the treatment time extended. The exudation of gallic acid was significantly inhibited by treatments A1 and B1 and high concentrations of root and soil extracts; high concentrations of root and soil extracts resulted in strong inhibition. When the concentration of the extracts was higher than that of A4 and B4, there was no significant difference in the degree of inhibition of gallic acid secretion among treatments.

2.5.4. Chlorogenic Acid

Chlorogenic acid is a depside formed by caffeic acid and quinic acid; it is a phenylpropanoid compound produced by the shikimate pathway during the aerobic respiration of plants. Under normal growth conditions, chlorogenic acid was the main substance secreted by the root system of *M. hupehensis* Rehd. seedlings during the early stage of growth (Figure 6a,b). Treatments A2 and B2 delayed the secretion of chlorogenic acid by the root system, and other treatments had no effect on chlorogenic acid secretion. Within 2 months, treatments A1, A2, and B3 significantly promoted the secretion of chlorogenic acid by roots, and the stimulatory effect first increased and then decreased with time; treatments A5, A6, and B6 significantly inhibited the secretion of chlorogenic acid by the roots. The amount of chlorogenic acid secreted by the roots of *M. hupehensis* Rehd. seedlings under each root extract treatment was higher than that under each soil extract treatment in the same low-concentration treatments, and the amount of chlorogenic acid secreted was lower in these treatments compared with the high-concentration soil extract treatments.

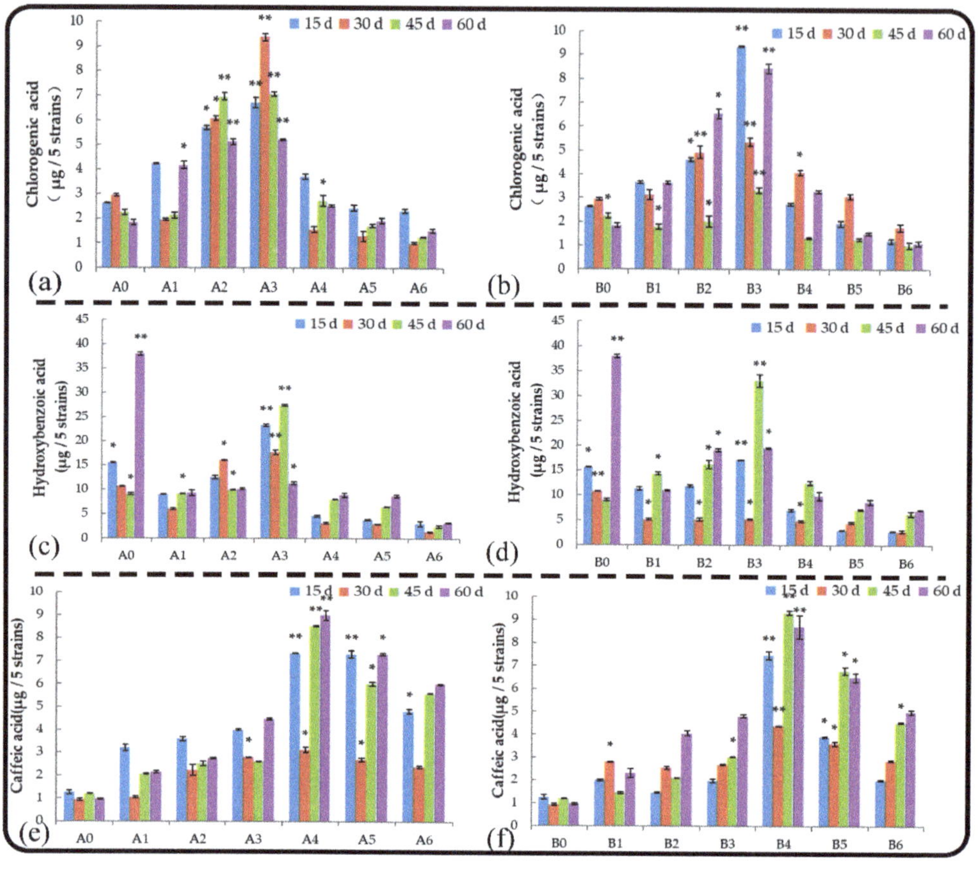

Figure 6. Effect of root and soil leaching liquor on the secretion of chlorogenic acid (**a**,**b**), hydroxybenzoic acid (**c**,**d**), and caffeic acid (**e**,**f**). Note: The displayed data are presented as mean ± SE (standard error) (n = 3). * Correlation is significant at the 0.05 level, ** correlation is significant at the 0.01 level by Duncan's new multiple-range test.

2.5.5. Para-Hydroxybenzoic Acid

Hydroxybenzoic acid is one of the main root exudates of *M. hupehensis* Rehd; it is also one of the main autotoxic substances secreted by strawberry roots, and it has a significant inhibitory effect on the growth of roots. Under normal growth conditions, p-hydroxybenzoic acid is the main substance secreted by the root system of *M. hupehensis* Rehd. seedlings during the late growth stage (Figure 6c,d), and no treatments affected p-hydroxybenzoic acid secretion. Treatment A3 significantly promoted the secretion of p-hydroxybenzoic acid in the first 45 d and then strongly inhibited its secretion. Within 2 months, root extracts of treatments A5 and A6 and soil extracts of all treatments (except treatment B3) significantly inhibited the secretion of p-hydroxybenzoic acid by the roots of seedlings, and the inhibitory effect of root extracts was stronger than that of soil extracts when they were applied at the same concentrations.

2.5.6. Caffeic Acid

Caffeic acid is one of the main allelochemicals secreted by plant roots, and it is mainly produced by the secondary metabolism of shikimic acid. Under normal growth conditions, caffeic acid is the main substance secreted by the roots of *M. hupehensis* Rehd. seedlings during the early growth stage, and treatments A1, A2, and B1 had no effect on caffeic acid secretion (Figure 6e,f); caffeic acid secretion by the roots was delayed in the other extract treatments. Root extracts and high concentrations of soil extracts promoted the secretion of caffeic acid from the roots of *M. hupehensis* Rehd. seedlings, and the magnitude of the stimulatory effect first increased and then decreased as the amount of leaching liquor applied increased.

2.5.7. Vanillin

Vanillin is one of the main allelochemicals secreted by roots; it can inhibit the growth of plant roots, destroy the structure of cell membranes, and inhibit seed germination. Under normal growth conditions, vanillin was the main substance secreted by the root system of *M. hupehensis* Rehd. seedlings during the late growth stage (Figure 7a,b). Treatment B1 promoted vanillin secretion by the root system, and other treatments had no effect on vanillin secretion. Within 2 months, treatments A4 and B4 significantly promoted vanillin secretion from the roots, and treatments A1 and B1 significantly inhibited vanillin secretion from the roots. Vanillin secretion was significantly lower when roots were treated with the same low concentrations of root extract compared with soil extract, and the opposite effect was observed when high concentrations of extracts were applied.

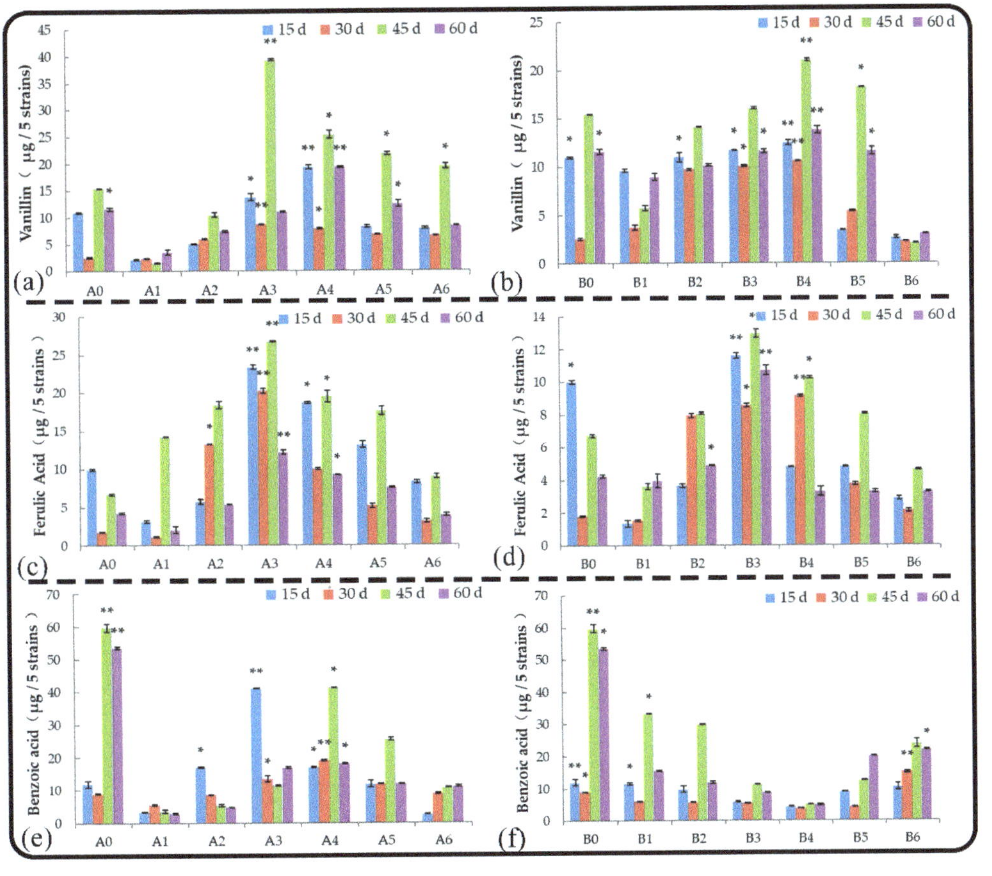

Figure 7. Effect of root and soil leaching liquor on the secretion of vanillin (**a**,**b**), ferulic acid (**c**,**d**), and benzoic acid (**e**,**f**). Note: The displayed data are presented as mean ± SE (standard error) (n = 3). * Correlation is significant at the 0.05 level, ** correlation is significant at the 0.01 level by Duncan's new multiple-range test.

2.5.8. Ferulic Acid

Ferulic acid is one of the derivatives of cinnamic acid. Ferulic acid was first detected in the seeds and leaves of plants. It is a phenolic acid that is widespread in plants. It combines with polysaccharides and proteins in the cell wall to form the skeleton of the cell wall. It is also one of the main allelochemicals secreted by roots. Root and soil extract treatments delayed the secretion of ferulic acid by the roots (Figure 7c,d). The effects of root and soil extracts on the secretion of ferulic acid from *M. hupehensis* Rehd. seedlings were consistent as the amount of leaching liquor applied increased. A stimulatory effect of leaching liquor on ferulic acid secretion was observed as the amount of leaching liquor applied increased; however, beyond a certain value, the stimulatory effect decreased, and ferulic acid secretion was observed. Treatments A3, A4, and A5 significantly promoted the secretion of ferulic acid in the roots, and treatment A3 had the most significant effect. The secretion of ferulic acid by the roots was three times higher in treatment A3 compared with the control. The effect of soil extract on the secretion of ferulic acid by the roots was relatively weak, and the effect of treatment B3 was significant compared with that of the control. The amount of ferulic acid secreted by the roots was 0.6 times higher in treatment B3 compared with the control.

2.5.9. Benzoic Acid

Benzoic acid is one of the main root exudates of *M. hupehensis* Rehd. It has significant allelopathic effects, destroys membrane permeability, causes root cell membrane dysfunction, and affects the absorption of nutrients by roots. Under normal growth conditions, benzoic acid was the main substance secreted by the root system of *M. hupehensis* Rehd. seedlings during the late growth stage (Figure 7e,f). Low-concentration root extract treatments advanced benzoic acid secretion, and the other treatments had no effect on benzoic acid secretion. In the later stage of the treatments, the root and soil extracts mainly inhibited the secretion of benzoic acid by seedlings, and the inhibition was strongest in treatments A1, A2, B3, and B4. The inhibition rates of treatments A1 and B4 were 74.75% and 75.66%, respectively, and there was no significant difference between them. In the root extract treatments, the amount of benzoic acid secreted by roots first increased and then decreased as the amount of root extract applied increased. In the soil extract treatments, the amount of benzoic acid secreted first decreased and then increased as the amount of soil extract applied increased, and this pattern was opposite the effect of root extracts. Benzoic acid secretion was significantly higher in A3 and A4 than in B3 and B4, respectively, and benzoic acid secretion was significantly lower in the root extract treatments than in the soil extract treatments at other concentrations.

2.5.10. Phloridzin

Phloridzin is a glycoside that is abundant in the bark, root system, and other organs of apple trees. A large amount of phloridzin is released into the soil via root secretion and residue decomposition, and this results in a decline in apple tree growth, yield, and quality. Phloridzin was the main substance secreted by the root system of *M. hupehensis* Rehd. seedlings during the late growth stage under normal growth conditions (Figure 8a,b), and all treatments had no effect on phloridzin secretion. The amount of phlorizin secreted first increased and then decreased as the concentration of leaching liquor applied increased. Root and soil extract treatments inhibited phloridzin secretion by seedlings. Phloridzin secretion by seedlings was higher when they were treated with root extract than with soil extract at the same low concentrations; the opposite pattern was observed in the high-concentration treatments.

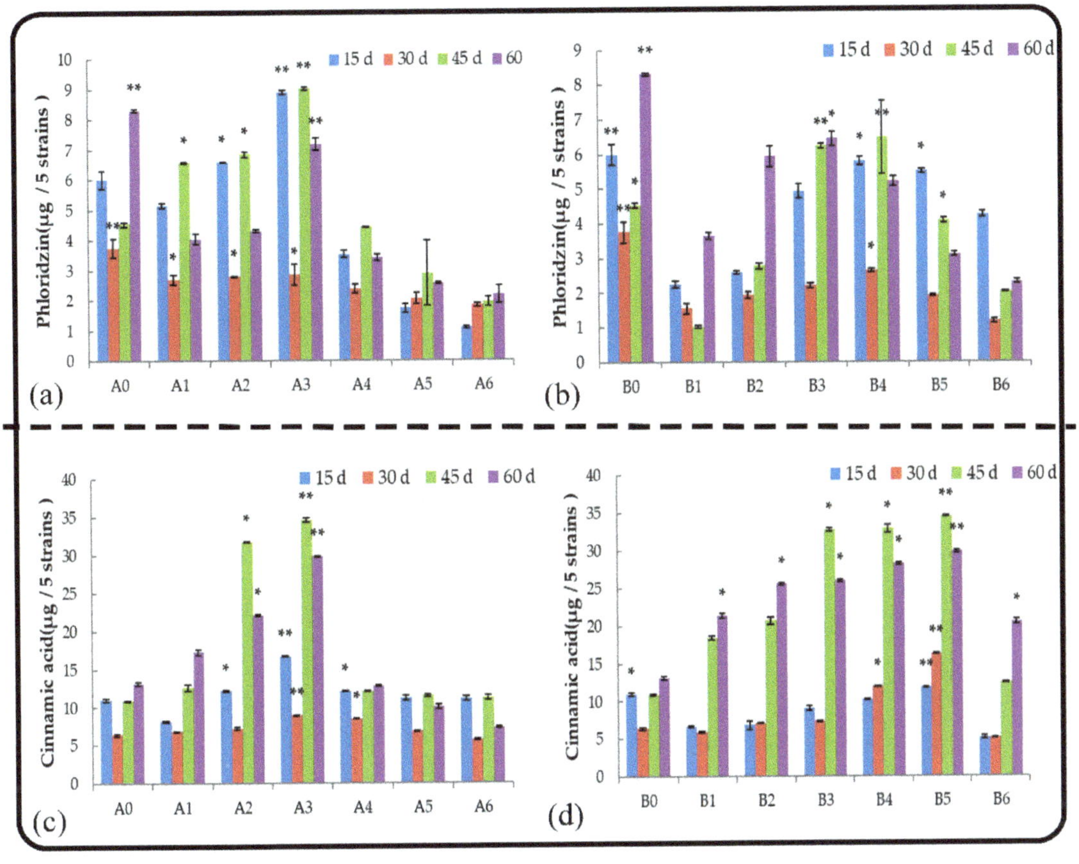

Figure 8. Effect of root and soil leaching liquor on the secretion of phloridzin (**a**,**b**) and cinnamic acid (**c**,**d**). Note: The displayed data are presented as mean ± SE (standard error) ($n = 3$). * Correlation is significant at the 0.05 level, ** correlation is significant at the 0.01 level by Duncan's new multiple-range test.

2.5.11. Cinnamic Acid

Cinnamic acid, which is produced by the deamination of phenylalanine, is the main root exudate of *M. hupehensis* Rehd. It affects plant growth by affecting ion absorption, water absorption, photosynthesis, and protein and DNA syntheses. Low-concentration root extract treatments and all soil extract treatments promoted the secretion of cinnamic acid by the roots (Figure 8c,d), and the stimulatory effect first increased and then decreased as the concentration of extract applied increased. At 45 d, A3 had the strongest stimulatory effect among root extract treatments, and cinnamic acid secretion was 2.17 times higher in A3 compared with that of the control. B5 had the strongest stimulatory effect in the treatment of soil extract, and cinnamic acid secretion was 2.15 times higher in B5 compared with that of the control. At the end of the treatments, low concentrations of root extracts promoted the secretion of cinnamic acid by the roots, and high concentrations of root extracts had a strong stimulatory effect on cinnamic acid secretion. All soil extract treatments significantly promoted the secretion of cinnamic acid. The stimulatory effect first increased and then decreased as the concentration applied increased. The cinnamic acid secretion of seedlings treated with soil extract was higher than that treated with root extract.

2.5.12. Correlation between the Growth of *M. hupehensis* Rehd. and Catechin, Phloretin, and Gallic Acid Secretions

Catechin, phloretin, and gallic acid are the main components of root exudates of *M. hupehensis* Rehd. seedlings. In this study, the correlations between the growth of *M. hupehensis* Rehd. seedlings and the amount of catechin, phloretin, and gallic acid secreted by roots was analyzed. Under normal growth conditions, the secretion of catechin was not related to plant height, ground diameter, and the chlorophyll content (Figures 9 and 10). There was a significant positive correlation in the secretion of these three root exudates between the root and soil extract treatments. Phloretin secretion was significantly correlated with plant height and basal diameter, and the correlation disappeared after treatment with root and soil extracts. The secretion of gallic acid was not correlated with plant height, basal diameter, and the chlorophyll content, but it was significantly positively correlated with plant height and chlorophyll after treatment with root extract and significantly positively correlated with seedling growth after treatment with soil extract.

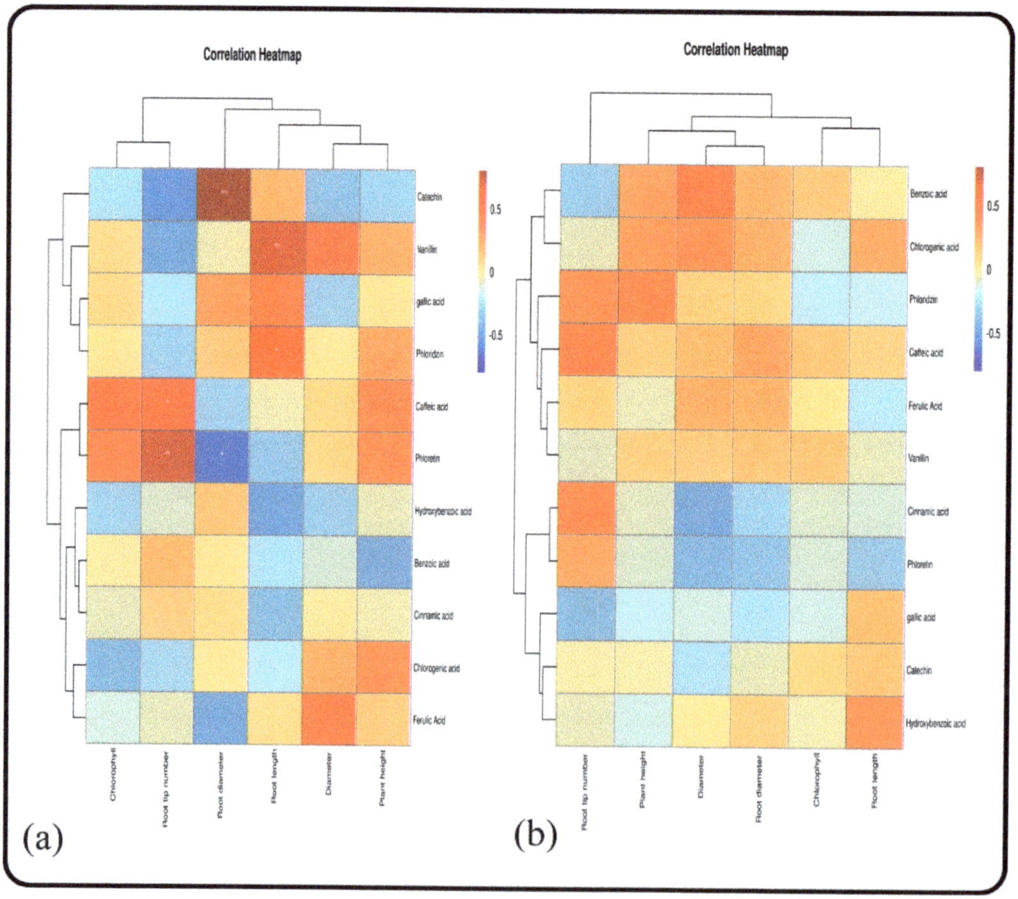

Figure 9. Correlations between seedling growth and phenolic acids of *M. hupehensis* Rehd. at different stages: 15 d (**a**) and 30 d (**b**). Note: * Correlation is significant at the 0.05 level, ** correlation is significant at the 0.01 level.

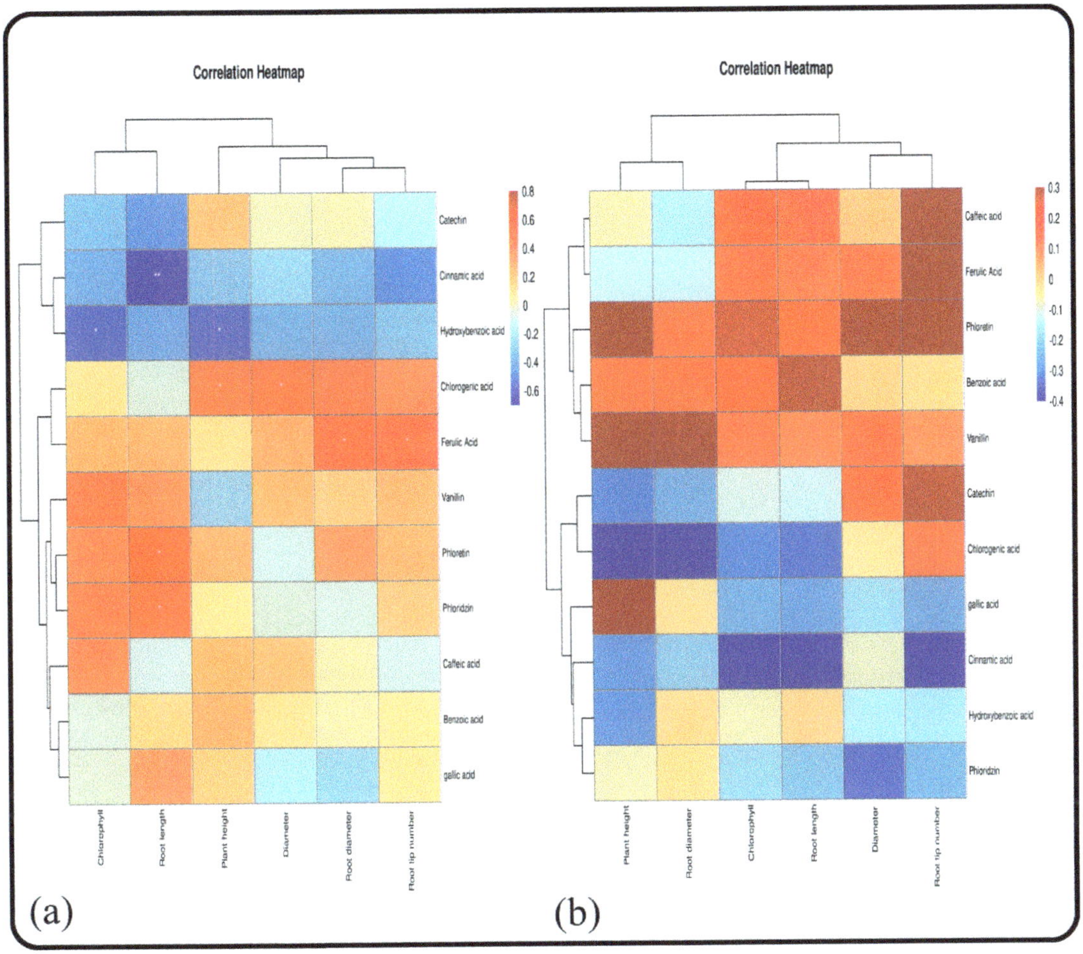

Figure 10. Correlations between seedling growth and the phenolic acids of *M. hupehensis* Rehd. at different stages: 45 d (**a**) and 60 d (**b**). Note: * Correlation is significant at the 0.05 level, ** correlation is significant at the 0.01 level.

3. Discussion

Phenolic acids, including chlorogenic acids, are major polyphenolic compounds in Jerusalem artichoke (*Helianthus tuberosus* L.). Phenolic acids are common allelochemicals in soil [20–22]. Our findings indicated that low concentrations of root and soil extracts could promote the growth of *M. hupehensis* Rehd. seedlings, and the degree of growth inhibition was positively correlated with the amount of extract applied. Growth inhibition first increased, decreased, and then increased as the treatment time extended. Allelopathy was positively correlated with the treatment concentration, and the strength of allelopathy first increased and then decreased as the treatment time extended. Low-concentration soil extract treatments had no significant effect on the ground diameter of seedlings and the synthesis of chlorophyll in leaves. The high-concentration root and soil extract treatments inhibited increases in ground diameter and the synthesis of porphyrin, accelerated the decomposition of chlorophyll, reduced the content of chlorophyll in leaves, and caused the leaves lose their green appearance [23]. The degree of inhibition was positively correlated with the concentration of leaching liquor applied. Changes in seedling height, net growth rate of ground diameter, and chlorophyll content were analyzed, and the seedling height and net growth rate of ground diameter were higher in the soil extract treatments than in the root extract treatments when the same concentrations of extract were applied. The content of phenolic acid allelochemicals in the root extract was higher than that in the soil extract, and phenolic acid and other allelochemicals have been shown to have significant effects on the aboveground growth of crops [24]. The chlorophyll content was lower in the soil extract treatments than in the root extract treatments when the same concentration of extract was applied.

Changes in the chlorophyll fluorescence of the leaves of *M. hupehensis* Rehd. seedlings in the different root and soil extract treatments were studied. When low concentrations of these extracts were applied, they increased the chlorophyll fluorescence of the leaves. However, the extracts did not have a stimulatory effect on the chlorophyll fluorescence when they were applied at high concentrations. The mechanism of the photosynthetic system was disrupted, the rate of photosynthetic electron transport decreased, the number of PSII open centers decreased, the proportion of light energy consumed in the form of heat dissipation increased, photosynthesis was hindered, and the growth of seedlings was inhibited.

The pollution of orchard soil is becoming increasingly widespread. There are several organic pollutants in soil aside from phenolic acid allelochemicals. We detected the content of PAHs in the soil, and the level of PAH pollution in the orchard soil was moderate (833.88 ng/G); the high ecological risks associated with PAH pollution can have adverse effects on crops. Ren et al. pointed out that allelochemicals can significantly inhibit the growth of cucumber roots [25]. The results of this study show that low concentrations of root extract can promote root elongation and thickening and stimulate root tip differentiation. However, above a certain concentration, root extracts can inhibit root growth and have allelopathic effects. Replant stress would thus have an adverse effect on the growth of crop roots, and this is consistent with the results of previous studies. During the experimental period, the root elongation of *M. hupehensis* Rehd. seedlings was significantly inhibited by the soil extract treatment, and the degree of inhibition was positively correlated with the concentration applied. Arabidopsis thaliana can absorb the three-ring PAH phenanthrene, and symptoms of phenanthrene stress, including root and seedling growth inhibition, fragrant hair tuft deformities, and root hair reductions, were observed [26]. Phenanthrene, pyrene, and 1,2,4-trichlorobenzene can inhibit the root elongation of taller plants (wheat, cabbage, and tomato), and there was a significant linear or logarithmic correlation between the concentration of leaching liquor applied and the inhibition rate of root elongation ($p = 0.05$) [27]. The results of these studies are consistent with the results of this experiment and confirm that the presence of PAHs in the soil has adverse effects on crops.

The root system is the first organ of plants exposed to harmful substances, and changes in root structure directly reflect the effects of harmful substances on plant growth and development [28]. Allelopathic effects on the roots of *M. hupehensis* Rehd. seedlings treated with root extract appeared earlier and were stronger compared with the shoots of seedlings. These findings indicated that phenolic acids that initially accumulated in the roots are not transported upward but accumulate in the roots, and the roots are damaged first. When the concentration exceeded a certain level, the phenolic acids accumulated in the roots and were transported to the shoot, which resulted in poor growth of the aboveground parts. Soil extracts had effects on the shoot and root growth of *M. hupehensis* Rehd. seedlings at the beginning of the treatment period, but the effect on root growth was stronger than that on shoot growth. This might stem from the fact that the root system absorbs PAHs in the soil and transports them upward; although the root system is affected by PAHs, the growth of the aboveground parts is also inhibited. However, the amount of PAHs transported upward is less than the amount of PAHs that the root system is exposed to, and this might explain why the root system is affected to a greater degree compared with the aboveground parts. The root parameters of seedlings treated with soil extract were lower than those of seedlings treated with root extract when soil and root extracts were applied at the same concentrations, and the effect of soil extract on the growth of seedling roots was greater.

A large number of microorganisms and PAHs occur in the soil in addition to phenolic acid allelochemicals, and these can have major effects on the development of the roots of seedlings and play an important role in mediating the effects of apple replant disease.

Recent studies have shown that plants produce more oxygen free radicals under stress conditions, and this aggravates the peroxidation of the plasma membrane and leads to the destruction of the membrane system [29]. The results of this study show that the allelopathic effects of root extract are closely related to the concentration of root extract applied. Changes in the POD activity of *M. hupehensis* Rehd. roots were similar among treatments: first increasing, decreasing, and then increasing. Low-concentration treatments can promote increases in POD and CAT activities, indicating that within a certain time and at certain concentrations, plants can activate their own defense mechanisms to resist stress [30]; intermediate concentrations are beneficial to the growth of plants. The activities of POD and CAT were not significantly affected by leaching liquor application. The activities of POD and CAT in roots of *M. hupehensis* Rehd. seedlings were significantly inhibited by the high-concentration treatments, the activities of enzymes were reduced, the protective mechanism in response to stress was ineffective, and the growth of crops was affected; these findings are consistent with the results of previous studies. Low-concentration treatments did not affect the growth of *M. hupehensis* Rehd. seedlings; these treatments even promoted increases in the activity of protective enzymes, root respiration, the growth of the root system, the synthesis of chlorophyll, and other physiological functions. The high-concentration treatments decreased the activity of root protective enzymes, weakened the molecular oxygen and H_2O_2-scavenging ability of the antioxidative defense system, and led to an imbalance in the production and elimination of active oxygen, the accumulation of oxygen free radicals, increases in the content of membrane peroxidation products, and disruption of the structure and function of the cell membrane. These effects alter the structure of chloroplasts and mitochondria and the content of hormones, and damages the ultrastructure of chloroplasts and mitochondria, which affects photosynthesis and inhibits the normal growth of plants [31,32].

The research on the phenolic acids secreted by crop roots under stress conditions has mainly focused on the effects of replant stress factors, such as heavy metals, nutrient deficiencies, and organic pollution [33]. In this experiment, the effects of root and soil extract stress on the secretion of phenolic acids from the roots of *M. hupehensis* Rehd. seedlings were studied. Eleven phenolic acids were detected in the root exudates of *M. hupehensis* Rehd. seedlings in all treatments, and the amount of phenolic acids secreted by seedlings varied among treatments. Replant stress significantly inhibited the secretion of p-hydroxybenzoic acid, benzoic acid, and phlorizin. The degree of inhibition of p-hydroxybenzoic acid secretion by the soil extract was greater than that of root extract when they were applied at the same concentrations. B3 and B4 had stronger inhibitory effects on benzoic acid than root extracts, and the opposite effect was observed for other concentrations of root extracts. The degree of inhibition of phlorizin associated with the application of low concentrations of soil extract was greater than that of the root extract, and the opposite pattern was observed when high concentrations were applied. P-hydroxybenzoic acid, benzoic acid, and phloridzin can affect plant growth by affecting cell membrane permeability, ion absorption, water absorption, photosynthesis, and protein and DNA syntheses [34–36]. Yin et al. showed that these three substances can significantly inhibit the growth of *M. hupehensis* Rehd. seedlings. This indicates that these three substances exist in large quantities in the root system and soil extract, which exerts feedback inhibition on root secretion [37].

Replant stress mainly promotes the secretion of caffeic acid and chlorogenic acid by the root system of *M. hupehensis* Rehd. These two phenolic acids are the main allelochemicals secreted by the root system of alfalfa [38] and can significantly inhibit the growth and development of soybean. The accumulation of caffeic acid and chlorogenic acid in soil further aggravated replant disease, had negative effects on the growth of crops, promoted the secretion of organic acids of crops, and increased the difficulty of overcoming the effects of replant disease. Vanillic acid reduces the absorption of 32P and methionine by soybean roots [39]. The inhibition of vanillin by low concentrations of soil extract was greater than that by root extract, and the opposite pattern was observed when high concentrations were applied. Treatments A1, A2, B1, and B2 promoted the growth of crop roots, significantly inhibited the secretion of vanillin by roots, reduced the accumulation of vanillin in soil, and alleviated the challenges of apple replant disease in orchards. All soil extract treatments and low-concentration root extract treatments significantly promoted the secretion of cinnamic acid. The stimulatory effect of root extracts was weaker than that of soil extracts when the same concentrations of root and soil extracts were applied. Cinnamic acid damages the cell membrane of the root epidermal cells of cucumber seedlings, affects the absorption of nutrients by roots, and inhibits the germination of crop seeds and the growth of seedlings.

Ferulic acid is a derivative of cinnamic acid. Treatments A1 and B6 promoted the secretion of cinnamic acid but inhibited the secretion of ferulic acid, indicating that these two treatments inhibited the conversion of cinnamic acid to ferulic acid.

Phloridzin lacks only one beta-D-glucopyranosyloxy group compared with Phloretin [40]. Replant stress inhibits the secretion of phloridzin by roots and promotes the secretion of phloridzin. This might stem from the fact that replant stress inhibits the addition of one beta-D-glucopyranosyloxy group to phloridzin or causes phlorizin to lose a beta-D-glucopyranose oxygen group to become phloretin; the specific conversion mechanism requires clarification. Under normal growth conditions, phloretin secretion was significantly positively correlated with plant height and significantly negatively correlated with ground diameter, indicating that phloretin affected the growth of *M. hupehensis* Rehd. seedlings; this suggests that allelochemicals secreted by the roots of *M. hupehensis* Rehd. seedlings significantly inhibited increases in ground diameter. The specific mechanism underlying this effect requires further study.

Gallic acid and catechin are the main components of the root exudates of *M. hupehensis* Rehd. Under normal growth conditions, the secretion of gallic acid and catechin was not correlated with plant height, basal diameter, and the chlorophyll content. After treatment with root and soil extracts, the secretion of gallic acid and catechin by roots was inhibited, but the secretion of these compounds was significantly positively correlated with seedling growth. Gallic acid and catechin in root exudates were allelochemicals of *M. hupehensis* Rehd. seedlings, but they did not have a negative effect on crop growth; by contrast, they might stimulate crop growth under replant disease conditions. The specific mechanism underlying this possibility requires further study.

4. Materials and Methods

4.1. Experimental Materials and Treatments

The experiment was conducted at the State Key Laboratory of Crop Biology at Shandong Agricultural University in Tai'an City, Shandong Province, China. *M. hupehensis* Rehd. (purchased from Shandong Horticultural Techniques & Services Co. Ltd., Tai'an, Shandong, China) was sterilized with 0.3% H_2O_2 for 20 min and then rinsed with sterile water. In January 2020, the seeds were buried in sand for 65 d. After the seeds were exposed, they were sown in sand culture in small flowerpots, with 10 seeds in each pot, and regularly irrigated with Hoagland nutrient solution. When the seedlings had 4–5 leaves, the seedlings were thinned, and 5 seedlings with relatively consistent growth were placed in each pot. When the seedlings were fully developed with 6–7 leaves, each seedling was treated with 1 mL of treatment solution. Root secretions were collected every 15 d in 3 biological replicates; root exudates were also collected on a regular basis.

4.2. Preparation of Root and Soil Extracts

The roots and soil were collected from a 20-year-old Fuji (Fuji/Malus × robusta (Carri CarriŠre) Rehder) replanted orchard in a suburb of Taian. The soil was brown loam, the soil bulk density was 1.31 G·cm^{-3}, and the pH was 5.61. The content of soil nutrients was as follows: 4.56 mg·kg^{-1} ammonium nitrogen, 7.38 mg·kg^{-1} nitrate nitrogen, 34.82 mg·kg^{-1} available phosphorus, 62.54 mg·kg^{-1} available potassium, and 7. 92 G·kg^{-1} organic matter.

According to differences in root diameter, the root system was divided into thick roots (diameters larger than 2 cm), middle roots (diameters 1–2 cm), and thin roots (diameters smaller than 1 cm). The weight ratio used in this experiment was thick heel: middle root:fine root = 1:1:1.The root system was mixed with equal amounts of water and shaken for at least 24 h; the samples were added to stock solution and stored at a low temperature (4 °C) for subsequent use. The stock solution was diluted and used as the solution to treat seedlings (Table S1).

For the preparation of soil extract, soil samples were collected in layers (0–15 cm, 15–30, and 30–45 cm) near the root system using the 4-point method and mixed in equal proportions. After the samples were air-dried, they were sieved through a 60-mesh sieve. The soil and water were mixed and shaken for 24 h; they were then filtered and stored at a low temperature (4 °C), and the stock solution was diluted and used to treat the seedlings (Table S1).

4.3. Preparation of Standard Solution

Preparation of PAHs standard solution: 16 types of USEPA priority PAH standard samples were purchased from O2si Smart Solutions, and each ml of acetonitrile contained 197.6 mg of Naphthalene (Nap, 2 rings), 198.1 mg of Dihydroacenaphthene (Acy, 2 rings), 198.1 mg of Acenaphthene (Ace, 2 rings), 198.1 mg of Dihydroacenaphthene (Acy, 2 rings), 198.1 mg of Dihydroacenaphthene (Acy, 2 rings), 198.1 mg of Dihydroac (2-ring) 199.1 mg of Fluorene (Flu, 2-ring), 198.3 mg of Phenanthrene (Phe, 3-ring), 198.1 mg of Anthracene (Ant, 3-ring), 198.2 mg of Fluoranthene (Fla, 3-ring), 198.1 mg of Pyrene (Pyr, 4-ring), 203.2 mg of Chrysene (Chr, 4-ring), 197.9 mg of Benzo [a] anthracene (Baa, 4-ring), 196.8 mg of Benzo [B] fluoranthene (Bbf, 4-ring), 197.5 mg of Benzo [K] fluoranthene (Bkf, 4-ring), 197.6 mg of Benzo [a] pyrene (Bap, 4-ring), 199.4 mg of Indeno [1,2,3, -cd] pyrene (Ilp, Dibenz [a, H] anthracene (Dba, 4-ring), and 197.6 mg of Benzo [G, H, I] pyrene (Bgp, 6-ring). Structures of the 16 PAHs are shown in Figure S1.

The preparation of phenolic acid standard solution was as follows: gallic acid, chlorogenic acid, catechin, caffeic acid, vanillin, benzoic acid, phloridzin, and phloretin were purchased from German Sigma Company; p-hydroxybenzoic acid was purchased from Beijing Xizhong Chemical Plant; ferulic acid was purchased from Shanghai No.1 Reagent Factory; and cinnamic acid was purchased from Tianjin Yuanhang Chemical Co., Ltd. (Tianjin, China). Gallic acid (1.3 mg), chlorogenic acid (1.5 mg), hydroxybenzoic acid (2.5 mg), catechin (1.5 mg), caffeic acid (3.3 mg), vanillin (3.0 mg), ferulic acid (1.5 mg), benzoic acid (4.1 mg), phloridzin (3.9 mg), cinnamic acid (2.7 mg), and phloretin (3.0 mg) were placed into a 10 mL volumetric flask, diluted with anhydrous methanol (chromatographic alcohol), and stored as the mother solution at a low temperature (4 °C) for subsequent use. The mixed standard sample was immediately prepared with 50 μL of each phenolic acid and 1 mL of methanol (chromatographic alcohol). The structures of 11 phenolic acids are shown in Figure S2.

4.4. Collection of Phenolic Acids Secreted by Roots

A total of 5 seedlings with similar growth outcomes were selected from the sand culture substrate by 5-point sampling, transferred to 250 mL of distilled water, pumped, and cultured at room temperature for 5–6 h under light, and the extract was collected. The extract was fully extracted with 200 mL of dichloromethane (analytical alcohol) 3 times, and the organic layers were combined, concentrated until dry in a rotary evaporator under reduced pressure, and reconstituted with 1.5 mL of methanol (chromatographic alcohol). The samples were placed in a refrigerator at −20 °C for 12 h to promote the precipitation of non-methanol-dissolved substances. After filtration through a 0.45 μm filter membrane, they were placed into 1.5 mL centrifuge tubes and stored in a refrigerator at −20 °C for subsequent use.

4.5. *Effect of Leaching Solution on Seedlings*

4.5.1. Effect on Seedling Growth and Development

The growth of *M. hupehensis* Rehd. seedlings was estimated every 15 d after the seedlings had 6–7 true leaves. The plant height and ground diameter were measured using vernier calipers; the chlorophyll content was measured using a SPAD-502 portable chlorophyll meter. A desktop scanner was to digitize images of the root system (NUScan 700, Tai'an, China), and morphological measurements were obtained using image analysis software (Deta-T scan, Delta-T Devices Lad, Cambridge, UK). Chlorophyll fluorescence was measured using a basic modulation fluorometer (WALZ Co., Dalian, China), and the maximum photochemical efficiency of photosystem II (PSII) F_v/F_m= $(F_m - {}_m/F_m)$ was measured. The photochemical quenching coefficient was determined using the following formula: qP = $(F_m{'} - F_s)/(F_m{'} - {}_m{'})$. The non-photochemical quenching coefficient was determined using the following formula: NPQ = $(F_m - F_m{'})/F_m{'}$.

4.5.2. Determination of Enzyme Activity

The activities of SOD [41], POD [42], and CAT [43] and the content of MDA [44] were determined following the methods of Yoshioka et al.

4.6. *Chromatographic Conditions*

4.6.1. Chromatographic Conditions for PAHs

The samples were analyzed by HPLC (UltiMate 3000 HPLC, Phenomenex EnviroSep-PP (125 mm× 4.6 mm) column) using the following conditions: column temperature 30 °C; injection volume 10 μL, mobile phase, acetonitrile, and water; gradient elution program, Table S1; UV detection, 254 nm; and fluorescence-detector wavelength program (Tables S2 and S3).

4.6.2. Phenolic Acid Chromatographic Conditions

The phenolic acids in the samples were determined using HPLC (Waters 515 HPLC, Kromasil C18 (125 mm × 4.6 mm) column). The determination conditions were as follows: column temperature, 28 °C; injection volume 20 μL, mobile phase, acetonitrile, and water; gradient elution procedure, Table S4; and dual-wavelength UV detector (waters), 280 nm (Table S4).

4.7. *Analysis Software*

SPSS13.0 was used to test the significance of differences in each index between treatments; the levels of significance were $p < 0.05$ and $p < 0.01$. The LSD method was used for multiple comparisons.

The allelopathy effect sensitivity index (RI) was estimated using Williamson's method [45]: RI = $1 - C/T$ (T ≥ C) or RI = $T/C - 1$ (T < C), where C is the control value and T is the treatment value. RI > 0 indicates a promoting effect and RI < 0 indicates an inhibiting effect. The absolute value of RI represents the intensity of allelopathy (autotoxicity).

5. Conclusions

Low-concentration root extract treatment (A1) significantly promoted the growth of *M. hupehensis* Rehd. seedlings, chlorophyll synthesis in leaves, the growth of seedling roots, and POD and CAT activities; inhibited the growth of ground diameter; and had no effect on SOD activity and fluorescence parameters. A6 had significant negative effects on all indices except the NPQ value.

Low-concentration soil extract treatment (B1) significantly promoted the growth of *M. hupehensis* Rehd. seedlings, chlorophyll synthesis in leaves, and the growth of seedling roots, and had no effect on the ground diameter; the activities of POD, SOD, and CAT; and fluorescence parameters. B5 and B6 increased NPQ, but had negative effects on all other indexes.

At the end of the treatment period, the root and soil extracts inhibited the secretion of gallic acid, p-hydroxybenzoic acid, benzoic acid, and phlorizin, but promoted the secretion of caffeic acid. Root extract treatment also inhibited catechin secretion and promoted the secretion of phloretin; soil extract treatment promoted the secretion of cinnamic acid. The secretion of the other phenolic acids varied depending on the concentration of leaching liquor applied.

Supplementary Materials: The following supporting information can be downloaded at: https://www.mdpi.com/article/10.3390/plants11212968/s1, Table S1: Process diagram; Table S2: Gradient eluting procedure; Table S3: Fluorescence detector's changes in procedures wavelength; Table S4: Gradient eluting procedure; Figure S1: 16 PAHs of EPA priority control; Figure S2: Structure of phenolics compounds; Figure S3: Standard sample 1 ((200 ng/g) in 254 nm ultraviolet detection), 2 ((20 ng/g) in different wavelengths fluorescence detection); Figure S4: Standard sample in 280 nm ultraviolet detection.

Author Contributions: Conceptualization, N.S., Y.M. and X.S.; methodology, N.S., Y.M. and X.S.; software, N.S.; validation, N.S., C.Y., Y.M. and X.S.; formal analysis, N.S.; investigation, N.S., C.Y. and X.Q.; resources, Y.L., M.S., Y.Z., X.C. (Xueli Cui), Y.Y. and R.W.; data curation, N.S.; writing—original draft preparation, N.S.; writing—review and editing, N.S., Y.H., X.C. (Xuesen Chen), Z.M., Y.M. and X.S.; visualization, N.S.; supervision, Y.M. and X.S.; project administration, Y.M. and X.S.; funding acquisition, X.S. All authors have read and agreed to the published version of the manuscript.

Funding: This research was funded by Shandong agricultural and rural Hall the fruit innovation team project of Shandong Province (CN), grant number SDAIT-06-07; Department of Science & Technolagy of Shandong Province, Natural Science Foundation of Shandong Province, grant number 32072520; Department of Science & Technolagy of Shandong Province, Natural Science Foundation of Shandong Province (CN), grant number ZR2020MC132.

Data Availability Statement: Data are presented in the article. All data and material are available from the manuscript.

Conflicts of Interest: The authors declare no conflict of interest.

Abbreviations

ARD: apple replant disease; F_v/F_m: PSII original light-energy-conversion efficiency; NPQ: non-photochemical quenching coefficient; qP: photochemical quenching coefficient; RI: indices of allelopathic effects; PAHs: polycyclic aromatic hydrocarbons; SOD: superoxide dismutase; POD: peroxidase; CAT: catalase; MDA: malondialdehyde.

References

1. Margaux, S.; Eva Lehndorff, A.W.; Wulf, A. In-field heterogeneity of apple replant disease: Relations to abiotic soil properties. *Sci. Hortic.* **2020**, *259*, 108809. [CrossRef]
2. Zhou, Z.; Zhu, Y.M.; Tian, Y.; Yao, J.L.; Bian, S.X.; Zhang, H.T.; Zhang, R.P.; Gao, Q.M.; Yan, Z.L. MdPR4, a Pathogenesis-Related Protein in Apple, Is Involved in Chitin Recognition and Resistance Response to Apple Replant Disease Pathogens. *J. Plant Physiol.* **2021**, *260*, 153390. [CrossRef] [PubMed]
3. Liu, X.X.; Xu, S.Z.; Wang, X.P.; Xin, L.; Wang, L.S.; Mao, Z.Q.; Chen, X.S.; Wu, S.J. MdBAK1 overexpression in apple enhanced resistance to replant disease as well as to the causative pathogen Fusarium oxysporum. *Plant Physiol. Biochem.* **2022**, *179*, 144–157. [CrossRef] [PubMed]
4. Tan, G.; Liu, Y.; Peng, S.; Yin, H.; Meng, D.; Tao, J.; Gu, Y.; Li, J.; Yang, S.; Xiao, N.; et al. Soil potentials to resist continuous cropping obstacle: Three field cases. *Environ. Res.* **2021**, *200*, 111319. [CrossRef]
5. Wang, H.Y.; Sheng, Y.F.; Jiang, W.T.; Pan, F.B.; Wang, M.; Chen, X.S.; Shen, X.; Yin, C.M.; Mao, Z.Q. The effects of crop rotation combinations on the soil quality of old apple orchard. *Hortic. Plant J.* **2022**, *8*, 1–10. [CrossRef]
6. Tewoldemedhin, Y.T.; Mazzola, M.; Labuschagne, I.; Adéle, M. A multi-phasic approach reveals that apple replant disease is caused by multiple biological agents, with some agents acting synergistically. *Soil Biol. Biochem.* **2011**, *43*, 1917–1927. [CrossRef]
7. He, C.N.; Gao, W.W.; Yang, J.X.; Bi, W.; Zhang, X.S.; Zhao, Y.J. Identification of autotoxic compounds from fibrous roots of *Panax quinquefolium* L. *Plant Soil* **2009**, *318*, 63–72. [CrossRef]

8. Tewoldemedhin, Y.T.; Mazzola, M.; Botha, W.J.; Spies, C.F.; McLeod, A. Characterization of fungi (Fusarium and Rhizoctonia) and oomycetes (Phytophthora and Pythium) associated with apple orchards in South Africa. *Eur. J. Plant Pathol.* **2009**, *130*, 215–229. [CrossRef]
9. Song, X.L.; Gao, X.D.; Zhao, X.N.; Wu, P.; Miles, D. Spatial distribution of soil oisture and fine roots in rain-fed apple orchards employing a Rainwater Collection and Infiltration (RWCI) system on the Loess Plateau of China. *Agric. Water Manag.* **2017**, *184*, 170–177. [CrossRef]
10. Elory, A. *The Rhizosphere*; Elroy, A.C., Bryan, T., Eds.; Springer: Berlin/Heidelberg, Germany; New York, NY, USA; Tokyo, Janpan, 1986; Volume 3, pp. 55–91. [CrossRef]
11. Marschner, H. *Mineral Nutrition of Higher Plants*, 2nd ed.; Academic Press: London, UK, 2012. [CrossRef]
12. Przybylska-Balcerek, A.; Szablewski, T.; Cegielska-Radziejewska, R.; Góral, T.; Kurasiak-Popowska, D.; Stuper-Szablewska, K. Assessment of Antimicrobial Properties of Phenolic Acid Extracts from Grain Infected with Fungi from the Genus Fusarium. *Molecules* **2022**, *27*, 1741. [CrossRef]
13. Bao, L.M.; Liu, Y.Y.; Ding, Y.F.; Shang, J.J.; Wei, Y.L.; Tan, Y.; Zi, F.T. Interactions Between Phenolic Acids and Microorganisms in Rhizospheric Soil From Continuous Cropping of Panax notoginseng. *Front. Microbiol.* **2022**, *13*, 791603. [CrossRef] [PubMed]
14. Liu, J.X.; Chang, Y.J.; Sun, L.H.; Du, F.F.; Cui, J.; Liu, X.J.; Li, N.W.; Wang, W.; Li, J.F.; Yao, D.R. Abundant Allelochemicals and the Inhibitory Mechanism of the Phenolic Acids in Water Dropwort for the Control of Microcystis aeruginosa Blooms. *Plants* **2021**, *10*, 2653. [CrossRef] [PubMed]
15. Politycka, B. Phenolics and the activities of phenylalanine ammonia-lyase, phenol-β-glucosyltransferase and β-glucosidase in cucumber roots as affected by phenolic allelochemicals. *Acta Physiol. Plant.* **1998**, *20*, 405–410. [CrossRef]
16. Callaway, R.M.; Aschehoug, E.T. lnvasive plant versus their new and old neighbors: A mechanism for exotic invasion. *Science* **2000**, *290*, 521–523. [CrossRef]
17. Patterson, D.T. Effects of allelopathic chemicals on growth and physiological responses of soybean (Glycine max). *Weed Sci.* **1981**, *29*, 53–59. [CrossRef]
18. Williams, A.; Langridge, H.; Straathof, A.L.; Fox, G.; Muhammadali, H.; Hollywood, K.A.; Xu, Y.; Goodacre, R.; de Vriesaf, F.T. Root functional traits explain root exudation rate and composition across a range of grassland species. *J. Ecol.* **2021**, *161*, 108391. [CrossRef]
19. Gadkari, P.V.; Balaraman, M. Catechins: Sources, extraction and encapsulation: A review. *Food Bioprod. Process.* **2015**, *93*, 122–138. [CrossRef]
20. Tian, G.; Bi, Y.; Sun, Z.; Zhang, L. Phenolic acids in the plow layer soil of strawberry fields and their effects on the occurrence of strawberry anthracnose. *Eur. J. Plant Pathol.* **2015**, *143*, 581–594. [CrossRef]
21. Muhammad, M.S.; Anne, B.F.; Knut, O.S. Phenolic Acids in Jerusalem Artichoke (*Helianthus tuberosus* L.): Plant Organ Dependent Antioxidant Activity and Optimized Extraction from Leaves. *Molecules* **2019**, *24*, 3296. [CrossRef]
22. Briana, A.O.; Clifford, P.R.; Brian, W.D.; Harry, H.S.; Steven, B.M.; Katherine, L.T. Phenolic acids released to soil during cereal rye cover crop decomposition. *Chemoecology* **2020**, *30*, 25–34. [CrossRef]
23. Pateron, R.P. Growth andspecific nodule activityofsoybean during application and recovery of leaf moisture stress. *Plant Physiol.* **1997**, *64*, 551–556. [CrossRef] [PubMed]
24. Li, Z.H.; Wang, Q.; Ruan, X.; Pan, C.D.; Jiang, D.A. Phenolics and plant allelopathy. *Molecules* **2010**, *15*, 8933–8952. [CrossRef] [PubMed]
25. Ren, K.; Hayat, S.; Qi, X.; Liu, T.; Cheng, Z. The garlic allelochemical DADS influences cucumber root growth involved in regulating hormone levels and modulating cell cycling. *Plant Physiol.* **2018**, *230*, 51–60. [CrossRef] [PubMed]
26. Lee, B.H.; Zhu, J.K. Phenotypic analysis of Arabidopsis mutants: Root elongation under salt/hormone-induced stress. *Cold Spring Harb. Protoc.* **2009**, *2009*, pdb.prot 4968. [CrossRef] [PubMed]
27. Song, Y.F.; Zhou, Q.X.; Xu, H.X. Eco-toxicological Effects of Phenanthrene, Pyrene and 1,2,4-Trichlorobenzene in Soils on the Inhibition of Root Elongation of Higher Plants. *Acta Ecol. Sin.* **2002**, *22*, 1945–1950. [CrossRef]
28. Hao, S.G.; Xue, J.Z.; Guo, D.L.; Wang, D.M. Earliest rooting system and root: Shoot ratio from a new Zosterophyllum plant. *New Phytol.* **2010**, *185*, 217–225. [CrossRef]
29. Caverzan, A.; Casassola, A.; Brammer, S.P. *Reactive Oxygen Species and Antioxidant Enzymes Involved in Plant Tolerance to Stress*; Shanker, A.K., Shanker, C., Eds.; IntechOpen: London, UK, 2016. [CrossRef]
30. Devi, E.L.; Kumar, S.; Singh, T.B.; Sharma, S.K.; Beemrote, A.; Devi, C.P.; Chongtham, S.K.; Singh, C.H.; Yumlembam, R.A.; Haribhushan, A.; et al. *Medicinal Plants and Environmental Challenges*, 1st ed.; Springer: Cham, Switzerland, 2017; pp. 359–413. [CrossRef]
31. Xu, J.M.; Ji, Z.Y.; Wang, C.L.; Xu, F.F.; Wang, F.J.; Zheng, Y.H.; Tang, Y.C.; Wei, Z.; Zhao, T.Y.; Zhao, K.J. Water-Soaked Spot1 Controls Chloroplast Development and Leaf Senescence via Regulating Reactive Oxygen Species Homeostasis in Rice. *Plant Sci.* **2022**, *13*, 918673. [CrossRef]
32. Chen, H.F.; Zhang, Q.; Lv, W.; Yu, X.Y.; Zhang, Z.H. Ethylene positively regulates Cd tolerance via reactive oxygen species scavenging and apoplastic transport barrier formation in rice. *Environ. Pollut.* **2022**, *302*, 119063. [CrossRef]
33. Li, H.; Xu, C.Y.; Han, L.; Li, C.Y.; Xiao, B.B.; Wang, H.; Yang, C.W. Extensive secretion of phenolic acids and fatty acids facilitates rhizosphere pH regulation in halophyte Puccinellia tenuiflora under alkali stress. *Physiol. Plant.* **2022**, *174*, e13678. [CrossRef]

34. Huang, C.Z.; Xu, L.; Sun, J.J.; Zhang, Z.H.; Fu, M.L.; Teng, H.Y.; Yi, K.K. Allelochemical p-hydroxybenzoic acid inhibits root growth via regulating ROS accumulation in cucumber (*Cucumis sativus* L.). *Integr. Agric.* **2020**, *19*, 518–527. [CrossRef]
35. Baziramakenga, R.; Simard, R.R.; Leroux, G.D. Effects of benzoic and cinnamic acids on growth, mineral composition, and chlorophyll content of soybean. *J. Chem. Ecol.* **1944**, *20*, 2821–2833. [CrossRef] [PubMed]
36. Duan, Y.A.; Zhao, L.; Jiang, W.T.; Chen, R.; Zhang, R.; Chen, X.; Yin, C.; Mao, Z. Biocontrol Potential of The Phloridin-Degrading Bacillus Licheniformis XNRB-3 against Apple Replant Disease. *Res. Sq.* 2021, *submitted*. [CrossRef]
37. Yin, C.M.; Duan, Y.N.; Xiang, L.; Wang, G.S.; Zhang, X.F.; Shen, X.; Chen, X.S.; Zhang, M.; Mao, Z.Q. Effects of phloridzin, phloretin and benzoic acid at the concentrations measured in soil on the root proteome of *Malus hupehensis* Rehd Seedlings. *Sci. Hortic.* **2018**, *228*, 10–17. [CrossRef]
38. Abdul-Rahman, A.A.; Habib, S.A. Allelopathic effect of alfalfa (*Medicago sativa*) on bladygrass (*Imperata cylindrical*). *Chem. Ecol.* **1989**, *15*, 2289–2300. [CrossRef]
39. Baziramakenga, R.; Leroux, G.D.; Simard, R.R. Effects of benzoic and cinnamic acids on membrane permeability of soybean roots. *Chem. Ecol.* **1995**, *21*, 1271–1285. [CrossRef]
40. Lyu, J.; Ning, J.; Meng, X.; Li, J.; Wang, S.Y.; Xiao, X.M.; Liu, Z.C.; Tang, Z.Q.; Yu, J.H. Exogenous silicon alleviates the adverse effects of cinnamic acid-induced autotoxicity stress on cucumber seedling growth. *Plant Sci.* **2022**, *13*, 968541. [CrossRef]
41. Beauchamp, C.; Fridovich, I. Superoxide dismutase: Improved assays and anassay applicable to acrylamide gels. *Anal. Biochem.* **1971**, *44*, 276–287. [CrossRef]
42. Smolinska, B.; Leszczynska, J. Photosynthetic pigments and peroxidase activity of *Lepidium sativum* L. during assisted Hg phytoextraction. *Environ. Sci. Pollut. Res. Int.* **2017**, *24*, 13384–13393. [CrossRef]
43. Aebi, H. Catalase. In *Methods of Enzymatic Analysis*; Bergmeyer, H.U., Ed.; Academic Press Inc.: Weinheim, Germany; New York, NY, USA, 1974; pp. 673–680. [CrossRef]
44. Yoshioka, T.; Kawada, K.; Shimada, T.; Mori, M. Lipid peroxidation in maternal and cord blood and protective mechanism against activated-oxygen toxicity in the blood. *Am. J. Obstet. Gynecol.* **1979**, *135*, 372–376. [CrossRef]
45. Williamson, G.B.; Richardson, D. Bioassays for allelopathy: Measuring treatment responses with independent controls. *J. Chem. Ecol.* **1988**, *14*, 181–187. [CrossRef]

plants

MDPI

Technical Note

Rhizoboxes as Rapid Tools for the Study of Root Systems of *Prunus* Seedlings

Ricardo A. Lesmes-Vesga [1], Liliana M. Cano [2], Mark A. Ritenour [1], Ali Sarkhosh [3], José X. Chaparro [3] and Lorenzo Rossi [1,*]

[1] Horticultural Sciences Department, Indian River Research and Education Center, Institute of Food and Agricultural Sciences, University of Florida, Fort Pierce, FL 34945, USA
[2] Plant Pathology Department, Indian River Research and Education Center, Institute of Food and Agricultural Sciences, University of Florida, Fort Pierce, FL 34945, USA
[3] Horticultural Sciences Department, Institute of Food and Agricultural Sciences, University of Florida, Gainesville, FL 32603, USA
* Correspondence: l.rossi@ufl.edu; Tel.: +1-772-577-7341

Abstract: Rootstocks are fundamental for peach production, and their architectural root traits determine their performance. Root-system architecture (RSA) analysis is one of the key factors involved in rootstock selection. However, there are few RSA studies on *Prunus* spp., mostly due to the tedious and time-consuming labor of measuring below-ground roots. A root-phenotyping experiment was developed to analyze the RSA of seedlings from 'Okinawa' and 'Guardian'™ peach rootstocks. The seedlings were established in rhizoboxes and their root systems scanned and architecturally analyzed. The root-system depth:width ratio (D:W) throughout the experiment, as well as the root morphological parameters, the depth rooting parameters, and the root angular spread were estimated. The 'Okinawa' exhibited greater root morphological traits, as well as the other parameters, confirming the relevance of the spatial disposition and growth pattern of the root system.

Keywords: peach; rootstock; rhizobox; root scanning; root system architecture; root angle

Citation: Lesmes-Vesga, R.A.; Cano, L.M.; Ritenour, M.A.; Sarkhosh, A.; Chaparro, J.X.; Rossi, L. Rhizoboxes as Rapid Tools for the Study of Root Systems of *Prunus* Seedlings. *Plants* **2022**, *11*, 2081. https://doi.org/10.3390/plants11162081

Academic Editor: Mariana Amato

Received: 30 June 2022
Accepted: 6 August 2022
Published: 9 August 2022

1. Introduction

The traits associated with root system architecture (RSA) can be employed as useful phenotyping tools for the breeding selection of plant materials, such as fruit-tree rootstocks. However, the study of underground root systems is more challenging than the study of the aerial part of the plant in several respects [1]. Soil is a complex medium that presents a more heterogeneous environment than the atmosphere. The variability of soil profiles in terms of physical, chemical, and biological conditions, as well as the pedological perturbation caused by planting and soil preparation, are variables to consider for RSA studies [2]. Currently, knowledge about root growth is relatively limited because of the complex dynamics of root systems compared to above-ground organs [3].

Rootstock selection plays a fundamental role in orchard management and successful fruit production. Some horticultural aspects to consider for rootstock breeding programs include soil adaptation, water and nutrient uptake, plant nutritional status, scion growth, pathogen resistance, and fruit quality [4,5]. 'Guardian'™, a commercial peach *Prunus persica* rootstock, was released in 1993 through the cooperative efforts of the USDA-ARS and Clemson University [6]. This rootstock provides longer tree life to overcome the problem of peach-tree short life (PTSL) [7]. In the last decade, 'Guardian'™ has been the predominant rootstock used for commercial peach production in the Southeastern United States, especially in South Carolina and Georgia. It is increasingly planted as a rootstock for almonds in California [8–10] with satisfactory field performance [11]. However, 'Guardian'™ is not a recommended rootstock for peach production in Florida due to its susceptibility to *Meloidogyne floridensis*. On the other hand, 'Okinawa', a rootstock introduced in the

United States in 1953 from Ryuku Islands (Japan), is one of the rootstocks utilized as genetic material in the University of Florida stone-fruit-breeding program for low-chill adaptation [12]. This low-chill rootstock is resistant to root-knot nematodes and exhibits good compatibility with several peach scion cultivars [13]. Despite these desirable features, 'Okinawa' rootstock is also not recommended for peach production in Florida because of its susceptibility to *M. floridensis* [14,15]. However, it is widely used outside of the United States, in countries such as Brazil [16] and Egypt [17].

Root architectural traits are fundamental for soil exploration, which influences the ability of plants to uptake below-ground resources [18]. Architectural traits are also strongly related to plant adaptation to sub-optimal conditions [19]. For instance, the most recognized contribution of RSA to water uptake is the ability to explore deeper soil layers. It is generally assumed that deeper root systems provide access to water stored deep in the soil [20]. Thus, deeper soil exploration contributes to drought resistance. Indeed, multiple studies provide evidence for the association between rooting depth and drought resistance in a panel of species, including trees [21]. However, the overall advantage of the root architectural traits must be interpreted in the context of the environment where the plants are established [19].

Among the geometric properties of root architectural traits, the root angle plays an important role in shaping the RSA of many species [19,22,23]. This is key, since it is one of the main traits contributing to rooting depth [24] and is strongly associated with the temporal and spatial acquisition efficiency of soil resources [25]. Moreover, the insertion angle of basal roots in dicotyledons determines the depth of root distribution; wider basal root angles determine shallower root systems [26,27]. For instance, a wider basal root angle in dicotyledons, such as the common bean (*Phaseolus vulgaris*), develops a shallower root system, enhancing topsoil foraging and phosphorous acquisition [19]. In contrast with monocotyledons such as rice (*Oryza sativa*), Kato et al. [28] demonstrated that the growth angle affects the vertical distribution and the rooting depth. A narrower angular spread indicates a deeper root system, which is advantageous for tolerance and adaptation to water-limited soil conditions [29]. Similarly, Manschadi et al. [19] demonstrated that in another monocotyledon, such as wheat (*Triticum aestivum*), drought-tolerant genotypes exhibit a narrower angular spread of seminal axes. Although the studies presented involved several species of economic importance, there is a lack of research evidence with regards to fruit-tree species, particularly rootstocks. Overall, the selection of the root angle may help to identify genotypes with a RSA more adapted to drought conditions, representing promising traits that can be exploited in crop-breeding programs [19].

Most plants repeat either their aerial or below-ground architectural units through reiteration, a morphogenetic process through which the organism duplicates its own elementary architecture entirely or partially [30]. Few attempts have been made to describe the genotypic variation in the root architecture of mature dicotyledon fruit-tree crops at the quantitative level; most of these have been on monocotyledon cereal crops [25]. Very few studies have been performed on *Prunus* spp., and in less detail. Through reiteration, the plant expresses its growth pattern repeatedly, from seedling to maturity [30]. Therefore, the analysis of the RSA at early developmental stages (seedlings) allows the measurement of root traits under controlled and relatively uniform conditions [31], especially in perennial trees such as *Prunus* spp. The quantification of root architectural traits becomes increasingly challenging as the plant grows. The root-insertion angles of seedlings may facilitate the use of root phenotyping as selection criteria in many cultivated species, since at early growth stages, this trait determines the development and performance of the mature root system. This has been demonstrated in the root angle of wheat genotypes, which developed either deeper or shallower root systems, depending on the root angle (narrower or wider) exhibited at the seedling growth stage [32,33].

Since most of the roots are below-ground and/or require a dark environment to grow naturally, root phenotyping requires special techniques. Rhizotrons (i.e., underground viewing chambers that offer a window of the root system) and, more recently, rhizoboxes (i.e., individual planting in thin translucent containers), are non-destructive techniques developed to perform direct and repeated measurements on the root system and high-quality imaging of the rhizosphere [34].

Rhizoboxes can present in many shapes and sizes, all utilizing translucent walls that allow the observation of the natural growth and development of an individual plant's roots. Typically, these rhizoboxes are flat to ensure the visualization of the root architecture and to allow for them to be scanned or pictured. Of course, this greatly diminishes the growing space for the roots, and therefore cannot be used as a direct comparison to field conditions; however, we are able to observe RSA traits and root characteristics from which we can extrapolate to form a better understanding of future developments in the field, or to identify particular traits of interest in breeding. According to Mašková and Klimeš [35], flat rhizoboxes did not affect plant growth in terms of total biomass, nor did they affect root-system-growth comparisons. This form of direct monitoring allows the non-invasive quantification and estimation of root-trait development, since it does not disturb the spatial disposition of roots. Rhizoboxes combine the controlled conditions of the laboratory with field-oriented research, even when the soil environment is artificial [36]. The imaging methods used in rhizoboxes help to study the development of root systems, as well as rhizosphere dynamics, in different growth media. This way, in rhizoboxes, the RSA can be analyzed holistically, and promising traits can be applied to breeding [37]. The main objective of this study was to analyze the root systems of peach rootstock seedlings with morphological and geometric approaches using rhizoboxes.

2. Results

2.1. Whole-Root System Depth: Width Ratio

Overall, the 'Guardian'™ whole-root system was significantly deeper (Figure 1a) and wider (Figure 1b) than the 'Okinawa' during the study. The depth:width ratio (D:W) of the 'Okinawa' was higher than that of the 'Guardian'™ each week for the duration of the experiment (Figure 1c); however, there were no significant differences throughout the 7-week period.

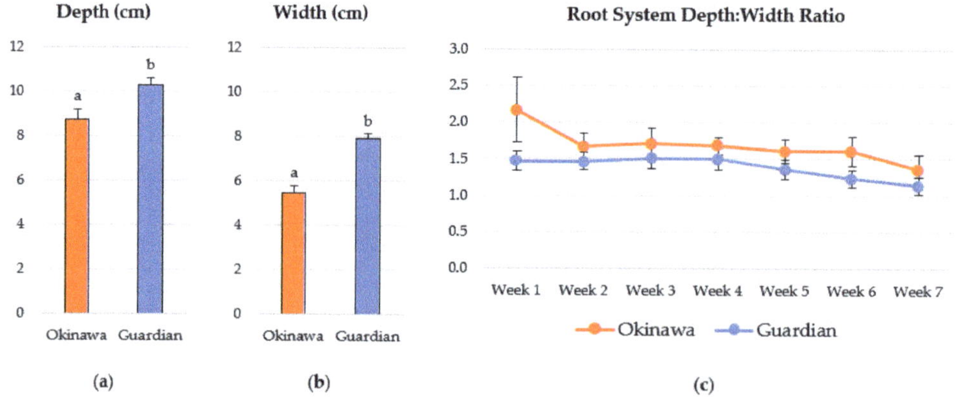

Figure 1. Average (**a**) depth and (**b**) width reached by the whole-root system of the 'Okinawa' and 'Guardian'™. (**c**) Depth:width ratio of the whole-root system of the 'Okinawa' and 'Guardian'™. Bars with different letters indicate significantly different values according to Tukey's (HSD) test ($p \leq 0.05$).

2.2. Root-Growth Parameters

The seedlings of the 'Okinawa' had significantly longer roots (Figure 2a) and a greater total-root-surface area (Figure 2b), root volume (Figure 2d), number of root tips (Figure 2e), and number of root forks (Figure 2f), compared with the 'Guardian'™ (789 cm). However, the average root diameter of the 'Okinawa' was significantly lower than that of the 'Guardian'™ (Figure 2c). The significant linear regressions between root-growth parameters in the two different cultivars are reported in the Supplemental Materials (Figure S1 and Table S1).

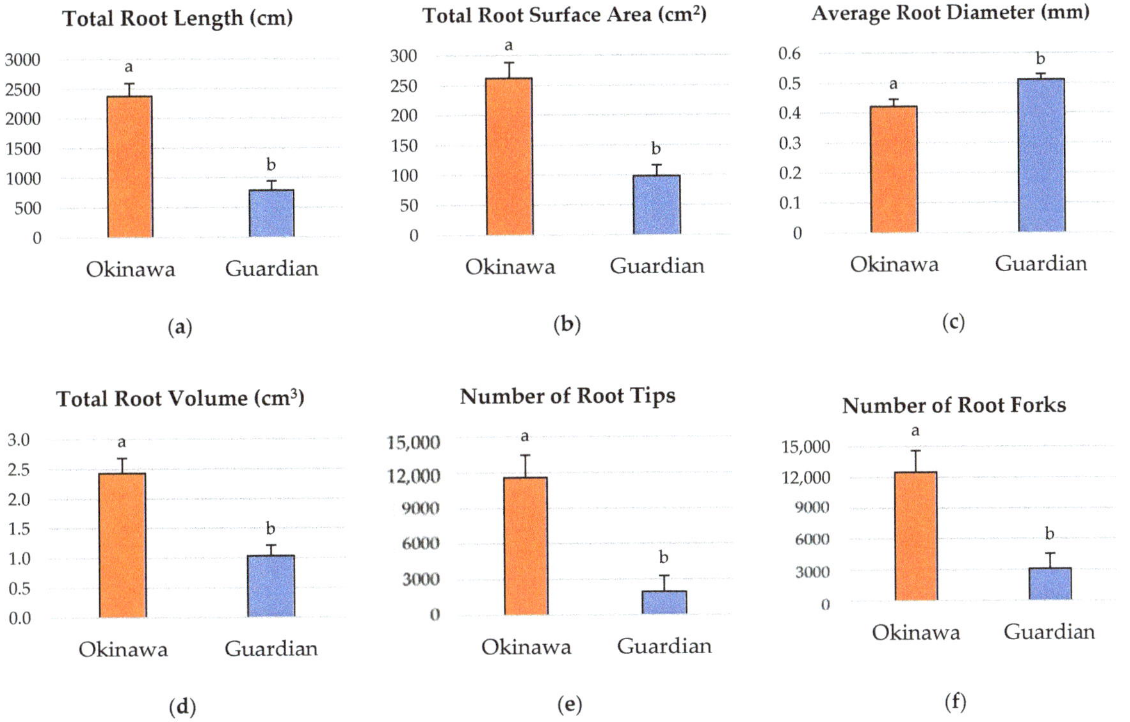

Figure 2. Comparisons of root morphological parameters in 'Okinawa' and 'Guardian'™ rootstocks for (**a**) total root length, (**b**) total-root-surface area, (**c**) average root diameter, (**d**) total root volume, (**e**) number of root tips, and (**f**) number of root forks. Bars with different letters indicate significantly different values according to Tukey's (HSD) test ($p \leq 0.05$).

2.3. *Total-Root-Length Distribution by Diameter Classes*

The total root length was divided into three diameter classes: very fine (≤0.5 mm), fine (>0.5 to ≤1.0 mm), and large (>0.5 to ≤1.0 mm). The very fine root length of the 'Okinawa' (1955.5 cm) was significantly greater than that of the 'Guardian'™ (611.1 cm). There were no significant differences in root length for the fine and large root diameters between the cultivars. (Table 1).

Table 1. Statistical analysis results of total root length (cm) distributed by root-diameter classes. The root-diameter classes are very fine (≤0.5 mm), fine (>0.5 to ≤1.0 mm), and large (>1.0 mm). Values not connected by the same letter are significantly different according to Tukey's (HSD) test ($p \leq 0.05$).

Root Length (cm) Distribution by Diameter Classes [a]				
Diameter Class	**Cultivar**	**Response**	**SE**	**Group**
Very Fine (≤0.5 mm)	'Okinawa'	1955.5	264.50	*a*
	'Guardian'™	611.1	74.69	*b*
Fine (>0.5 to ≤1.0 mm)	'Okinawa'	366.5	76.10	*bc*
	'Guardian'™	154.7	27.15	*c*
Large (>1.0 mm)	'Okinawa'	53.7	7.01	*c*
	'Guardian'™	23.1	4.88	*c*

[a] where SE = standard error.

2.4. Root-Depth Pattern and Root-Depth Index

Root horizon A of the 'Okinawa' represented a significantly larger portion of the total root length (44.2%) compared to horizons C (14.8%) and D (15.1%), but was not significantly different from horizon B (25.9%). Furthermore, there were no significant differences in root length between horizons B, C, and D (Figure 3a). Similarly, root horizon A of the 'Guardian'™ presented a root-length percentage (63.36%) that was significantly higher than that of horizons B (28.4%), C (9.8%), and D (7.12%). Moreover, horizon B was significantly higher than horizons C and D in this cultivar for this parameter, whereas horizons C and D were not significantly different (Figure 3b). Finally, the root-depth index (RDI) of the 'Okinawa' (15.09) was significantly higher than that of the 'Guardian'™ (9.58) (Figure 3c).

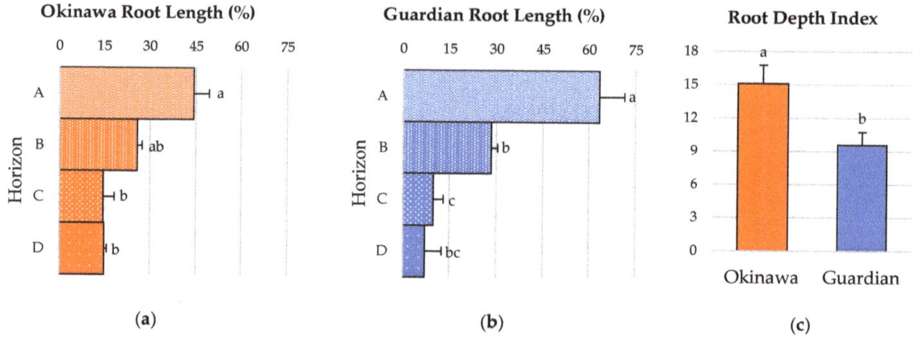

Figure 3. (**a**) Root-distribution pattern (RDP) in terms of root-length percentage of (**a**) 'Okinawa' and (**b**) 'Guardian'™ rootstocks. (**c**) Root-depth index (RDI) of 'Okinawa' and 'Guardian'™ rootstocks. Bars with different letters indicate significantly different values according to Tukey's (HSD) test ($p \leq 0.05$).

2.5. Root Spreading Angle

The total root length was not significantly different in the shallower (0–25°), shallow (25–45°), or deeper spreading angles (65–90°). However, the total root length was significantly different in the deep (45–65°) spreading angle between the 'Okinawa' (303.35 cm) and the 'Guardian'™ (168.08 cm). On the other hand, each cultivar presented the same pattern between spreading-angle ranges; the shallower and shallow spreading angles were not significantly different from each other, but significant differences were observed for the deep and deeper spreading angles (Table 2).

Table 2. Statistical analysis results of the root spreading angle from the central portion of the top horizon (A) of 'Okinawa' and 'Guardian'™ rootstocks. The RSG was estimated in terms of the total root length distributed between angular sections: shallower (0–25°), shallow (25–45°), deep (45–65°), and deeper (65–90°). Values not connected by the same letter are significantly different according to Tukey's (HSD) test ($p \leq 0.05$).

Total Root Length (cm) by Spreading Angle from the Cultivar–Angle Interaction [a]				
Angle	**Cultivar**	**Response**	**SE**	**Group**
Shallower (0–25°)	'Okinawa'	17.89	4.51	*a*
	'Guardian'™	7.15	3.88	*a*
Shallow (25–45°)	'Okinawa'	103.57	16.84	*ab*
	'Guardian'™	36.49	12.30	*a*
Deep (45–65°)	'Okinawa'	303.35	42.92	*c*
	'Guardian'™	168.08	22.31	*b*
Deeper (65–90°)	'Okinawa'	460.17	29.89	*d*
	'Guardian'™	382.70	22.72	*cd*

[a] where SE = standard error.

3. Discussion

There were significant statistical differences ($p \leq 0.05$) between the rootstock cultivars for most of the RSA parameters analyzed. The estimation of the whole-root-system depth:width ratio (D:W) was proposed, from an allometric perspective, to assess the evolution of the proportion of depth vs. width, as a useful indicator of the root-system growth and architecture for future studies. The average depth and width reached by the whole-root system measured weekly in the 'Guardian'™ seedlings were significantly higher than in the 'Okinawa'. Conversely, the D:W showed no significant differences between cultivars or weeks. Moreover, the D:W decreased as the plants grew, with the ratio in the 'Okinawa' higher than that in the 'Guardian'™, with significant differences. Furthermore, the D:W of the 'Okinawa' tended to be more stable than that of the 'Guardian'™, with the D:W of the 'Guardian'™ decreasing sharply earlier (week 4) than in the 'Okinawa' (week 7).

Significant differences were also observed between the root-growth parameters of both rootstock cultivars. Except for the average root diameter, all the root-growth parameters of the 'Okinawa' seedlings were significantly higher than those of the 'Guardian'™. The linear regressions performed (reported in Figure S1 and Table S1) showed similar significant patterns except for the root diameter. These results contrasted with the average of the whole-root-system depth and width reached by the 'Okinawa' seedlings, which were significantly lower than those of the 'Guardian'™. Therefore, it can be inferred that the depth and width reached by the root system alone is a poor descriptor of the actual root-system growth.

Despite both cultivars belonging to the same species, *P. persica*, these results confirm the contrast that can be found between rootstock cultivars. It is noteworthy that no previous studies on the root-system architecture (RSA) of the 'Okinawa' rootstock nor the 'Guardian'™ rootstock have been published. Although there are reports that mention the 'Guardian'™ as a rootstock that imparts excellent scion vigor and productivity [38,39], they do not describe its RSA. Similarly, this also occurs with the 'Okinawa' rootstock. The RSA features of these rootstocks confirm the findings from parallel studies that mention the 'Okinawa' as a rootstock that transmits high vigor to the scion cultivar [16]. In Northern Thailand, it was reported that 'Okinawa' induced good scion performance and reached the highest tree growth compared to other rootstocks, such as 'Khunwang', 'White Angkhang', 'Red Angkhang', and 'Flordaguard' [40]. This is consistent with the study by Kucukyumuk and Erdal [41], which indicates that vigorous fruit rootstocks have large root systems.

The morphology of the root system is as important as the actual soil volume explored by the roots [42]. The significantly greater length of the very fine roots ($\leq$0.5 mm) of the 'Okinawa' is a significant morphological trait, since knowledge of root diameters can provide important information about the root penetration and exploration potential in relation to the soil-pore size. Fine roots (diameter < 1 or 2 mm) are believed to play an important role in water and mineral uptake in higher plants [43,44]. However, it is necessary to keep in mind that this does not guarantee the absorption ability of the root system, since very fine roots may be suberized or even lignified [45].

In addition to the growth and morphological parameters, the spatial arrangement and distribution of the roots is another key aspect of the RSA. The D:W ratio estimation of the root systems studied in this chapter revealed little information and few contrasts between the two rootstocks. Nevertheless, these results agreed with the root-distribution patterns (RDPs) of both cultivars; the 'Okinawa' exhibited a more deeply distributed root system than the 'Guardian'™. Most of the total root lengths of the 'Guardian'™ seedlings were concentrated in the upper level, whereas in the 'Okinawa' seedlings, they were more deeply distributed. In addition, the percentage of the root length distributed at deeper horizons was almost twice as high in the 'Okinawa' than in the 'Guardian'™, where more than half of the total root length was in the top horizon (A). These results were confirmed by the superior root depth index (RDI) of the 'Okinawa', reflecting the deeper root system of this cultivar. The differences in root spreading angle (RSG) between the cultivars indicate that the root system of the 'Okinawa' tends to grow not only deeper due to its RDP and RDI, but also downwards. This trait is key for soil exploration in the pursuit of resources [25,32]. Solari et al. [46] confirmed that tree vigor is positively related to rootstock hydraulic properties. Moreover, such RSA traits can explain the differences in nutrient concentrations among rootstocks. In general, the root system architecture is different in *Prunus* from one rootstock cultivar to another, influencing their nutrient-uptake ability [41], which is evidenced by the nutritional status of the scion [47,48]. Rootstocks have a significant influence on leaf-mineral composition. Thus, according to Kumar et al. [49], the higher nutrient concentration in scion leaves is induced by invigorating rootstocks. The same authors reported similar findings in citrus rootstocks, indicating a significant positive correlation between the nitrogen (N), Ca, K, and Mg concentrations in leaves with the total root length, surface area, root volume, and number of tips. This was found by Shahkoomahally et al. [4] in the 'UFSun' peach on 'Okinawa', which showed greater concentrations of calcium, potassium, and magnesium in leaves than scions from the same cultivar grafted on other rootstocks, inducing a tree nutritional balance that was superior to those of other commercial rootstocks, maintaining the highest concentrations of macronutrients throughout the year. Given its better mineral-uptake efficiency, this study suggests the use of 'Okinawa' for sandy soils. Our results are comparable with those in the study by Ahmed et al. [17], where scions grafted on 'Okinawa' in sandy soils achieved superior percentages of initial and final fruit set. Similarly, it has been reported that 'Chimarrita' peach on 'Okinawa' reached higher levels of Brix, L-ascorbic acid, and carotenoids [50] than other rootstocks. Furthermore, the 'Okinawa Clone 1' rootstock induced greater Brix levels in 'Maciel' peach [51].

Since this study was performed on young plants, it can be considered too early to infer the potential performance of these rootstocks ('Okinawa' and 'Guardian'™); however, it is possible that the plant express the growth pattern shown in its seedling stage in maturity through reiteration [30]. This reiteration ability has been observed in oil-palm (*Elaeis guineensis* Jacq.) root systems [52]. It has been found that reiteration occurs in the entire root system of other stone fruit species, such as plum (*Prunus cerasifera*) [53], implying stone fruit species are not exceptions to this trait.

In summary, a larger root system may influence fruit quality beyond scion nutritional status. The measurement of the D:W of the root system alone is not a good indicator of the actual root-system size; however, the ratio can suggest how the root system grows, which is as relevant as the root-system morphology.

4. Materials and Methods

4.1. Plant Material, Seeds Stratification and Germination

Seedlings of two peach rootstock cultivars of *P. persica* used in commercial peach production in the United States ('Okinawa' and 'Guardian'™) were propagated from seeds provided by the Fruit Tree Breeding and Genetics Laboratory of the University of Florida in Gainesville, FL, USA. 'Okinawa' is a rootstock cultivar selected from landrace seed imported from Ryuku, Japan, and is also highly homozygous. 'Guardian'™ is a seed-propagated rootstock that is highly homozygous and uniform in performance.

The seeds were submerged in water, and those that floated were discarded. The remaining seeds were left in water for 96 h for stratification, and the water was renewed every 24 h. After 96 h, the seeds were submerged in a solution of water with Captan fungicide at 0.15% (*w/v*) for another 24 h. After this period, the solution was drained and the seeds were stored in a plastic bag with perlite, previously moistened with the same fungicide solution. The seed bags were stored at 4–8 °C until germination for a period of 6 weeks.

After germination, the seeds were sown in plastic trays (720700C SureRoots®; T.O. Plastics, Inc.; Clearwater, MN, USA) that consisted of star-shaped deep cell plugs (12.7 cm depth × 5.1 cm top width) containing a mixture (1:1) of potting-mix sphagnum (Jolly Gardener® Pro-Line C/20 Growing Mix; Jolly Gardener Products, Inc.; Poland Spring, ME, USA) and coarse perlite (Specialty Vermiculite Corp.; Pompano Beach, FL, USA). The substrate mixture was previously autoclaved at 121 °C for 90 min. The germination trays were maintained within a plastic-covered greenhouse located in Fort Pierce, FL, USA, at 27°25′34.2″ N–80°24′34.0″ W, covered with a shade cloth at 70% of shading, and watered manually. The seedlings were grown under greenhouse conditions. The average temperature was 20–30 °C during the day and 15–25 °C at night. Relative humidity (RH) was an average of 70–80% and photoperiod followed a natural phase of roughly 12–12 h (day–night).

4.2. Seedling Growing Conditions

After 10 days, at the emergence stage, the seeds that exhibited a 3–5-centimeter-long healthy and straight radicle were selected and sown in rhizoboxes (40 cm × 40 cm × 2.5 cm) (Figure 4a). One seed was transplanted per rhizobox. The rhizoboxes consisted of two clear glass panes (40 cm × 40 cm) attached to both faces of a three-sided plastic frame with a perforated bottom to allow water filtration. The plates and plastic frames were previously disinfected with a solution of sodium hypochlorite (1.5% v/v) in water for 30 min. The rhizoboxes were filled with a mixture of potting-mix sphagnum:coarse perlite (3:1) from the same manufacturers as those used for germination trays. Furthermore, the substrate was blended with the controlled-release (3 months) fertilizer, Osmocote® Plus (15-9-12) (A.M. Leonard, Inc., Piqua, OH, USA), at 0.5% (v/v).

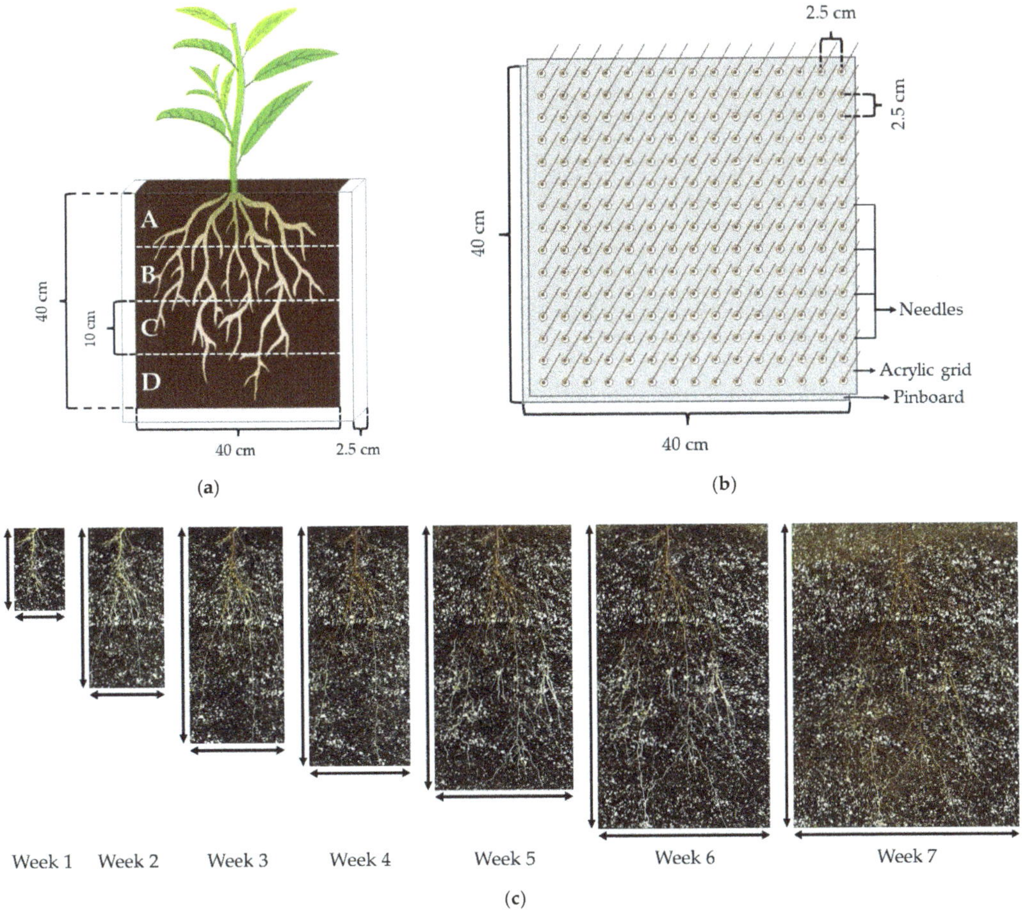

Figure 4. *Cont.*

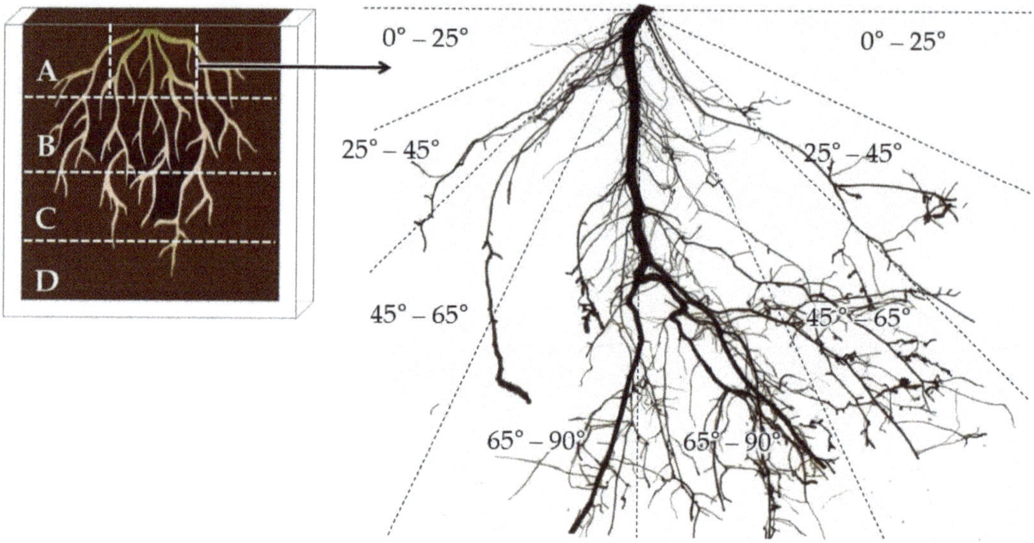

(d)

Figure 4. (**a**) Scheme of the rhizoboxes used to observe the seedling root system of 'Okinawa' and 'Guardian'™ rootstocks (1 seedling/rhizobox). The dashed white lines indicate the four horizons (A, B, C, and D) into which the root systems were split for scanning with their respective dimensions. (**b**) Diagram of the pinboard used for root-system extractions from the rhizoboxes. The acrylic grid was fitted with the needles of the pinboard. (**c**) Images captured once a week for 7 weeks from a rhizobox with the root system of one seedling. The maximum depth (↕) and width (↔) reached by the whole-root system, from which whole-root system D:W was estimated. (**d**) Representation of the central portion of the top root horizon (A) from which the root spreading angle (RSG) was estimated. RSG was estimated by measuring the root length (cm) within each angular section: shallower (0–25°), shallow (25–45°), deep (45–65°), and deeper (65–90°).

The emergent seeds were transplanted next to one of the glass panes of the corresponding rhizobox. Both panes of each rhizobox were covered with aluminum foil to exclude light and keep the roots under dark conditions. The rhizoboxes with the seeds were set in black plastic containers with bottom-draining holes, standing on wooden racks reclined at 30°. The experiment was maintained under laboratory conditions, with 23 °C and 65% relative humidity (RH. Complementary light-emitting diode (LED) bulbs and high-pressure sodium (HPS) lamps were adapted to keep the photoperiod at 16/8 h (day/night). The seedlings were watered manually throughout the experiment daily.

4.3. Root Systems Scanning

The root systems were scanned from the rhizoboxes once a week for 7 weeks with a flatbed scanner EPSON Expression 10000XL (EPSON America, Inc., Los Alamitos, CA, USA). Full color images were captured in Tagged Image Format File (TIFF) at resolution of 400 dots per inch (dpi). After this period, the root systems of the seedlings were extracted for measurement by removing one of the glass panes from the corresponding rhizobox. A pinboard was used to remove each plantlet, maintaining the spatial distribution of the roots. The pinboard consisted of an acrylic panel with 2.5-centimeter-long nails separated by 2.5 cm × 2.5 cm. Another acrylic panel with holes that corresponded with the nails was matched as a grid on the pinboard. The pinboard-grid was inserted into the rhizoboxes at the sampling moment (Figure 4b). The substrate was washed off the roots inserted in the pinboard by gently spraying with water.

After the root systems were extracted from the rhizoboxes, these were split into 4 horizons (A, B, C, and D, from top to bottom) of 10 cm (Figure 4a). The roots from each horizon were scanned with a flatbed scanner EPSON Perfection V800/V850 (EPSON America, Inc., USA) by placing them in a Plexiglas tray (20 cm × 30 cm × 1.5 cm). The plexiglass tray contained water to untangle the roots and minimize their overlapping. The images were captured in TIFF at 600 dpi resolution. Subsequently, the central portion of the top horizon (A) within an area of 10 cm × 10 cm was scanned with the EPSON Perfection V800/V850.

4.4. Image-Analysis Software and Measurements

From the images obtained every week with the EPSON Expression 10000XL scanner, the depth (D) and width (W) reached by the whole-root system were measured, and the depth:width ratio (D:W) was estimated. The images obtained with the EPSON Perfection V800/V850 scanner were analyzed using the root-image-analysis software, WinRHIZO™ Pro (Regent Instruments Inc., Quebec City, QC, Canada) (Figure 4c). The measurements of the root-growth parameters were total root length (cm), total root-surface area (cm^2), average root diameter (mm), total root volume (cm^3), number of root tips, and number of root forks. Additionally, the root length was estimated for each root diameter class, following the criteria applied by Caruso et al. [54]: very fine ($\leq$0.5 mm), fine (from >0.5 mm to <1.0 mm), and large (>1.0 mm).

Based on the total root length from each horizon, the root-distribution pattern (RDP) and the root-depth index (RDI) were calculated, following the protocol proposed by Oyanagi et al. [33,55]. The RDP was estimated by calculating the percentages of the root length of each 10-centimeter horizon from the whole-root system. Subsequently, the percentage values were multiplied by the value of the middle depth (cm) of the corresponding horizon (5 for A, 15 for B, 25 for C, and 35 for D) and divided by 100.

To study the root spreading angles of each genotype, a slightly modified version of the protocol followed by Ramalingam et al. [56] was applied. The scanned images from the central portion of horizon A were split into four angular sections projected from the base of the seedling: shallower (0–25°), shallow (25–45°), deep (45–65°), and deeper (65–90°) (Figure 4d). Subsequently, the root length (cm) was measured from each angular section.

4.5. Experimental Design and Statistical Analysis

The experiment was arranged in a completely randomized design with two cultivar treatments and sixteen single-tree replications per genotype. Each tree was an experimental unit. The data collected from the described RAS parameters were compared among cultivars by analysis of variance (ANOVA). The data were processed using RStudio software (2019), and a Tukey Honest Significance Difference (HSD) test was used to compare the means when the differences between treatments were significant ($p \leq 0.05$).

5. Conclusions

Studying the architectural root traits of rootstocks provides higher detail on tree growth and development. To date, most peach rootstocks have been studied only for their field performance and productivity, which offers the industry a fundamental but short-term outlook. However, very little attention has been given to the RSA traits that can benefit the long-term breeding selection processes for stone-fruit rootstocks. The use of rhizoboxes allows the study of these root traits in controlled environments. The experiment conducted showed the potential of the use of rhizoboxes as a rapid tool for the study of the root systems of *Prunus* seedlings. However, field comparisons between the data obtained from rhizotrons, destructive analysis, and rhizoboxes are needed to further support the use of rhizoboxes as reliable tools in fruit-tree-species RSA studies.

Supplementary Materials: The following supporting information can be downloaded at: https://www.mdpi.com/article/10.3390/plants11162081/s1, Figure S1: Linear regressions of root morphological parameters in 'Okinawa' and 'Guardian'™ rootstocks for total root length, total root surface area, average root diameter, total root volume, number of root tips, and number of root forks. Table S1: Linear regressions R^2 values of root morphological parameters in 'Okinawa' and 'Guardian'™ rootstocks for total root length, total root surface area, average root diameter, total root volume, number of root tips, and number of root forks.

Author Contributions: Conceptualization: R.A.L.-V., L.R., M.A.R. and J.X.C.; methodology: R.A.L.-V.; data collection and statistical analysis: R.A.L.-V.; writing—original draft preparation: R.A.L.-V.; writing—review and editing: L.R., A.S., L.M.C., R.A.L.-V. and J.X.C.; funding: L.R. and J.X.C. All authors have read and agreed to the published version of the manuscript.

Funding: This research was funded by United States Department of Agriculture (USDA), National Institute of Food and Agriculture, Hatch project #FLA-IRC-005743.

Institutional Review Board Statement: Not applicable.

Informed Consent Statement: Not applicable.

Data Availability Statement: The data presented in this study are available within this article. Raw data are available upon request from the corresponding author (l.rossi@ufl.edu).

Acknowledgments: We thank John-Paul Fox for language editing and assistance with the overall paper submission.

Conflicts of Interest: The authors declare no conflict of interest.

References

1. Danjon, F.; Stokes, A.; Bakker, M.R. Root Systems of Woody Plants. In *Plant Roots: The Hidden Half*, 4th ed.; Eshel, A., Beeckman, T., Eds.; CRC Press: Boca Raton, FL, USA, 2013; pp. 435–460, ISBN 978-1-4398-4649-0.
2. Moore, J.; Tombleson, J.; Turner, J.; van der Colff, M. Wind Effects on Juvenile Trees: A Review with Special Reference to Toppling of Radiata Pine Growing in New Zealand. *For. Int. J. For. Res.* **2008**, *81*, 377–387. [CrossRef]
3. Basile, B.; Bryla, D.R.; Salsman, M.L.; Marsal, J.; Cirillo, C.; Johnson, R.S.; DeJong, T.M. Growth Patterns and Morphology of Fine Roots of Size-Controlling and Invigorating Peach Rootstocks. *Tree Physiol.* **2007**, *27*, 231–241. [CrossRef] [PubMed]
4. Shahkoomahally, S.; Chang, Y.; Brecht, J.K.; Chaparro, J.X.; Sarkhosh, A. Influence of Rootstocks on Fruit Physical and Chemical Properties of Peach Cv. UFSun. *Food Sci. Nutr.* **2021**, *9*, 401–413. [CrossRef] [PubMed]
5. Chatzissavvidis, C.A.; Therios, I.N.; Molassiotis, A.N. Seasonal Variation of Nutritional Status of Olive Plants as Affected by Boron Concentration in Nutrient Solution. *J. Plant Nutr.* **2005**, *28*, 309–321. [CrossRef]
6. Beckman, T.G.; Chaparro, J.X. Peach Rootstock Development for the Southeastern United States. In Proceedings of the VIII International Peach Symposium, Matera, Italy, 20 May 2015; International Society for Horticultural Science (ISHS): Leuven, Belgium, 2015; pp. 171–178.
7. Okie, W.R.; Beckman, T.G.; Nyczepir, A.P.; Reighard, G.L.; Newall, W.C.; Zehr, E.I. BY520-9, a Peach Rootstock for the Southeastern United States That Increases Scion Longevity. *HortScience* **1994**, *29*, 705–706. [CrossRef]
8. Fisk, C.L.; Tu, C.; Ritchie, D.F.; Parker, M.L.; Reighard, G.L. Rootstock Effect on Soil Ecology in a Young Peach Orchard. In Proceedings of the XI International Symposium on Integrating Canopy, Rootstock and Environmental Physiology in Orchard Systems, Bologna, Italy, 5 December 2018; International Society for Horticultural Science (ISHS): Leuven, Belgium, 2018; pp. 265–272.
9. McGee, T.; Shahid, M.A.; Beckman, T.G.; Chaparro, J.X.; Schaffer, B.; Sarkhosh, A. Physiological and Biochemical Characterization of Six Prunus Rootstocks in Response to Flooding. *Environ. Exp. Bot.* **2021**, *183*, 104368. [CrossRef]
10. San, B.; Li, Z.; Hu, Q.; Reighard, G.L.; Luo, H. Adventitious Shoot Regeneration from in Vitro Cultured Leaf Explants of Peach Rootstock Guardian®Is Significantly Enhanced by Silver Thiosulfate. *Plant Cell Tissue Organ Cult. PCTOC* **2015**, *120*, 757–765. [CrossRef]
11. Wilkins, B.S.; Ebel, R.C.; Dozier, W.A.; Pitts, J.; Eakes, D.J.; Himelrick, D.G.; Beckman, T.G.; Nyczepir, A.P. Field Performance of 'Guardian'™ Peach Rootstock Selections. *HortScience* **2002**, *37*, 1049–1052. [CrossRef]
12. Maquilan, M.A.; Sarkhosh, A.; Dickson, D.W. Peach Root-Knot Nematode. *EDIS* **2018**, *2018*, HS1320. [CrossRef]
13. Sharpe, R.H. *Okinawa Peach Shows Promising Resistance to Root-Knot Nematodes*; University of Florida, Agricultural Experiment Stations Bulletin: Gainesville, FL, USA, 1957; pp. 320–322.
14. Nyczepir, A.P.; Beckman, T.G. Host Status of "Guardian" Peach Rootstock to *Meloidogyne* Sp. and *M. Javanica*. *HortScience* **2000**, *35*, 772. [CrossRef]
15. Sarkhosh, A.; Olmstead, M.A.; Miller, E.P.; Andersen, P.C.; Williamson, J.G. Growing Plums in Florida. *EDIS* **2018**, HS895.

16. Souza, A.; Spinelli, V.; de Souza, R.; Smiderle, O.; Bianchi, V. Optimization of Germination and Initial Quality of Seedlings of *Prunus Persica* Tree Rootstocks. *J. Seed Sci.* **2017**, *39*, 166–173. [CrossRef]
17. Ahmed, E.A.; El-Habashy, S.; Maklad, M.F. Trend of Vegetative Growth and Fruiting of Some Peach Cultivars Budded on "Okinawa" and "Nemaguard" Rootstocks. *Middle East J. Agric. Res.* **2017**, *6*, 1346–1358.
18. Lobet, G.; Hachez, C.; Chaumont, F.; Javaux, M.; Draye, X. Root Water Uptake and Water Flow in the Soil–Root Domain. In *Plant Roots: The Hidden Half*, 4th ed.; Eshel, A., Beeckman, T., Eds.; CRC Press: Boca Raton, FL, USA, 2013; pp. 355–372, ISBN 978-1-4398-4649-0.
19. Manschadi, A.M.; Hammer, G.L.; Christopher, J.T.; deVoil, P. Genotypic Variation in Seedling Root Architectural Traits and Implications for Drought Adaptation in Wheat (*Triticum Aestivum* L.). *Plant Soil* **2008**, *303*, 115–129. [CrossRef]
20. King, J.; Gay, A.; Sylvester-Bradley, R.; Bingham, I.; Foulkes, J.; Gregory, P.; Robinson, D. Modelling Cereal Root Systems for Water and Nitrogen Capture: Towards an Economic Optimum. *Ann. Bot.* **2003**, *91*, 383–390. [CrossRef]
21. Pinheiro, H.A.; DaMatta, F.M.; Chaves, A.R.; Loureiro, M.E.; Dicatti, C. Drought Tolerance Is Associated with Rooting Depth and Stomatal Control of Water Use in Clones of Coffea Canephora. *Ann. Bot.* **2005**, *96*, 101–108. [CrossRef]
22. Atger, C.; Edelin, C. Preliminary Data on the Comparative Architecture of Roots and Crowns. *Can. J. Bot.* **1994**, *72*, 963–975. [CrossRef]
23. Pandey, R.; Chinnusamy, V.; Rathod, G.; Paul, V.; Jain, N. Evaluation of Root Growth and Architecture. In Proceedings of the Manual of ICAR Sponsored Training Programme on Physiological Techniques to Analyze the Impact of Climate Change on Crop Plants, New Delhi, India, 16–25 January 2017; ICAR Indian Agricultural Research Institute (IARI): New Delhi, India, 2017; pp. 16–25.
24. Hammer, G.L.; Dong, Z.; McLean, G.; Doherty, A.; Messina, C.; Schussler, J.; Zinselmeier, C.; Paszkiewicz, S.; Cooper, M. Can Changes in Canopy and/or Root System Architecture Explain Historical Maize Yield Trends in the U.S. Corn Belt? *Crop Sci.* **2009**, *49*, 299–312. [CrossRef]
25. Manschadi, A.M.; Manske, G.; Vlek, P. Root Architecture and Resource Acquisition: Wheat as a Model Plant. In *Plant Roots: The Hidden Half*, 4th ed.; Eshel, A., Beeckman, T., Eds.; CRC Press: Boca Raton, FL, USA, 2013; pp. 319–336, ISBN 978-1-4398-4649-0.
26. Bonser, A.M.; Lynch, J.P.; Snapp, S. Effect of Phosphorus Deficiency on Growth Angle of Basal Roots in *Phaseolus Vulgaris*. *New Phytol.* **1996**, *132*, 281–288. [CrossRef]
27. Liao, H.; Rubio, G.; Yan, X.; Cao, A.; Brown, K.M.; Lynch, J.P. Effect of Phosphorus Availability on Basal Root Shallowness in Common Bean. *Plant Soil* **2001**, *232*, 69–79. [CrossRef]
28. Kato, Y.; Abe, J.; Kamoshita, A.; Yamagishi, J. Genotypic Variation in Root Growth Angle in Rice (*Oryza sativa* L.) and Its Association with Deep Root Development in Upland Fields with Different Water Regimes. *Plant Soil* **2006**, *287*, 117–129. [CrossRef]
29. Nakamoto, T.; Oyanagi, A. The Direction of Growth of Seminal Roots of *Triticum Aestivum* L. and Experimental Modification. *Ann. Bot.* **1994**, *73*, 363–367. [CrossRef]
30. Barthélémy, D.; Caraglio, Y. Plant Architecture: A Dynamic, Multilevel and Comprehensive Approach to Plant Form, Structure and Ontogeny. *Ann. Bot.* **2007**, *99*, 375–407. [CrossRef] [PubMed]
31. Thangthong, N.; Jogloy, S.; Pensuk, V.; Kesmala, T.; Vorasoot, N. Distribution Patterns of Peanut Roots under Different Durations of Early Season Drought Stress. *Field Crops Res.* **2016**, *198*, 40–49. [CrossRef]
32. Manschadi, A.M.; Christopher, J.; deVoil, P.; Hammer, G. The Role of Root Architectural Traits in Adaptation of Wheat to Water-Limited Environments. *Funct. Plant Biol.* **2006**, *33*, 823–837. [CrossRef]
33. Oyanagi, A. Gravitropic Response Growth Angle and Vertical Distribution of Roots of Wheat (*Triticum aestivum* L.). *Plant Soil* **1994**, *165*, 323–326. [CrossRef]
34. Rewald, B.; Ephrath, J.E. Minirhizotron Techniques. In *Plant Roots: The Hidden Half*, 4th ed.; Eshel, A., Beeckman, T., Eds.; CRC Press: Boca Raton, FL, USA, 2013; pp. 728–743, ISBN 978-1-4398-4649-0.
35. Mašková, T.; Klimeš, A. The Effect of Rhizoboxes on Plant Growth and Root: Shoot Biomass Partitioning. *Front. Plant Sci.* **2020**, *10*, 1693. [CrossRef]
36. Polomski, J.; Kuhn, N. Root Research Methods. In *Plant Roots: The Hidden Half*, 3rd ed.; Waisel, Y., Eshel, A., Kafkafi, U., Eds.; CRC Press: New York, NY, USA, 2002; pp. 295–321, ISBN 978-0-8247-0631-9.
37. Ephrath, J.E.; Klein, T.; Sharp, R.E.; Lazarovitch, N. Exposing the Hidden Half: Root Research at the Forefront of Science. *Plant Soil* **2020**, *447*, 1–5. [CrossRef]
38. Blaauw, B.; Brannen, P.; Lockwood, D.; Schhnabel, G.; Ritchie, D. *Southeastern Peach, Nectarine, and Plum Pest Management and Culture Guide*; University of Georgia, Cooperative Extension Services: Athens, GA, USA, 2020.
39. Reighard, G.L.; Beckman, T.G.; Belding, R.; Black, B.; Byers, P.; Cline, J.; Cowgill, W.; Godin, R.; Johnson, R.; Kamas, J.; et al. Six-Year Performance of 14 Prunus Rootstocks at 11 Sites in the 2001 NC-140 Peach Trial. *J. Am. Pomol. Soc.* **2011**, *65*, 26–41.
40. Wongtanet, D.; Boonprakob, U. Effect of Rootstocks on Growth of Peaches in the Highland of Northern Thailand. In Proceedings of the VIII International Symposium on Temperate Zone Fruits in the Tropics and Subtropics, Florianopolis, Brazil, 31 August 2010; International Society for Horticultural Science (ISHS): Leuven, Belgium, 2010; pp. 327–332.
41. Kucukyumuk, Z.; Erdal, I. Rootstock and Cultivar Effect on Mineral Nutrition, Seasonal Nutrient Variation and Correlations among Leaf, Flower and Fruit Nutrient Concentrations in Apple Trees. *Bulg. J. Agric. Sci.* **2011**, *17*, 633–641.

42. Psarras, G.; Merwin, I.A. Water Stress Affects Rhizosphere Respiration Rates and Root Morphology of Young 'Mutsu' Apple Trees on M.9 and MM.111 Rootstocks. *J. Am. Soc. Hortic. Sci. Jashs* **2000**, *125*, 588–595. [CrossRef]
43. An, H.; Luo, F.; Wu, T.; Wang, Y.; Xu, X.; Zhang, X.; Han, Z. Dwarfing Effect of Apple Rootstocks Is Intimately Associated with Low Number of Fine Roots. *HortScience* **2017**, *52*, 503–512. [CrossRef]
44. Eissenstat, D.M. Costs and Benefits of Constructing Roots of Small Diameter. *J. Plant Nutr.* **1992**, *15*, 763–782. [CrossRef]
45. Böhm, W. *Methods of Studying Root Systems*, 3rd ed.; Springer: Berlin/Heidelberg, Germany, 1979; Volume 16.
46. Solari, L.I.; Pernice, F.; DeJong, T.M. The Relationship of Hydraulic Conductance to Root System Characteristics of Peach (*Prunus Persica*) Rootstocks. *Physiol. Plant.* **2006**, *128*, 324–333. [CrossRef]
47. Mayer, N.; Ueno, B.; Silva, V. Leaf Nutrient Content of Peach on Five Rootstocks. *Rev. Bras. Frutic.* **2015**, *37*, 1045–1052. [CrossRef]
48. Mestre, L.; Reig, G.; Betrán, J.A.; Pinochet, J.; Moreno, M.Á. Influence of Peach–Almond Hybrids and Plum-Based Rootstocks on Mineral Nutrition and Yield Characteristics of 'Big Top' Nectarine in Replant and Heavy-Calcareous Soil Conditions. *Sci. Hortic.* **2015**, *192*, 475–481. [CrossRef]
49. Kumar, S.; Awasthi, O.P.; Dubey, A.K.; Pandey, R.; Sharma, V.K.; Mishra, A.K.; Sharma, R.M. Root Morphology and the Effect of Rootstocks on Leaf Nutrient Acquisition of Kinnow Mandarin (*Citrus Nobilis* Loureiro × *Citrus Reticulata* Blanco). *J. Hortic. Sci. Biotechnol.* **2018**, *93*, 100–106. [CrossRef]
50. Barreto, C.; Moreno, M.; Pricila, S.; Andrade, S.; Cesar, V.; Marcelo, B.; José, C. Growth, Yield and Fruit Quality of "Chimarrita" Peach Trees Grafted on Different Rootstocks. *Afr. J. Agric. Res.* **2017**, *12*, 2933–2939. [CrossRef]
51. Almeida, C.; Souza, A.; Argenta, J.; Fachinello, J.; João, B. The Effect of Rootstocks on the Vigor, Yield, and Fruit Quality of "Maciel" Peach Trees. *Rev. Ciênc. Agrar. Amaz. J. Agric. Environ. Sci.* **2015**, *58*, 301–307. [CrossRef]
52. Jourdan, C.; Rey, H. Architecture and Development of the Oil-Palm (Elaeis Guineensis Jacq.) Root System. *Plant Soil* **1997**, *189*, 33–48. [CrossRef]
53. Vercambre, G.; Pagès, L.; Doussan, C.; Habib, R. Architectural Analysis and Synthesis of the Plum Tree Root System in an Orchard Using a Quantitative Modelling Approach. *Plant Soil* **2003**, *251*, 1–11. [CrossRef]
54. Caruso, T.; Mafrica, R.; Bruno, M.; Vescio, R.; Sorgonà, A. Root Architectural Traits of Rooted Cuttings of Two Fig Cultivars: Treatments with Arbuscular Mycorrhizal Fungi Formulation. *Sci. Hortic.* **2021**, *283*, 110083. [CrossRef]
55. Oyanagi, A.; Nakamoto, T.; Wada, M. Relationship between Root Growth Angle of Seedlings and Vertical Distribution of Roots in the Field in Wheat Cultivars. *Jpn. J. Crop Sci.* **1993**, *62*, 565–570. [CrossRef]
56. Ramalingam, P.; Kamoshita, A.; Deshmukh, V.; Yaginuma, S.; Uga, Y. Association between Root Growth Angle and Root Length Density of a Near-Isogenic Line of IR64 Rice with DEEPER ROOTING 1 under Different Levels of Soil Compaction. *Plant Prod. Sci.* **2017**, *20*, 162–175. [CrossRef]

MDPI
St. Alban-Anlage 66
4052 Basel
Switzerland
www.mdpi.com

Plants Editorial Office
E-mail: plants@mdpi.com
www.mdpi.com/journal/plants

www.ingramcontent.com/pod-product-compliance
Lightning Source LLC
Chambersburg PA
CBHW061502210926
51850CB00008B/1799
9783725810178